STEELMAKING: 1850–1900

K. C. Barraclough

The Institute of Metals
1990

Book Number 458

Published in 1990 by The Institute of Metals
1 Carlton House Terrace, London SW1Y 5DB

and
The Institute of Metals
North American Publications Center
Old Post Road, Brookfield VT 05036
U S A

I S B N 0-901462-71-3

Printed and bound in Great Britain by Biddles Ltd,
Guildford and King's Lynn.

Contents

Preface

The author's two previous volumes, under the general title "Steelmaking Before Bessemer" derived from a thesis presented to Sheffield University. They dealt particularly with the processes which historically had provided the basis of Sheffield's reputation as the steelmaking centre of the world before Bessemer even contemplated blowing air through the liquid product of the blast furnace. These earlier processes were relatively small scale operations which provided superior materials for special applications and at relatively high cost. The detail had proved difficult to find; it relied on much hitherto unpublished documentation and on the diaries of continental visitors and was the result of over thirty years spare-time research by a practising steelmaker.

This present volume is different in many ways. In the first instance, it is a commissioned work from a retired steelmaker. It also deals specifically and in detail with bulk steel production; it is true that the earlier volume on Crucible Steel, in order to chart the reasons for the change in direction of the use of the crucible process after about 1865, did devote some forty pages to a summary entitled "The Need for More Steel". This same theme is now extended in considerable detail and forms the entire subject of the present volume. It covers the steelmaking activities in the half century which witnessed the rise in the production of steel world wide from some mere fifty thousand tons to just under thirty million tons and a reduction in cost by around four fifths. It discusses, in other words, the impact of the inventions of Bessemer, Siemens and Thomas and the introduction of what came to be termed "bulk steel".

The background research for this volume quickly revealed a completely different pattern of available evidence. No longer has the information to be derived as a result of detective work on unpublished manuscripts; after 1850 standard steelmaking texts begin to appear and these proliferate as the years go by; in addition, the various Learned Societies become more and more interested in the topic and their Transactions and Proceedings become valuable sources; with the publication of the Journal of the Iron and Steel Institute from 1869 onwards, a mine of information becomes available.

In order to provide a self contained study, it has been thought fit to provide a general background to steel and steelmaking, including some of the essential chemistry to the various steelmaking processes, and, in addition, covering the various means by which steel could possibly be produced in reasonable quantities and at reasonable cost.

Subsequently, the study becomes a survey of the growth of bulk steelmaking. In the first instance, there is a discussion of the work of Bessemer and others in the development of the process generally known by his name, followed by a survey of the growth of the process in Britain, Europe and America to around 1880. A similar treatment is then provided of the work of Siemens and others in the development of the Open Hearth process. This is logically followed by the story of Gilchrist Thomas and what may be called the Solution of the Phosphorus Problem and a survey of the growth of the Basic Bessemer and the Basic Open Hearth processes. The study concludes with a discussion of the developments in the final decade of the nineteenth century.

By far the best known of the early treatises dealing with steelmaking is *Metallurgy: Iron and Steel,* by Dr. John Percy. This was published in 1864, at a time when the Bessemer process had become firmly established but when the Open Hearth processes devised by Siemens were in early stages of development. It was clearly the intention of Dr. Percy to provide a revised version of his text, as witnessed by the massive collection of information represented by the so-called "Percy Papers", still held by the Institute of Metals. Indeed, as late as 1885, in his Presidential Address to the Iron and Steel Institute, he remarked that "the treatise was published twenty one years ago and, since that date, I need hardly remind my audience, immense progress has been made in the metallurgy of iron and steel. May I venture to announce that, during many years, I have been engaged, though not uninterruptedly, in preparing a new edition of the treatise in question and, in so doing, I have found so much to correct, so much to excise and so much to add that I believe the labour of revision will exceed that of rewriting the whole volume". Time, unfortunately, ran out for Dr Percy and posterity was denied what would have been a most historical contribution. It seems appropriate, therefore, that the present volume, dealing as it does with the advances in steelmaking which he would have covered, had his project been completed, should be going to press on the centenary of the death of Dr. John Percy.

Acknowledgements

The work of consulting the various steelmaking texts and journals has been rendered much more difficult by the disbanding of the several excellent steelworks libraries in the Sheffield area over the past few years. It is, however, gratifying to note that some, at least, of these valuable collections have been gathered together at the Kelham Island Industrial Museum; in addition, this library has acquired many of the earlier patent specifications. The facilities for study which have been most kindly provided there by the Keeper at Kelham Island, Peter Smithurst, have been invaluable and are gratefully acknowledged; without them, the author's research could have been much more difficult and much less productive.

Likewise, acknowledgement is due to Maureen Anderson of the Science and Commerce Section, to Martin Olive of the Local History Section and to Dr. Postles of the Archive Section of the Sheffield City Libraries for their courteous and efficient service in providing source material.

A most comprehensive, and little known, source of information on activities during the later 1870s has proved to be the collection of papers put out by Alexander Holley "for the use of his clients"; by the kind collaboration of Martha Goodway, metallurgist at the Smithsonian Institution in Washington, copies of these have been obtained from the Library of Congress. The author wishes to place on record his gratitude for this invaluable assistance.

Gratitude is also due to the late Wilhelm Ekman of Stockholm for helpful discussions and the provision of background information on early Swedish developments; also to Kerstin Fernheden of Jernkontoret and to Magnus Bjorklund of Sandvik Steel for other items of Swedish developments. Likewise, acknowledgement is also given of the assistance rendered by Friedrich Toussaint of Verein Deutsher Eisenhuttenleute and to the Historisches Archiv of Fried Krupp for German information. Thanks are also due to Professor R. B. Gordon of Yale University for the provision of copies of various papers from American journals not readily accessible in Sheffield.

The advent of the word processor since the author's previous texts were committed to paper has removed the need for secretarial assistance which was so vital previously in preparing a presentable manuscript. The necessity of an independent check of the text, however, at numerous stages, has remained and the author would be sadly remiss not to acknowledge the valuable assistance rendered by his wife in this respect; the final presentation is considerably "refined" as a result and the product more free from "extraneous inclusions".

1 Introduction to Steel and Steelmaking

Historical background

The growth of human endeavour is universally recognised by the periods referred to as the Stone Age, the Bronze Age and the Iron Age. The time scale for these periods varies according to location – indeed, some isolated areas of the world may still be seen as belonging to the Stone Age – but the Iron Age must be considered to have originated around 1500 B.C. somewhere in the Near East. The significance of this statement is that, although the iron-rich minerals had been well recognised and had been used as pigments by Stone Age Man and as gemstones during the Bronze Age, the winning of metallic iron from its ores did not happen until the much rarer metal, copper, alloyed with an even rarer metal, tin, had been produced for almost two thousand years.

The origin of the smelting of iron ores is shrouded in mystery. Iron does occur as part of many copper ores and, after roasting the ore, iron oxide would have been a useful flux for siliceous impurities. Certainly, fayalite (a silicate of iron) is present in copper slags produced in antiquity and it could well be that some such slag had reacted with excess charcoal in the hearth of a copper smelting furnace to give a spongy bloom of metallic iron particles. Some observant smith might then have tried hammering such a bloom, so laying the foundations of a technique very different from any prior smelting operation.

The melting point of pure iron is 1535°C; the most impure cast iron (with about 4% carbon, together with other impurities) requires a temperature of around 1200°C to render it free running, so that the copper smelting process could not produce liquid iron. In addition, even the purest iron is somewhat difficult to smith without reheating it. The completely new technique so developed is now generally referred to as the **bloomery process**: the smelting of the iron ore by heating it with charcoal in a furnace blown by bellows, followed by the extraction from the hearth of the hot bloom – a spongy mass of solid iron particles surrounded by and intermingled with slag – and hammering it to give a solid mass of metal and at the same time expelling the bulk of the slag. It was a process which, with continuing minor modifications, mostly concerned with increasing the scale of operations, was to survive for over three thousand years.

The first people renowned for their prowess in the making of iron were the Hittites and tradition has it that their superiority in the battlefield – and that of the Assyrians who inherited their skills – was due to the superiority of their weapons over those of their contemporaries. This clearly presents a paradox, since iron as produced normally from the bloomery is a relatively soft material and a sword or cutting weapon made from such iron would be a poor thing compared with the conventional bronze weapon of the time; indeed, there are stories from as late a period as the Icelandic sagas of warriors who found it necessary to straighten their swords under their feet after using them for a powerful thrust. A **steel** sword would have been a different matter. Since the difference between iron and steel rests on the seemingly minor matter of introducing a small amount of carbon into the metal, followed by quenching and tempering, it is interesting to speculate how much of the art was known to the Hittites. The evidence is confusing; there is an Egyptian axe from about 900 B.C. which has an edge with 0.9% carbon, fully hardened; an adze from Ras Shamrah dating from about 500 B.C. has a steel edge forge welded to an iron backing – a procedure used as an economy measure in tool manufacture well into the twentieth century; chariot tyres from Llyn Cerrig Bach in Anglesey, dating from 50 B.C. or thereabouts, were made up by welding small batches of steel, all differing somewhat in carbon content (between the limits of 0.74% to 0.96%) but uniform within themselves. This, and other evidence, leads to the conclusion that the bloomery process, quite early in the Iron Age, was consciously and deliberately modified, with adjustment of the ratio of charcoal to ore, to provide a higher carbon material, so providing steel from the bloomery, a material which came to be known as **natural steel**.

As time went on, the bloomery operations grew in scale; the original small bowl shaped furnace, worked with the natural draught derived from its situation on an open hillside, began to be fitted with hand- or foot-operated bellows; it had protective walls built around it to concentrate the heat; eventually the mid-European high shaft bloomery was provided with bellows actuated by a water wheel. As the efficiency of the furnace was increased, the hearth temperature tended to rise and it was only a small step from such a bloomery to a **blast furnace**. The difference between the two lies in the formation of a liquid product from the blast furnace as distinct from the spongy bloom from the bloomery; the composition of the two metals, however, is vastly different. As the temperature rises, the solid iron particles have an opportunity to absorb carbon from the charcoal in the hearth and this significantly lowers the melting point, such that some 4% carbon provides an alloy which is liquid at 1200°C and can be run out from the furnace and then cast into **cast iron** or **pig iron**, a brittle, intractable, unforgeable material.

With the advent of the blast furnace, the production of malleable iron, similar to the bloomery iron, and of steel, had to take different routes. The bloomery provided a **direct process**; with the blast furnace as intermediary, there was the **indirect process** to be considered. The original process to find general application was the **finery**; here the cast iron was remelted in a hearth full of charcoal and air blown in from bellows. The carbon in the metal was in this way burned out; this caused the melting point of the metal to rise and, at the temperature employed, it first became pasty and eventually changed to a spongy mass of virtually carbon-free iron particles with entrained slag – essentially similar to the product from the bloomery – which could be lifted out and hammered to give **wrought iron**, as distinct from the original cast iron.

To obtain **steel** from the finery, a nicety of judgement was needed to conclude the oxidation process whilst there was still an appropriate amount of carbon present. Nevertheless, within a relatively short period after the introduction of the blast furnace in continental Europe, there grew up numerous processes, such as the Styrian, the Carinthian, the Westphalian, the Siegen, the Paal, the Tyrolese, the Rivois and the Dauphiné methods, all in effect variants of a central theme, producing a metal with sufficient carbon for it to harden on quenching, generally with a somewhat segregated structure as regards carbon distribution and still with entrained slag streaks similar to those found in the wrought iron. Confusingly, this material was generally known as **natural steel**; in Britain, where it was imported during the seventeenth and eighteenth centuries for the manufacture of cutlery and edge tools, it tended to be called **German steel**.

After the Cort inventions of the 1780s, the use of the finery (and its partner, the **chafery**, where the bloom was reheated and reforged into the form required by the user) was superseded by the **puddling process**, really a cheaper and more productive method of doing the same thing, and by the use of a rolling mill in place of the chafery. Eventually, the puddling process was to be modified to allow steel to be produced; this development comes within the scope of this present volume.

The methods involving partial decarburisation of the cast iron to produce steel, however, were uncertain and depended heavily on the skill of the operator. The product was invariably sorted after manufacture as to its suitability for use; in France, for instance, the Rivois process provided **acier** (steel), **fer fort** (strong iron, or steely iron, suited to making agricultural implements) and **fer doux** (soft iron or wrought iron) and a production of over 50% as "acier" was considered good. The other approach was to burn out all the carbon, thus providing wrought iron or bar iron and then to carburise this under more controlled conditions by heating it, packed in charcoal in sealed chests, to a bright red heat for up to a week. During this period, carbon would diffuse into the solid iron and convert it into **blister steel**. This was the **cementation process**,[1] cementation being the alchemist's term for a reaction taking place in the solid state. Although its origins were central European, it became

[1] An extended study of the Cementation Process is to be found in the Author's prior publication "Steelmaking Before Bessemer": Vol. 1 is subtitled "Blister Steel - The Birth of an Industry" (London, 1984)

more and more the British process. The bars of blister steel were broken into short lengths, these put together in a bundle known as a **faggot** which was then reheated as a single entity and the whole then forge welded to provide **shear steel**. Again there is confusion, since the structure of such a material, with bands of high and low carbon content alternating and with the slag from the original wrought iron still present, is very similar to the continental **natural steel** from finery operations; to add even further to the confusion, such material, although produced in Britain, could also be known as **German steel**!

Such a material provided excellent knife blades and general edge tools. For more delicate instruments and particularly for clock and watch springs its segregated structure proved detrimental and it was the result of trials carried out by a clockmaker, Benjamin Huntsman, that **cast steel** – no steel had ever been melted previously in Western Europe – was introduced. This was provided by melting broken pieces of blister steel in closed crucibles surrounded by an incandescent bed of burning coke, this being the **crucible process**.2 The combination of the cementation furnace and the crucible furnace formed the basis of the concentration of steelmaking in the Sheffield area during the first half of the nineteenth century, especially after the improvement in communications brought about by the introduction of the railways in the area after 1838.

As it so happens, there are estimates of the total European steel production for the year 1843 from a French source, which can be summarised as follows:3

Producing Country	Natural Steel tonnes	Cementation Steel tonnes	Total Product tonnes
Great Britain	Nil	20500	20500
Austrian Monarchy	12800	Nil	12800
German Federation	7100	100	7200
France	3370	3770	7140
Russia	530	2680	3210
Sweden and Norway	2000	900	2900
Spain	200	100	300
Italian Peninsula	100	100	200
Total	**26100**	**28150**	**54250**

[2] The Crucible Process forms the subject of the second volume of " Steelmaking Before Bessemer". This is subtitled "Crucible Steel —The Growth of Technology" (London, 1984).

[3] F. le Play: "Memoire sur la Fabrication de l'Acier en Yorkshire", Annales des Mines, Serie 4 me, Tome III (1843), p. 692.

In passing, the report from which the above figures are taken, points out that 1843 was not a good year for Sheffield production and that the capacity in the town could have provided some 33,000 tonnes of blister steel, with crucible furnaces capable of melting 15,500 tonnes annually. A further point of interest is that the American market was largely supplied from the Sheffield area, the domestic output in U.S.A. being only 6,078 tons per annum in 1847; of this a mere 44 tons was crucible steel.[4]

This then was the position of the steelmaking industry at the commencement of the period under survey.

Over the next fifty years there were revolutionary changes in steelmaking. As a result of intensive investigations into the possibilities of producing steel more cheaply and on a larger scale there came what was to become known as "bulk steelmaking". The first breakthrough was the production of **puddled steel**. This was followed by the development by Bessemer of the so-called **Pneumatic Process** and the somewhat later work by Siemens which led to the **Open Hearth Process**. As developed, both these suffered from the limitation that they were only applicable to materials low in phosphorus. Then came the work of Gilchrist Thomas, which overcame the phosphorus problem, introduced **basic steelmaking** and opened up virtually all the iron ore supplies of the world to steelmaking. This rendered the production of puddled steel obsolete. It is worthy of comment that, with the exception of puddled steel, these revolutionary inventions were all British in origin. This probably is not surprising, since Britain led the world in ferrous metallurgy during this period; paradoxically, however, the final innovation proved to be the root cause of the rapid overtaking of Britain as a steelmaking nation, first by the United States of America and then, just before the end of the century, by Germany.

It is the background to these innovations and their technical and commercial development which forms the subject matter of this study.

Basic metallurgy

Up to 1850, it had been perfectly logical and straightforward to define steel as an alloy of iron with a suitable proportion of carbon, nominally from 0.5% to 1.5%, this material being capable of showing a significant increase in hardness on quenching in water from a bright red heat. In this way it was distinguishable from wrought iron, which normally contained so little carbon that the quenching operation showed little change. The coming of the bulk steelmaking processes brought about the introduction of what came to be known as **mild steel** with a much lower carbon content, usually

[4] A report of the Philadelphia Convention of Ironmasters in 1847, quoted by J. M. Swank: " History of the Manufacture of Iron in All Ages", 2nd ed., (Philadelphia, 1892), p.385.

from 0.10% to 0.20%, used largely as a more satisfactory substitute for wrought iron, which showed only a minor degree of response to quenching. Nevertheless, by far the major bulk of steel produced between 1850 and 1900 could still be categorised as an alloy of iron and carbon – or as **carbon steel**. It was only after about 1870 that any commercial steels were produced with intentional amounts of other additional or alloying metals, the purpose of whose addition was to modify the characteristics of the steel so as to render it more useful under certain conditions of service. This defines the group of materials known as **alloy steel**; the definition must be such as to exclude the proportions of elements other than iron and carbon which are present as incidental impurities in materials made by commercial processes and also the elements such as silicon and, particularly, manganese which were added as essential aids to the production of sound material in such processes. The alloy steels, up to the end of the period under review were only produced in small quantities and mainly by the crucible process; they will not, therefore, be discussed in detail here.[5]

The phenomenon of hardening on quenching from a bright red heat (generally a temperature of 900–1000°C) is almost unique in metallurgy, most metals tending to be rendered softer, rather than harder, on comparable treatment. The extreme hardness of steel after quenching can be reduced progressively by reheating it to temperatures from 150–650°C, this procedure being known as tempering. In general, the combination of quenching combined with a suitable tempering can provide a wide range of properties. This arises since the increase in hardness – or tensile strength, since the two are complementary – is accompanied by a reduction in both toughness and ductility, as illustrated in Fig. 1, which gives typical values for a sample of steel with around 0.5% carbon, the hardest value being obtained by quenching from 920°C and then tempering at 150°C; raising the tempering temperature gradually reduces the hardness but at the same time restores toughness and ductility.

The hardness or strength can also be modified by decreasing the severity of the quenching medium. Water is the most effective medium; substitution of oil for water reduces the intensity of the quench and thus the ultimate level of hardness obtained. A still less violent quench may be obtained by cooling the heated steel in an air blast, a process usually known as normalising, whilst cooling the steel slowly in a furnace from, say, 900 to 600°C, is known as annealing and produces the softest state for the steel,

[5] A general survey of the rise of alloy steel manufacture is to be found in Chapter 4 of "Steelmaking Before Bessemer", Vol 2. It finds its place there as a later adaptation of the Huntsman Crucible process to meet the needs of the latter part of the nineteenth century. It could be considered as equally relevant to the present study and it is, therefore, recommended to the interested reader.

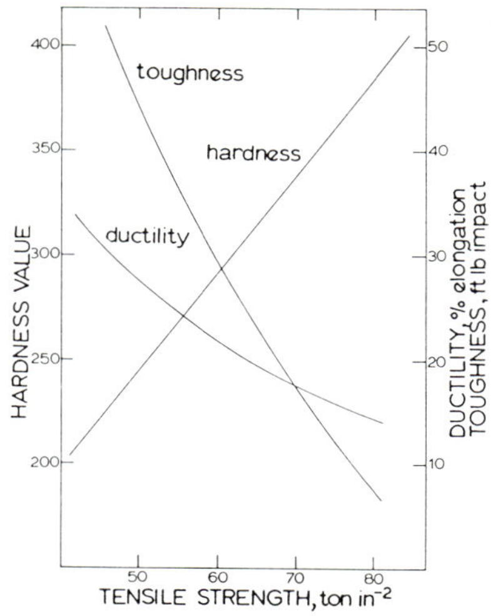

Fig 1. Steel: Relationship between tensile strength, ductility, hardness and toughness. The relationship between hardness and tensile strength remains reasonably constant for any material. The relationship shown between ductility, toughness and tensile strength is typical for a medium carbon steel in small bar size in the hardened and tempered condition.

in which the level of hardness is little higher than that of iron, despite the amount of carbon present. A general idea of the relative effectiveness of these various heat treatments may be obtained from Fig. 2, which also shows the effect of the amount of carbon present on the level of hardness obtained; it should be appreciated that these curves are only indicative of general trends since, in practice, a scatterband of properties will be registered.

This curve shows that there is no sharp cut dividing line between "iron" which does not quench-harden, and "steel" which does. It was only as a consequence of the clear-cut dividing line between carbon contents of wrought iron (generally below 0.10% carbon) and steel (generally produced by the diffusion of carbon into the bar iron and rarely containing less than 0.5% carbon) that such a distinction existed prior to 1850; the coming of mild steel from the Bessemer and Open Hearth processes within the next twenty years rendered the old definition inadequate and, indeed, there was much philosophical discussion around 1875 as to the manner in which steel might be defined.[6]

[6] There was so much controversy concerning this issue that an International Committee was set up in 1875 to consider the subject and many distinguished metallurgists deliberated the matter , including I. Lowthian Bell of England, Hermann Wedding of Germany, Peter Ritter von Tunner from Leoben in Austria, Richard Akerman of Sweden, L. Gruner of France and A. L. Holley of the U.S.A. Many schemes of nomenclature were published but there was no final agreement. See, for example, L. Gruner: "Note sur les Sens des Mots FER et ACIER", Revue Universelle, Vol 40 (1876), pp 437-440: H.M. Howe: "The Nomenclature of Iron", Trans. Amer. Inst. Mining Engineers , Vol. 5 (1876-77), pp 515-528; A. L. Holley: "What is Steel?" Iron, Vol. 7, pp. 102-103 (22.1.1876) and pp. 165-166 (5.2.1876): R. Akerman (trans. from Jernkontorets Annaler): "The Proper Definition of Steel", Iron, Vol. 7, p. 738 (10.6. 1876).

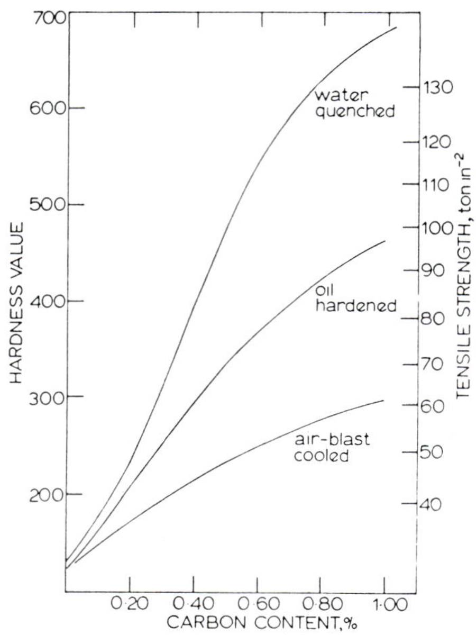

Fig. 2. Relationship between hardness and carbon content for small section bars in carbon steel in various hardened conditions. The figures are typical of the values likely to be obtained by heating small bar samples of steel to a temperature of between 900 and 1000°C and subsequently cooling to room temperature by the treatment indicated.

Pure iron, as stated earlier, melts at the relatively high temperature of 1535°C. Carbon steels, however, show two differences from iron in their behaviour at high temperatures. In the first place, they do not melt at a specific temperature but pass from solid to liquid and vice versa over a temperature range. This situation is demonstrated in Fig. 3; it is to be appreciated that this diagram refers to pure iron-carbon alloys and thus gives temperatures which are somewhat higher than those applicable to commercial alloys, since these generally contain small amounts of other impurities which themselves have the effect of depressing the melting points still further. It will be noted from this diagram that there is one further material, apart from pure iron, which melts and freezes at a single temperature, this being the alloy with 4.3% carbon. The cast irons normally contain from 3% to 4% carbon; such alloys all start to melt at 1130°C, as indeed does any alloy with over 1.7% carbon, but are not completely molten until some higher temperature is reached. For example, a pure alloy with 3.3% carbon is completely molten at 1260°C and exists as a mushy mixture of liquid and crystals between 1130°C and 1260°C. All materials with less than 1.7% carbon – this group covering wrought iron, mild steel and the other carbon

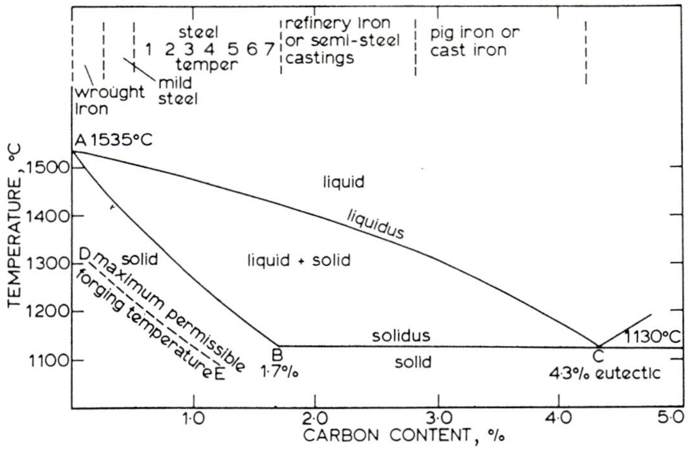

Fig. 3. Effect of carbon content on melting and freezing point of iron. Pure iron melts and freezes at 1535°C. An alloy of iron with 4.3% carbon melts and freezes at 1130°C; this is a "eutectic" alloy. All other iron-carbon alloys melt and freeze over a temperature range. This is determined by the line AC, which is known as the **liquidus,** and the line AB continued along BC, which is known as the **solidus.** On heating, when the solidus line is reached, partial melting begins, continuing as the temperature rises, and culminating in the alloy becoming completely liquid when the liquidus line is reached. The reverse occurs on cooling. The maximum forging temperature lies below the solidus line, typically as indicated by the line DE. Alloys with more than 1.7% carbon are virtually unforgeable. Also indicated are the classes of material defined by carbon content. The steel tempers refer to the old crucible steel production; the range between wrought iron and these steel tempers, here indicated as mild steel, covers a group of materials which became of increasing importance after the introduction of the bulk steelmaking processes.

steels – begin to melt along the line AB and are completely molten at the appropriate point shown by AC, the former line being termed the **solidus**. and the latter the **liquidus**. A steel with 1.5% carbon, suitable for the production of razors, will start to melt at 1160°C and will be completely molten at 1430°C; conversely, on solidifying from a liquid melt, crystals will begin to form at 1430°C but the mass will not be completely solid until the temperature has fallen to 1160°C. A pure "mild steel" with 0.25% carbon, however, would start to solidify at 1510°C and be completely solid at 1430°C.

There is, however, a further line to be considered: that which indicates the maximum temperature at which forging may be carried out successfully without rupturing the material. This is related to the solidus and is typically indicated by the line DE, lying some 100° to 150°C below it. This obviously implies that the maximum forging temperature falls with rising carbon content. On the other hand, the resistance to deformation (or the hot strength) increases as the temperature falls, as indicated in Fig. 4. As with properties at room temperature, an increase in strength at high temperature is accompanied by an increased tendency to loss in ductility and thus to difficulty in forging without rupture. It should be clear, therefore, that there is a maximum carbon content above which the material becomes unforgeable; in practical terms this must be considered to have been reached at around 1.5% to 1.7% carbon. Conversely, the lowest carbon material – wrought iron or, later, the mildest of mild steels – is eminently forgeable. At the highest temperatures, certainly up to 1350°C, pieces of wrought iron could be forge-welded together under a hammer; thus it was that such temperatures were referred to as welding heat. Again, at much lower temperatures, referred to as dull red heat, wrought iron was still ductile under the hammer.

Chemistry of Steelmaking

The chemistry of steelmaking becomes much more important after 1850 with the introduction of the bulk steelmaking processes.

In the older Continental processes for the production of **natural steel** from cast iron, the blowing of air into the molten metal on a charcoal hearth, with a certain amount of slag present, brought about removal of manganese, silicon and then carbon from the metal. The current of air had two functions. First of all it would have some oxidising effect on the molten iron, producing a certain amount of what was then termed "protoxide of iron" – ferrous oxide (FeO):

$$2Fe + O_2 = 2FeO$$

Secondly, and probably just as importantly, it would stir the mixture of metal and slag together, maintaining freshly changing contact surfaces between the two. The iron oxide would augment the quantity already present in the slag until it became sufficiently oxidising to react with, first, the manganese, then the silicon:

$$Mn + FeO = MnO + Fe.$$
$$Si + 2FeO = SiO_2 + 2Fe;$$

These metallic oxides would be absorbed into the slag. The silicon content of the cast iron was generally quite low and the manganese content relatively high; this produced a slag which was low in silica and high in manganese oxide. The silica, by and large, would be absorbed by excess of the iron oxide and manganese oxide present, producing silicates in the slag:

$$2FeO + SiO_2 = Fe_2SiO_4;$$
$$2MnO + SiO_2 = Mn_2SiO_4$$

The remaining iron and manganese oxides would remain available for the oxidation of the carbon:

$$C + FeO = CO + Fe,$$
$$C + MnO = CO + Mn.$$

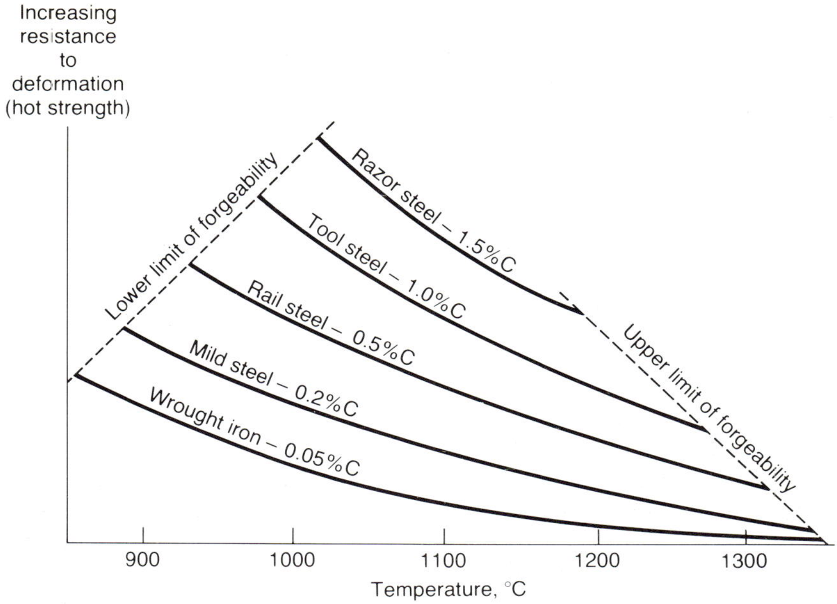

Fig. 4. Typical hot strength comparison for various types of steel at forging temperatures. The hot strength (or resistance to deformation) of steel falls as the temperature rises. The level of carbon in the steel has an effect on the hot strength, the higher the carbon content the higher the hot strength at any particular temperature. As indicated in Fig. 3, the addition of carbon lowers the melting point and, in particular, the solidus line; this lowers the maximum temperature at which the material may be successfully forged. In addition, as the strength rises, brittleness tends to increase, as indicated in Fig. l; the result is the raising of the temperature at which forging has to be discontinued. The combination of these two factors leads to a more restricted range over which the material may be forged or rolled as the carbon content increases.

The carbon monoxide would push up through the slag in bubbles which would then burn on top of the bath with a blue flame:

$$2CO + O_2 = 2CO_2.$$

The high manganese content of the slag was important. In the first place, replacement of FeO by MnO in the slag reduced the oxidation potential of the slag as far as the removal of the carbon was concerned; this meant that there was a better control of the process as far as stopping the operation when the desired carbon content had been reached. Equally importantly, however, it reduced the melting point of the slag somewhat, so that at the lower forging temperature necessary for steel as compared with wrought iron the slag was still fluid and could be expelled more effectively by the hammering operation.

This overall sequence of events was still valid for the later production of **puddled steel**. Many of the cast irons available, however, were not high in manganese and it was found of benefit to add what was known as "Schafhautl's Mixture" – a mixture of manganese ore (braunite or brownstone) with common salt – to the slag in the steel puddling furnace to raise the manganese oxide content.[7]

A further point to be raised in connection with the puddling process, however, concerns the removal of phosphorus. Phosphorus present in the cast iron will tend to be oxidised:

$$5FeO + 2P = P_2O_5 + 5Fe$$

Provided the phosphorus pentoxide can be held in combination with a suitable base, it can be prevented from reverting to the elementary state. Whether or not this happens depends on the composition of the slag present. If the amount of silica present is sufficiently low for it all to be held as a silicate, leaving a significant excess of iron and manganese oxides, the nett effect is a reduction in phosphorus content in the final product:

$$3FeO + P_2O_5 = Fe_3P_2O_8$$

In general, the very rich oxide slag which was obtained in the puddling process ensured that there was a significant removal of phosphorus from the metal. Thus it was, as will be noted later in its proper context, that the incorporation of Staffordshire puddling cinder as flux in the Cumberland blast furnaces contaminated the metal with phosphorus; operating on the good haematite ores of the region it should have been almost phosphorus-free. It should be made quite clear that the puddling operation carried out on high silicon cast irons could provide a slag which was not favourable to phosphorus removal; thus it was that some operators installed a refinery hearth, in which the pig iron was melted and most of the silicon oxidised out, prior to casting slabs which were then charged to the puddling furnaces.

Both these processes were operated in furnaces working at a temperature of around 1350°C. At this temperature the cast iron charge would melt quite readily; during the working, however, as the carbon content was reduced, the metal would begin to solidify and eventually a semi-solid mass of metal particles together with entrained slag could pulled together in the hearth and eventually taken from the furnace for forging and consolidating.

The **Bessemer Process** utilised much the same chemistry but produced a liquid steel. The blowing of air through the molten metal produced iron oxide in a very similar manner, building up gradually a volume of oxidising slag. The next stage of the process was the oxidation of the manganese and then the silicon by reaction with iron oxide in the slag. The iron oxide was constantly being renewed by further combustion and equally constantly being reduced in oxidising the manganese and silicon. These reactions were

[7] C. Schafhautl : British Patent 6837, 13 May 1835

exothermic, releasing a quantity of heat sufficient to raise the temperature of the metal quite significantly; in this way, it was kept molten, in spite of the fact that it had probably been charged at 1300°C and that it would need to be around 1600°C at the end of the process to enable it to be cast successfully into ingots. The commencement of the oxidation of the carbon was accompanied by the increase of flame, due to the combustion of the carbon monoxide to carbon dioxide by the excess air. When the flame died down, the carbon had virtually been removed. It was at this stage that the operations could go wrong. Too long a blow could produce further oxide of iron which, in the absence of silicon, manganese and carbon, would begin to precipitate within the metal itself. Such overoxidised metal, when solid, would prove to be "rotten" – it would break up on forging, showing intergranular cracks. The remedy was to add to the metal some material which would react with the iron oxide present; the early solution was to add spiegeleisen (a cast iron containing from 5 to 15% of manganese, referred to by Mushet as "triple compound" – of iron, carbon and manganese). This was precisely the same blast furnace product as the Continental workers had been refining for the making of their "natural steel" for centuries. The reactions which occurred on the addition of the spiegeleisen have all been given above – reactions of iron oxide with the three elements. With sufficient additions, the overoxidation of the metal was corrected and it then behaved perfectly well under the hammer. The lining of the Bessemer converter, however, had its part to play. Originally this was lined with "ganister", a refractory rock occurring in the Sheffield neighbourhood, which is virtually pure silica; the lining, therefore, had an **acid** character. This was compatible with the slag produced in the Bessemer converter, which contained a relatively high proportion of silica and in this respect differed from the puddling slag. Any oxidation of the phosphorus in the metal charge in the original Bessemer converter was nullified by there being no free metal oxide with which it could combine. The original Bessemer process, with its acid lining, therefore, was incapable of removing phosphorus and its use, perforce, had to be limited to the refining of cast irons low in phosphorus.

The **Open Hearth Process**, developed by Siemens, had the same function as the Bessemer process but achieved it by different means. The cast iron, usually mixed with a quantity of suitable scrap or puddled blooms, was melted on the hearth of the furnace by the burning of producer gas and air within the chamber above the hearth. To the slag which formed over the molten metal were gradually added charges of iron ore. In this way the iron oxide content of the slag was raised and eventually the manganese and the silicon in the charge would be removed by the now familiar reactions with iron oxide. Then would commence the oxidation of the carbon, gradually at first and then gathering momentum until the evolution of carbon monoxide became so violent that the contents of the furnace were apparently "boiling". At a suitable stage, bearing in mind that the end in view was to stop the boil with sufficient carbon still left in the metal, the additions of iron ore would be replaced by limestone. In the slag, the limestone would decompose and the calcium oxide would release iron oxide from the iron silicate present:

$$CaCO_3 = CaO + CO_2;$$
$$Fe_2SiO_4 + 2CaO = Ca_2\ SiO_4 + 2FeO$$

This had the eventual effect of reducing the iron oxide content of the slag and thus its oxidising potential, thus slowing down the removal of carbon. When examination of samples of metal and slag from the bath indicated to the experienced eye that the correct carbon level had been arrived at, an addition of spiegeleisen to the bath, followed by tapping the metal into a ladle would provide sound ingots. Again, the original Open Hearth process was operated in a silica lined furnace, later to be referred to as the **Acid Open Hearth Process**; again, the raw materials had to be selected as being low in phosphorus. There were, however, interesting and significant differences between the Bessemer and the Open Hearth processes. In the first place, the reactions were much slower in the Open Hearth and, therefore, were much more under control; in the second place, the yield of metal from the Bessemer converter was lower than the amount of cast iron charged, since the impurities were removed by combustion to their oxides. In the Open Hearth process, however, the chemically equivalent amount of iron was produced by the removal of carbon, silicon and manganese and this gave a nett gain in weight at the end of the operation, the extra iron, of course, being derived from the iron ore added. This ore, obviously, had to be high grade haematite.

The phosphorus problem took some twenty years to solve. Meanwhile, Britain maintained an advantage over its Continental rivals due to the extensive deposits of high grade haematite ore in Furness and Cumberland. Work on the continent showed that serviceable steel could, in fact, be produced with phosphorus contents of up to 0.25%, provided the carbon content was not higher than 0.18% and the manganese content was of the order of 0.6–1.0%.[8] The real answer, however, came with the work of Thomas leading to the introduction of **basic steelmaking**. This, in the first instance, was applied to the Bessemer process by changing the lining of the converter to one of burnt dolomite (a mixture of lime and magnesia) and by modifying the procedure to include the addition of lime as a slag making

[8] A.L. Holley: "The Effect of Manganese, Carbon, Phosphorus and Other Ingredients on Iron and Steel", Private Publication, 2nd Supplement to No. 2, New York, 20.1.1876.

constituent, together with an "afterblow", that is, continuing the blowing of air for a sufficient period after the carbon flame dropped to oxidise the phosphorus present. In this manner was the **Basic Bessemer process** introduced. Under these operating conditions, the reversal of the phosphorus to the metal, as was usual in the acid Bessemer process, was prevented. The oxidation of the phosphorus to the pentoxide:

$$2P + 5FeO = P_2O_5 + 5Fe$$

took place in a "basic" or limey environment and the lime reacted with the oxidised phosphorus:

$$3CaO + P_2O_5 = Ca_3P_2O_8$$

thus being firmly held and removed from the metal. The bulk of the slag had to be run off before any finishing additions were made to prevent reaction of carbon or silicon with the calcium phosphate, since this would return some phosphorus to the metal:

$$Ca_3P_2O_8 + 5C = 2P + 5CO + 3CaO;$$
$$2Ca_3P_2O_8 + 5Si + 4CaO = 4P + 5Ca_2SiO_4.$$

The excess oxygen in the overblown metal had then to be removed by the addition of spiegeleisen or some such material, as noted for the acid process; here, however, the deliberate overblow implied that extra additions were necessary.

The application of the new technique to the Siemens furnace followed shortly afterwards and thus arose the **Basic Open Hearth process**, which was to provide the bulk of the world's steel in the first half of the twentieth century. The technique had to be slightly different from that in the Acid Open Hearth since the slag control by adding limestone was obviously no longer available. The technique developed was to "block" the boil by addition of spiegeleisen or some such alloy in sufficient quantity and then to tap the metal into the ladle as quickly as possible. Any final adjustments to composition would be made by additions to the metal stream on tapping, before the slag came over. These would be of spiegeleisen or ferrosilicon and ferromanganese and, if necessary, carbon as graphite or, more usually, as anthracite. The heat available in the molten metal was sufficient to melt in these additions and the swirl of metal sufficient to mix them adequately.

As a postscript, it should be made quite clear that the metallurgy of steel and the chemistry of steelmaking are far more complex than has been indicated here; an appreciation of the essential features from this simplified treatment will, however, be of assistance in the further reading of this study.

2 The Development of New Steelmaking Processes

By 1850, steel production in this country, which at that time led the world, was certainly not more than 50,000 tons per annum, with the complementary production of wrought iron running at rather more than 2 million tons per annum. The price differential was wide, steel selling at around £50–£60 per ton whilst wrought iron was available at £10–£15 per ton. Steel, therefore was used only where necessary, for cutlery, razors, metal cutting tools and files, surgical instruments and springs. For general engineering applications and, in particular, for the growing market in connection with the railway boom – for rails, wheels and tyres, for buffers and locomotive frames and boilers – there was an urgent need for a stronger material than wrought iron for which some modest increase in price would be acceptable. The crucible process could not be expected to meet this need, either from the standpoint of economics or production capacity. In this situation, the search for substitute steelmaking processes was intensified, evidenced from a survey of the various patent applications over the quarter century from 1850 to 1875. Many of the proposed methods, with hindsight, may be seen to have had little hope of success; others which appeared to have had a bright future were doomed to oblivion; some filled a temporary need and were then superseded. Even the successful ones had a long period of gestation. The story of Bessemer and Siemens will be expanded later, but it is worth commenting here that it took six years for Bessemer to put the process he described at Cheltenham in 1856 into a fully viable commercial form; similarly, the Siemens Open Hearth Process, pioneered in France in 1864 and patented in Britain in 1867, was only producing some 70,000 tons of steel per annum in 1873, probably just over half as much as the small scale crucible process was then producing in Sheffield. At this date, the Bessemer process was responsible for about 500,000 tons of British steel output, but this was at a time when wrought iron production was running at nearly six times this figure.

Methods involving Direct Reduction of Iron Ore

A number of attempts were made over the years for the direct production of steel from iron ore. Paradoxically, probably the most successful of these was the ancient bloomery, which was perpetuated in the form of the

Catalan Hearth (also known as the Corsican Hearth).[1] This consisted of a furnace structure with a vertical back wall, through which a tuyere pointed downwards into a hearth which had a sloping front wall, the depth of the hearth being from 16 to 24 in., the distance from front wall to back wall from 20 to 32 in., with a length varying from 20 to 40 in. The corners of the hearth were well rounded, giving in general an elliptical plan. The furnace could be used at will for the production of either soft iron or steel; the only difference in the furnace was that the tuyere angle was some 30–40° to the vertical for the production of iron and only about 10–15° for steel production. The type of ore favoured appears to have been the brown haematite or spathic ore, rich in manganese, roasted and broken into small pieces. A sieving of the ore gave a coarse powder and the more lumpy material.

In order to charge the furnace, the bottom of the hearth was covered by a "brasque" of charcoal dust, well rammed down. On the side around the tuyere, where some charcoal would still be burning from the previous heat, a further supply of charcoal was piled up. On the opposite side, known as the "contrevent", a wall was made with the ore, piling it up so that it covered the tap hole. The ore was then covered with charcoal, on top of which a mixture of charcoal with the coarsely powdered ore and some wetted slag from previous operations was beaten down. When charged, the blast was gently applied and slowly increased, so that the full power took about an hour to achieve. As the charcoal was burned around the tuyere, fresh supplies were added so that there was always a support for the opposing bank of ore; in this way the ore was reduced and ready for fusion. Working for iron, the slag was run out after about two hours; working for steel, the slag was retained longer, so that the reduced iron could absorb carbon and sink below the slag, and in this manner be protected from decarburisation by the air blast. Eventually, the remaining ore and charcoal would be brought together for final fusion, which resulted in the formation of an agglomeration of particles of solid iron or steel below the slag, which was worked together with a crowbar to provide rough balls of metal. These were removed with tongs and hammered into blooms, each of which was then cut in two pieces, one being gently reheated whilst the other was forged into a bar, followed by the second. A typical smelting operation would last about six hours; the weight of product per smelt obviously varied according to the size of the furnace and was generally between 100 and 300 lb. The consumption of charcoal was approximately equal in weight to that of the ore; approximately 300 lb. of rich ore, containing 55–60% iron, needed 300 lb. of charcoal to produce 100 lb. of steel; the recovery of the iron from the ore

[1] Further details on the use of the Catalan Hearth may be found in H. C. Landrin: "A Treatise on Steel" (trans. A. A. Fesquet), Philadelphia, 1868, pp. 142–148 and in J. Percy: "Metallurgy: Iron and Steel", London, 1864, pp. 278-285.

was of the order of only 60% and the very fluid slag contained around 30% iron. The product of the Catalan Forge, particularly in the Pyrenees region, was reputed to be of very high quality. Clearly, productivity was restricted.

Mushet[2] had proposed a direct production of steel from iron ore, using the crucible process, as early as 1800:

> . . . when I meet with or procure iron stones or iron ores sufficiently rich and free from foreign mixtures I save the time and expense necessary for the conversion of such iron stone, first into cast and pig iron and afterwards, by a tedious and expensive process, accompanied by a great waste of material, into bar iron. . . . Such iron stone, being previously roasted or torrified . . . may be substituted [in the crucible] for the bar iron . . . and the result will be cast steel, if a proper quantity of charcoal, charcoal dust . . . or any substance or things containing carbonaceous matter has been used.

This use of the crucible furnace has its counterpart in the proposals set out by Samuel Lucas in 1854[3] and by Newton in 1856.[4] In both these, ore was placed with charcoal in the chests of cementation furnaces, normally used for the carburisation of Swedish bar iron into blister steel, the latter being melted in crucibles and the metal then teemed to give steel ingots. Lucas placed alternate layers of iron bars and iron ore, separated from each other by a layer of charcoal containing some ore of manganese, in the chests of the cementation furnaces. After operating the furnace in the usual manner, the pieces of ore could either be forged as steel or, preferably, melted down in crucibles to produce cast steel. Newton took the ore and roasted and crushed it, subjected it to a magnetic separation for removal of impurities, then placed the magnetic portion in layers, separated by charcoal, in the cementation chests. After firing to a white heat for a suitable period and allowing to cool, the charge was crushed and again magnetically separated, the magnetic portion then being melted in crucibles and cast into ingots.

It seems that some unspecified direct process was also operated on a limited scale by Riley at the Dowlais works, since Dr. Percy reported that "excellent steel for chisels, etc., of which I have seen specimens, was occasionally obtained, but it was not found possible to ensure uniform results".[5]

Direct reduction of iron ore to provide an iron powder , which could then be utilised for making steel, short circuiting the blast furnace and the iron refinery, appeared to have its attractions in the middle years of the nineteenth century. Most prominent among the workers in this field were the members of the Chenot family. The earliest patent was taken out in

2 D. Mushet, British Patent 2447, 13 November 1800.

3 S. Lucas, British Patent 1730, 7 August 1854.

4 W. E. Newton, British Patent 851, 8 April 1856.

5 Percy, op. cit., p. 765.

1846,[6] with a further six between 1854 and 1856.[7] A picture of the process may be deduced from a study of these patents.

In the first instance, "suitable ores" had to be obtained; the preferred varieties were the spathic carbonate ores, relatively rich in manganese. The ores were then roasted and crushed, followed by separation of the earthy matter from the "ferriferous" substances. Originally carried out by gravity methods, magnetic separators were eventually specified. These were highly developed units for their time, employing either permanent magnets or electromagnets; Chenot preferred the latter as they were more powerful and gave a cleaner separation. Subsequently, if the mineral was powdery, it was compressed into small blocks with the appropriate amount of charcoal and 3% resin as binder; otherwise, the small size ore was simply mixed with charcoal. The ratio of charcoal was carefully calculated; for a typical mineral which at this stage contained 73% iron, the charcoal requirement was 1.4 times the volume of the mineral or 270 kg of charcoal to 1000 kg of enriched ore.

At this stage it was charged to the retort. A typical form had a height of 40 feet above ground level; the central firebrick retort was surrounded by a gallery of flues down each side, the internal part of the retort being 20 in. wide and 60 to 80 in. long. The exhaust gases could be utilised either for the roasting of further ore or in waste heat boilers. Below the retort was a cooling chamber, since it was necessary to cool the sponge before allowing it to come into contact with the air to prevent its spontaneous reoxidation. This was a double walled water cooled box, running the whole length of the retort. Its base was closed by a series of iron stopper rods pushed through holes in the supporting frame. A bogie ran below, with a movable base, and the charge was eventually allowed to fall out by removing a number of the push rods. It is not made clear how the process was first started up, but in its normal semi-continuous process, the new charge would be put in on top of the previous charge which, now reduced, would have shrunk somewhat. The top sealed in this manner, the reduced charge would be let into the cooling chamber and the fires relit to reduce the new charge. Three days later, they would again be allowed to go out; the previous charge, now fully cooled, would be let out through the bottom and the newly reduced charge let into the cooling chamber and a new charge put in at the top. Three days later the process would be repeated in all its stages. In six days a single furnace could produce 1,100 kg of sponge from 1,500 kg of ore and 400 kg of charcoal, with a similar yield every subsequent three days until the campaign was stopped. Some 1,300 kg of coal was required for the above quantities for external heating. The sponge was then broken up and again magnetically separated, this time to remove the excess charcoal, which could be reused.

The sponge was then available for use in various ways. It could be melted

[6] A. Chenot, Brltish Patent 11515, 31 December 1846.

[7] C. A. B. Chenot: British Patents 246, 1 February 1854 and 658, 20 March 1854; A. L. S. Chenot and E. C. A. Chenot: British Patents 1587–1590, 7 July 1856.

down in crucibles with a suitable proportion of cast iron to provide steel; the method preferred by the inventor, however, was to mix the sponge with carbonaceous matter in the right proportions, together with a resin binder and a small quantity of either manganese ore or crushed spiegeleisen, to compress this mixture into small blocks and to melt these direct in crucibles to provide steel. Chenot claimed that steel made in this way was the equal of that made by "the house of Huntsman" and figures are quoted for the costs of forged bars in cast steel[8] which indicate very similar figures to those quoted by le Play for standard Sheffield practice some 17 years earlier.[9] The Sheffield firm of Moss and Gamble tried out the Chenot material in 1864–65, apparently with no real success. Owing to the bulk of the compressed blocks compared with the more usual metal charges, the quantity of liquid steel in the crucible had to be lowered, with adverse effects on the economics. Elsewhere, however, there are declarations that its use led to a superior product.[10]

The Chenot process does not appear to have had the success predicted for it.[11] Nevertheless, there were many efforts made to improve the process. Snelus, a leading British metallurgist, suggested a vertical circular retort.[12] Here the powdered ore was to be fed to a hopper at the top of the furnace, from which it passed in a steady trickle through a heated column of reducing gas, before being collected in a chamber filled with spent gas to prevent reoxidation. Blair, in America, devoted ten years to his attempts, first with a horizontal retort and then with a vertical one, eventually using a lime addition and handling 200 tons of ore a week in a single 36 foot high furnace.[13] The patent literature of 1850–1880 contains many variants of such processes.[14]

What is not usually appreciated, however, is that the bulk of Siemens'

8 E. Grateau: Memoire sur la Fabrication de l'Acier Fondu par le Procédé Chenot", Revue Universelle, Vol 6 (1859), pp. 56-57 and attached table.

9 F. le Play: "Memoire sur la Fabrication de l'Acier en Yorkshire", Annales des Mines, 4me. Serie, Tome III (1843), p. 668.

10 J. S. Jeans: "Steel: Its History, Manufacture, Properties and Uses", London, 1880, p. 41.

11 Grateau (Loc.cit) considered that the methods of Bessemer and Uchatius (and others) had been proved unworkable; on the other hand, he felt that Chenot's process presented the happiest alliance of science and practice, which had broadened the horizon of industry by presenting a reliable means of obtaining common steel of good quality. This, of course, was written during the period in which Bessemer was striving to overcome his initial problems.

12 G. Snelus, British Patent 616, 20 August 1869.

13 H. M. Howe: "The Metallurgy of Steel", 2nd. edition, New York, 1892, pp. 278–280.

14 Among the most interesting are the following: Isaac Rogers proposed a rotating cylindrical furnace along which a mixture of ore and charcoal was driven by a screw, the reduced product passing into a reverberatory hearth for conversion to steel (British Patents 629, 20 March 1855 and 262, 31 January 1861); Frederick Gurlt had a shaft type furnace, producer gas fired, in which the gas composition was controlled so as to be carburising (British Patent 1679, 16 July 1856) and it is interesting to note that this process was in operation in Bilbao as late as 1884 (Howe, op. cit., p. 275); a patent taken out by Bonneville on behalf of J. Absterdam somewhat later used a coal fired retort containing a mixture of ore and carbonaceous matter, hydrocarbon oil being introduced during the process into the retort and the gases taken off as 'illuminating gas', the retort subsequently being allowed to go quite cold before taking out the reduced ore (British Patent 186, 26 January 1869).

experiments, at the time when he was developing the Open Hearth furnace, were, in reality, a search for a direct steelmaking process based on the pre-reduction of ore to sponge, prior to converting the sponge to steel. He made two furnace designs to this end, one using a vertical retort above the Open Hearth and the other using a horizontal rotating retort feeding the furnace. The earlier design[15] had a pair of vertical cast iron retorts above the steelmaking furnace, the surrounding space being heated by the flame from the furnace itself. About 28 kg of charcoal were charged through each retort hopper and then the remaining space filled with ore. Producer gas was injected through pipes in the centres of the hoppers and this deoxidised the ore which had been already heated to redness. About half a ton of pig iron had meanwhile been charged to the furnace; on melting, it began to dissolve the bottom end of the column of reduced ore. Regular feeding of ore through the top of the hoppers would continue for three or four hours; feeding would then cease and a clay-coated cast iron plate, suspended on wires, would be placed on top of the remaining charge, sinking with the charge and eventually sealing the mouth of the retort, allowing the next charge to begin to be added. Meanwhile the contents of the furnace would be brought to the desired composition, spiegeleisen added and the metal tapped. A somewhat harsh comment on this concept came from an American metallurgist quarter of a century later:[16]

> Today we wonder that a man of Siemens' genius and judgment could have seriously entertained so crude a project even twenty-one years ago. To maintain these hoppers, exposed thus in an Open Hearth furnace, to heat these thick bodies of ore through and to deoxidise them at the necessarily low temperature in any reasonable time, to keep the Open Hearth furnace waiting while the charge of ore was descending – well, well! Today's folly is wiser than yesterday's wisdom.

In 1867, it has to be remembered, bulk steelmaking was still in its infancy. By the time Howe made this judgement, the basic steelmaking processes had led to a phenomenal growth in world output and the Open Hearth furnace, by then rapidly overhauling the Bessemer process as the foremost means of steel production, was surely sufficient tribute to Siemens' genius, leaving aside the use of his regenerative furnace in crucible steelmaking, in the glass industry and elsewhere. Moreover, Siemens had a firm grasp on the principles involved, even if his engineering was at times over ambitious. His second version of a direct steelmaking process[17] is a most elegant conception and, since it predated the currently accepted process by almost 100 years, is worthy of comment. It consisted of a rotary reducing furnace

[15] C. W. Siemens, British Patent 2395, 1867 (final specification 18 February 1868).
[16] Howe, op. cit., p. 283.
[17] C. W. Siemens, British Patent 1892, 10 June 1868.

feeding an Open Hearth furnace. That it worked, there seems to be no doubt. In the event, however, improvements in blast furnace economy and the growing availability of steel and iron scrap combined to make the direct process an unnecessary complication and the normal Open Hearth process took over.[18]

The direct steelmaking processes and the production of a reduced iron product from iron ore, other than by the conventional blast furnace route, had, therefore, no great impact on steelmaking during this period; that they were ever conceived, however, is indicative both of the growing appreciation of steelmaking technology and of the increasing need for steel.

Methods involving the Melting of Cast and Wrought iron

The traditional Huntsman crucible process consisted of taking blister steel bars from the cementation furnace, breaking them into short lengths, charging them to a crucible, melting them down and finally pouring the liquid steel into an ingot mould.[19] This was evolved as a viable steelmaking process before there was any clear idea of the role of carbon in differentiating between iron and steel; this matter was only elucidated in the period just prior to the French Revolution. Almost immediately, the need for carburising the bar iron in the cementation furnace prior to melting was queried, actually in the same Mushet patent of 1800 already quoted.[2] The new proposal was to charge bar iron and charcoal to the crucible, thus making economies in production cost. The proposal foundered, however, since the bar iron had to be molten before the carbon would be absorbed and this involved running the furnace at a higher temperature, since the carbon present in the blister steel lowered the melting point significantly. This led to an unacceptable degree of crucible erosion, with consequent loss of metal, and also considerable furnace damage, causing the procedure to be abandoned; it was, however,

[18] The works set up by Charles Tennant at Hallside near Glasgow were originally intended to utilise "Blue Billy". Iron pyrites was initially roasted to release the sulphur dioxide, to be converted subsequently to sulphuric acid in Tennant's chemical works. The residue was further treated to recover the copper which it contained. The balance was the Blue Billy, rich in iron oxide, and huge mounds of it had accumulated. William Siemens examined the material and recommended his direct reduction process. Furnaces were installed, in accordance with his second patent, and the process set in action. In addition, however, it was found that the furnaces worked very well on the pig and ore process, using the Blue Billy in place of ore. Eventually it was found more convenient and more economical to install conventional Open Hearth furnaces; the plant set up in this way by 1873 was to produce Open Hearth steel for the next hundred years.

[19] An extended discussion of the development of the Huntsman process may be found in the author's prior publication "Steelmaking Before Bessemer", Vol 2, subtitled "Crucible Steel the Growth of Technology", London, 1984.

taken up again at a later date,[20] when the Siemens gas-fired furnace became available in the late 1860s. The required furnace temperatures could then be achieved and held without any problem; moreover, the use of crucibles in which a proportion of graphite was incorporated with the clay (the so-called "plumbago" crucibles) provided an answer to the erosion problem. So it was that this procedure became a standard operation in the River Don Works of Naylor-Vickers, the largest crucible melting plant in the world in 1867, built without any provision for the making of blister steel as raw material.

The proposal put forward by William Vickers in 1839,[21] involving the melting of bar iron with cast iron, was much more practical under the prevailing conditions. The cast iron would start to melt at a temperature as low as 1150°C; as it became progressively more fluid, it would gradually absorb the bar iron and the whole would not require any excess temperature over that required by blister steel. Obviously, steels of different carbon contents could be produced at will by varying the proportions of bar iron and cast iron. There was still a problem, however. The crucible steelmaker, from the very early days, had relied on Swedish or Russian bar iron as his initial raw material for the making of blister steel; this had been found to be the only way of ensuring quality of product. Essentially, this was due to the extremely low phosphorus content of the Scandinavian and Russian ores; in addition, the smelting of the ores was carried out using charcoal, which eliminated sulphur contamination. The general supplies of cast iron commercially available in this country were from phosphoric ores smelted with coke and their use as part of the crucible charge would inevitably down-grade the steel. The Swedish authorities had imposed strict control on exports, since it was quite clear that the use of available resources to produce the bar iron was economically correct, as this enhanced the selling value of the cast iron by a factor of from three to six, dependent on the forge location and the type of ore used. With increasing demand, however, charcoal resources began to prove insufficient and, in 1854, the export of Swedish cast iron, low in both sulphur and phosphorus, was permitted,[22] allowing the Vickers proposals to become viable. Whilst the traditional method of melting selected blister steel bar remained almost obligatory for the provision of steel for fine tools, razors, taps and dies, turning tools, gun cylinders, pen nib sheet and watch springs – some would say as a matter of prejudice – there is no doubt that the melting together of Swedish cast iron with Swedish bar iron soon grew into a perfectly accceptable method

20 Report on the River Don Steelworks, Engineering, 25 October 1867, pp. 383-385.
21 W. Vickers, British Patent 8129, 26 August 1839.
22 Riksdags Bilaga, 19 December 1854.

for the production of crucible steel; merchant firms such as Daniel Doncaster provided ready made up mixtures of the two for the smaller Sheffield firms.[23]

There were two further developments of the concept of melting together bar iron and cast iron, both of which owed their inception to William Siemens. In the first instance, he applied the regenerative gas fired furnace developed by his brother Frederick and himself[24] to the melting of steel in crucibles – the so-called **gas fired crucible furnace**. It is not clear when the first furnace of this type was installed; there was a Siemens furnace at the Sheffield works of Marriott and Atkinson as early as 1858 but it is probable that this was a billet reheating furnace and not a melting furnace.[25] The advantages of such a furnace, fired with producer gas, were that there was no difficulty in maintaining the temperature required for the melting of the contents of the crucibles; the chambers could be made sufficiently large to hold from 10 to 30 crucibles, as against the normal two per hole for the coke furnace; in addition, the possibility of continuous round the clock working existed. Siemens described the furnace and its operations in detail in a paper to the Chemical Society in 1868,[26] claiming that fuel costs were greatly reduced and indicating that the furnace could provide 15 to 20 weeks's continuous use between lining repairs. At this date, gas firing was regularly applied to at least one third of the crucible steel output from the River Don Works of Naylor, Vickers and Company.[27] Meanwhile, in America, such furnaces were beginning to be built to serve the relatively young crucible steel output of Pittsburgh. The first gas fired crucible furnaces were at the works of Anderson, Woods and Company and were commissioned in 1868;[28] by 1876 there were twenty four units in the Pittsburgh works, of the standard 24-pot type, against a residual total of 299 coke fired 2-pot holes.[29] At this stage, therefore, about half the Pittsburgh

23 A typical example from the Doncaster Archives, dated 1880, offers mixtures of Swedish Bar Iron together with "Warner's Carbonizing Iron" (this being a Swedish cast iron with about 3.8% carbon), the ratio of bar iron to cast iron being adjusted according to the desired carbon content in the resulting steel (for example 29 lbs to 20 lbs for a steel with 1.5% carbon or 40 lbs to 9 lbs for 0.75% carbon), each mix having a 1 lb addition of Spiegeleisen to bring it to a 50 lbs charge. The price charged was 10s. per cwt. (£10 per ton).

24 F. Siemens, British Patent 2861, 2 December 1861.

25 C. W. Siemens: "On a Regenerative Gas Furnace", Proc. Institute Mechanical Engineers, 1862, p. 42.

26 C. W. Siemens: "On the Regenerative Gas Furnace as applied to the Manufacture of Cast Steel", J. Chemical Society, 1868, pp. 279-310. See also Fig. 20, p. 138.

27 W. Hackney: "The Manufacture of Steel", Proc. Institute Civil Engineers, xlii (1874–5), pp 2-68.

28 W. F. Durfee: "Development of American Industries since Columbus", Popular Science Monthly, xxxix (1892), p. 736.

29 G. H. Thurston: "Pittsburgh and Allegheny in the Centennial Year", Pittsburgh, 1876, p. 196.

crucible steel was made in gas fired furnaces. Tradition, or prejudice, ensured that Sheffield never reached as high a proportion. Nevertheless some of the larger firms with reputation for quality did install such furnaces; Sanderson Brothers put up a new shop with a gas-fired furnace in 1872,[30] Wm. Jessop and Sons installed a modified Siemens furnace, designed by Dawson, Robinson and Pope in 1897[31] and Thomas Firth and Sons built a new tool steel works at Tinsley around a "New Form" Siemens furnace.[32]

From the use of such a furnace, which consisted of a melting chamber with regenerators at each end, on whose hearth numerous crucibles were placed, to the idea of melting steel on the hearth itself and tapping the molten metal in bulk into a ladle for casting, was a relatively simple progression. The first attempt at this seems to have been at Tow Law in County Durham in 1862.[33] Here Charles Attwood was melting cast iron and wrought iron together, in appropriate proportions to give the desired composition of steel. He was operating on the brown Weardale ones, which were low in phosphorus and also high in manganese. His proportions of cast iron to wrought iron could vary between 6% for the mildest steels to 30–40% for hard steels. He also added a certain amount of cullet [broken glass] with his metal so that, when molten, it would be covered with a layer of more or less inert material.

This very early trial, in which the use of the hearth itself was merely an experimental adaptation of a crucible melting furnace, was followed two years later by the pioneering work of Pierre and Emile Martin at Sireuil in France, in a furnace designed for the purpose by Siemens. Originally, they melted cast iron on the open hearth and then melted in puddled or scrap iron, or scrap steel or turnings, the additions preferably preheated. They then ran off half the charge, melted in more cast iron and more scrap or puddled iron. An important feature was the use of a protective slag cover made by adding blast furnace slag and lime or alumina as necessary to fluidise it. In such a manner the process could be kept going continuously for a week or more.[34]

This procedure was soon to be followed by a batch procedure[35] in which a charge of cast iron was melted down in the hearth and blooms of puddled steel,[36] together with scrap from previous heats as available, fed into it at

30 Hackney op. cit., p. 11.

31 A chance survival is a list of wages to be paid for the working of this 30-pot gas melting furnace at the works of William Jessop (Private Collection).

32 This was built in 1907 and a record book for this year has survived (Private Collection).

33 C. Attwood, British Patent 1473, 15 May 1862.

34 R. A. Brooman, on behalf of Pierre and Emile Martin, British Patent 2031, 15 August 1864.

35 R. A. Brooman, on behalf of Pierre and Emile Martin, British Patent 2137, 18 August 1865. Fuller details of the procedures utilised by the Martin Brothers can be found in the later discussion of the development of the Open Hearth methods (See Chapter 7).

36 Please refer below for information on Puddled Steel.

such a rate that they readily dissolved. When the whole charge was melted, a sample was withdrawn to ascertain its "temper", and a small further charge of preheated cast iron added to adjust the "exact grain" of the steel, which was then run out over the lip of the furnace into moulds.

In this way was developed the **Siemens Martin process**, also known as the **pig and scrap process**, a method which was to have considerable application in Europe over the next fifteen years. It will be obvious that it drew on steel puddling procedures as well as on experience of melting in crucibles. It should be made clear that, as originally developed, the addition of iron ore to the furnace (other than a passing notice elsewhere in the patent to the possible use of a small addition as a flux to assist in dealing with any small pieces of metal which may be caught in the "scum") was not an integral feature of the process. This was purely a melting operation, just as the crucible process, whether the charge was of blister steel or of a mixture of cast iron and wrought iron, was purely a melting operation, with the aim of producing a homogeneous melt of the desired carbon content without any deliberate steps being taken to change the chemical composition of the total charge.

The Siemens Martin process, however, is a first significant step towards producing a reasonable quality steel, essentially similar to that which previously could be provided by the use of medium quality materials in the crucible process, but in larger quantities and at a reasonable cost. It was eventually to be superseded by what became known as the **Siemens Open Hearth process**, also known as the **pig and ore process**, in which the carbon content of a molten bath of cast iron would be reduced by additions of iron ore to the slag and the period of supersession saw a gradual blending of the two processes. This leads to some confusion in the interpretaton of the available evidence, as the discussion of the Open Hearth process will clearly show, but this does not in any way detract from the importance of the early work at Sireuil in the history of steelmaking.

Methods involving Partial Decarburisation of Cast iron without the production of Molten Steel

From the time of the introduction of the blast furnace there had been developed several local variants of a similar process, by which the cast iron, melted in a hearth, was oxidised by an air blast; this reduced the carbon to an intermediate level, providing a material with steely characteristics, rather than oxidation to the ultimate production of soft wrought iron. On the European mainland this type of procedure was particularly prevalent and, as has already been shown,[37] some 75% of the continental steel output in 1843 was produced in this way. The products could all be referred to as **natural steel**, generally a somewhat segregated material, containing alternating bands of high and low carbon material, not dissimilar in character to the **shear steel**. This latter was the result of forge-welding together bundles of pieces of blister steel, a process derived in the Newcastle area and thence transmitted to Sheffield in the late eighteenth century; since the blister steel was invariably more highly carburised on the outside, with a softer central material, this procedure also produced a banded structure. This was no disadvantage in the making of knife blades, since the higher carbon, more brittle areas needed to provide the cutting edge, were interspersed with softer, more ductile regions; correctly hardened and tempered, such a blade combined a marked degree of flexibility with exceptionally good cutting characteristics. Shear steel was also known in Sheffield in the early days as **Newcastle steel** and as **German steel**; the former showed its place of origin, the latter its similarity to the Continental materials. The most noted centres of the Continental production were Styria and Carinthia and in Austria, Siegen in Germany and the Dauphiné in France; all these operated procedures which differed only in minor detail, essentially based on the old finery and chafery processes for the conversion of cast iron into wrought iron.

The production of wrought iron had been revolutionised at the end of the eighteenth century by the introduction of the puddling process pioneered by Henry Cort.[38] So far, the fuel used in refining cast iron had been charcoal, since the use of coal gave rise to the phenomenon known as hot shortness, that is, brittleness on forging. This was due to the absorption of sulphur from the coal by the metal. Cort designed a furnace in which the fuel was burned on a grate separated from the working chamber by a bridge, with the heat from the combustion reflected down on to the metal by a reverberatory roof, the sulphur in the fuel already having been oxidised into a harmless form. Cort's process was improved upon in detail but, by and large, provided the means by which Britain became the world leader in iron production; indeed, the process was still operated up to about twenty five years ago.

There are innumerable descriptions of the process and drawings of furnaces in the literature.[39] Fig. 5 shows the furnaces at the Dowlais Works

37 le Play: op. cit., p. 692

38 H. Cort, British Patent 1420, 13 February 1784.

39 The most recent account is that of W. K. V. Gale: "The Black Country Iron Industry", London, 1966, pp. 125-132. Contemporary accounts are those of W. Truran: "The Iron Manufacture of Great Britain", London, 1855, pp. 130-146; B. Baylis "On Puddling - by a Practical Puddler", London, 1866, 56 pp. (Published privately) and the Welsh Trade Union leader, William Lewis, under the name Lewys Afan:"The Tinman's Companion", Pontypridd, 1876, pp. 121–154.

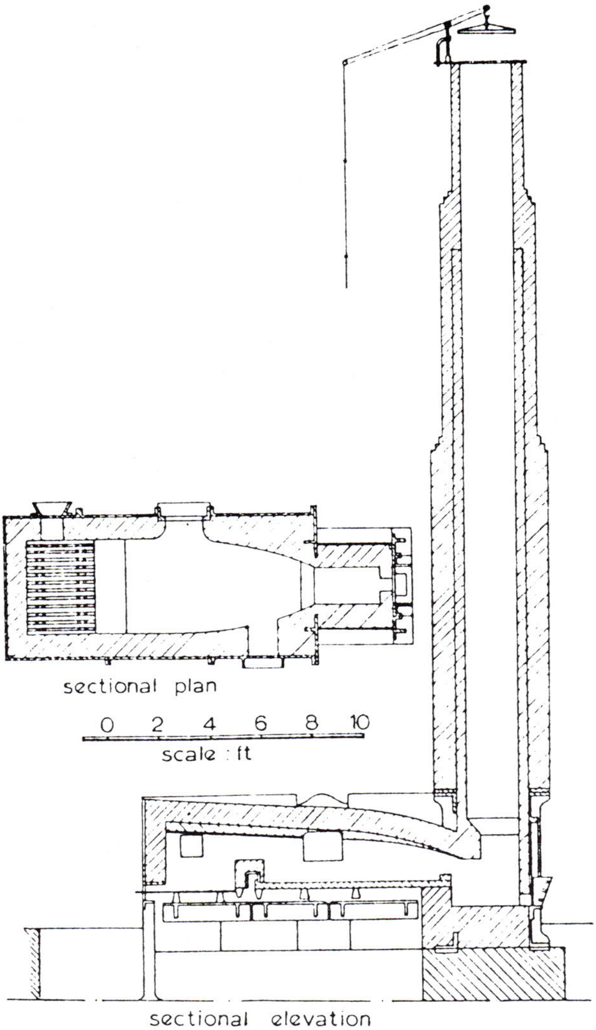

Fig. 5. The Puddling Furnace. This drawing of a typical puddling furnace is taken from W. Truran: "The Iron Manufacture of Great Britain", London, 1855, Plate 9. The essentials of the puddling furnace were a firegrate (left) separated from the working chamber by a bridge, so that the sulphur in the fuel could be oxidised before the flames had any opportunity of coming into contact with the metal in the hearth (to the right of the bridge). The heat reverberated on to the hearth from the furnace roof. An adequate draught was provided by the tall chimney (right), provided with a damper arrangement for control of the draught to enable the temperature to be adjusted, as necessary, during the various stages of the operation. There were over 8,000 puddling furnaces of this type in Great Britain in the 1870s.

in South Wales in 1855. Here can be seen the firegrate, the bridge and the working chamber. The hearth was lined with iron ore mixed with the roasted cinder from previous operations; this mixture was rich in iron oxide and relatively free from siliceous matter. On to the hot hearth would be charged some 300 to 500 lb. of cast iron, which would then be melted down under the action of heat transmitted from the coal fire on the adjoining firegrate. During the melting period a fluid, highly oxidising slag would begin to form and would eventually cover the molten metal. By **rabbling**, or working the metal and slag together with a rabble – a wrought iron bar whose end carried a stubby wrought iron plate welded at right angles[40] – much of the silicon, manganese and phosphorus from the cast iron would be oxidised, but very little carbon would be removed. On continued rabbling, however, bubbles of carbon monoxide would begin to rise through the slag[41] and the reaction would gather momentum, eventually the whole mass seething, with slag running out over the furnace sill. Then, quite suddenly, the reaction would slacken; moving the rabble through the bath would become more difficult, since the metal was becoming spongy rather than liquid. This was due to the removal of carbon, which brought about a rise in melting point; the metal had now '**come to nature**' and could be collected together into **balls** within the still fluid slag. These balls could then be lifted out from the furnace with tongs; they consisted of a loose agglomeration of iron particles with entrained slag. The were taken to the **shingling hammer**, originally water-wheel driven but later steam powered, where the bulk of the slag would be forced out and the metal formed into a solid **bloom**. This bloom would be charged to a furnace for reheating prior to being rolled down into bar, whilst the succeeding balls were taken from the furnace and shingled in turn. This was obviously a process in which physical strength was combined with skill, together with experience in regulating the furnace.

Cort had originally considered that he might use his process to provide steel as well as iron. It should be remembered, however, that he was working at a time when the clear distinction between iron and steel had not yet been satisfactorily elucidated. It slowly became clear that the conditions in the puddling furnace could be modified so that the full decarburisation of the metal was prevented but this took many years and it was, in fact, not until 1850 that a workable

[40] Furnace tools were inevitably made in earlier times from wrought iron as being the readily available blacksmiths material; it had the advantage of having a higher melting point than steel as well as being much cheaper. More recently, the wrought iron has been replaced by the softest "mild steel" for this purpose and for so-called "wrought ironwork" for domestic and other uses.

[41] Where they came to the surface they would burn with a blue flame to carbon dioxide; their appearance led to them being sometimes referred to as "puddler's candles".

process was developed in Germany and subsequently patented in Britain.[42] From this patent and from other later evidence[43] it is clear that the type of cast iron used had to be carefully chosen. It was preferably high in manganese, smelted from spathic or brown haematite type of ore. The melting down of the cast iron had to be carried out as quickly as possible. The slag had to be rich in manganese oxide, since such a slag had a less fiercely oxidising tendency than the corresponding iron rich slag and, moreover, it remained more fluid down to lower temperatures; to achieve this, the addition of manganese oxide together with sea salt was a normal procedure.[44] During the rabbling period, it was necessary to have a smoky flame within the furnace. Finally, when the metal had come to nature, the removal of the balls for shingling had to be done as quickly as possible to avoid further decarburisation.

Among the first to take up the 1850 patent was the Low Moor Company of Bradford. Herr Fehland, the German representative sent to assist them, reported on the initial trials as follows:[45]

> The steel working at Low Moor began on 13th October 1851. The puddling furnaces had preheating hearths which were very small and terrifically hot, so I did not find it necessary to alter anything, in contrast to what had to be done by me at all the works visited earlier or later. As in iron puddling at the time here, 300 lb. of pig iron were put in and, on 16th October, from 6.20 am to 5.30 pm. ten heats were made and a total of 2,796 lb. of steel balls were taken out, which corresponds to a waste of about 6.8%. The raw iron was charged into the preheater as soon as the metal in the furnace had balled up and shown grain. The steel was pressed together like butter and welded extremely well.

William Clay, of the Mersey Iron and Steel Company, also took up the patent and reported his experiences in 1857:[46]

> From the first commencement there has been found no difficulty in heating, forging or rolling this steel into any form or shape, as it has been made into plates, bars, angles, rivets, rails, railway points and forgings of all kinds with

[42] E. Riepe, British Patent 12950, 29 January 1850. This patent was taken out on behalf of Lohage, Bremme and Lehrkind of Unna in Germany; despite this, steel puddling was often referred to in Britain as "Riepe's Process".

[43] W. Crookes and E. Rohrig: "A Practical Treatise on Metallurgy: III. Iron and Steel", London, 1870, pp. 70-81.

[44] C. Schafhautl, British Patent 6837, 13 May 1835; see also details quoted in Crookes and Rohrig (43, above).

[45] H. Fehland: "Geschichtliches uber die Puddelstahl Fabrikation", Stahl und Eisen, 1886, p. 228. (Translation by courtesy of Otto Hirston, Esq.)

[46] W. Clay:"On the Manufacture of Puddled or Wrought Steel, with an Account of Some of the Uses to which it has been Applied", Journal Society of Arts, VI (1857–8), pp. 140-148.

> perfect ease and success and, ever since the manufacture was commenced at the Mersey Iron and Steel Works, this steel has been used for almost anything that was required to be of a strong and durable nature. . . . It is somewhat worthy of remark that, although the process is so novel and, apparently, of so delicate a nature, yet, with the specifications as my only guide, having never before heard of or seen the operation, it succeeded perfectly in the first trial which was made, and produced so excellent a steel that, after working about 100 tons, it has hardly been surpassed. I have used pig iron of all descriptions, North Welsh, South Welsh, Staffordshire and Scotch, with the same result, viz., the production of excellent steel.

He went on to describe its various uses: ordnance, naval chains, boiler plates, girders and bridges were all made satisfactorily out of **puddled steel**. There are no production figures, other than the reference to the original 100 tons, but clearly this was no small scale trial. He concluded his paper as follows:

> I do not for a moment anticipate that the steel manufactured by this process will supplant the best description of steel, but I feel confident that it must come largely into use for the most ordinary purposes, where cast steel, from its great cost, cannot be used. Indeed, if I might indulge somewhat in prophecy, I would express the belief that, in a few years, the manufacture of this wrought steel will have become as important a branch of our national industry as that of iron now is.

This puts the material well in context in the engineering field as a superior type of wrought iron with higher strength and, accordingly, important weight saving characteristics. It could be produced in quantity – there were thousands of puddling furnaces operating – at a reasonable price, somewhat higher than wrought iron, since it took longer with a somewhat lower yield than wrought iron in its making, but much lower in cost than crucible steel. Clay's prophecy was, in part, correct. Eventually the material which displaced wrought iron was **mild steel**, produced by the Bessemer or Siemens processes but, at the time he spoke, this was some years ahead and in the interim it is clear that puddled steel had an important part to play.

It is easier to perceive the development on the Continent. In 1854, the province of Westphalia (the centre of German steelmaking) produced 5,957 tons of natural steel, 3,878 tons of puddled steel and 2,610 tons of cast steel. The comment was that this showed an increase of 1,263 tons of puddled steel over the previous year.[47] That the production of puddled steel continued to be of importance in Germany is attested from a report of operations at the Krupp Works in Essen in 1866[48] from which it is clear that the bulk of

[47] Delvaux de Fenffe: "De la Fabrication de l'Acier Puddlé en Allemagne", Revue Universelle, vol. 1 (1857), p. 59.

[48] C. B. B., letter to The Times, Thursday, 8 September 1866; see also F. C. G. Muller: "Krupp's Steel Works", London, 1898, described as "an authorised translation from the German", pp. 33-35.

the charges to the crucibles used to produce the ingots for all the Krupp ordnance forgings were of puddled steel, carefully sorted into four grades with increasing carbon content according to the appearance of the bar fracture, all made from high manganese cast iron.

The puddled steel process was introduced into the Loire district in France in 1855 by German workers; previously the area had produced steel mainly by cementation and crucible processes. Official figures[49] for the Loire district, in metric quintals, are as follows:

	1853	1854	1855	1856	1857	1858	1859
Acier forgé	Nil	Nil	13,000	21,136	65,595	74,741	72,160
Acier cementé	80,430	86,300	74,331	37,698	42,840	32,468	30,260
Acier fondu	58,545	63,545	48,609	37,252	43,091	39,267	26,290

These figures indicate that, within four years of its introduction, puddled steel had overtaken the total by the established routes. At the same time, the puddled steel from the Loire only cost 80% as much as natural steel from the Isère and only 65% as much as Styrian or Carinthian steel.[50] The relative importance of puddled steel in France is corroborated by the number of papers which appeared in 1859;[51] the most definitive work, however, was a complete treatise published in 1861,[52] giving full working instructions for production of both wrought iron and puddled steel. Later information on French puddled steel production shows that there were 51 steel puddling furnaces (as against 995 iron puddling furnaces) in 1877; there were also five open fire refineries for the production of natural steel. The total output of natural and puddled steels was quoted as 20,373 tons in the year. Ten years later, there were only 35 steel puddling furnace and five open fire refineries; their combined output then was 12,532 tons. These statistics were accompanied by a comment that the output of puddled steel

49 Ministère de l'Agriculture, du Commerce et des Travaux Publiques: "Statistique de l'Industrie Minerale . . . 1853–1859", Paris, 1861, pp. 484–499.

50 L. E. Gruner and C. Lan: "L'Etat Present de la Metallurgie du Fer en Angleterre", Paris, 1862, p. 743.

51 Among them were M. Janoyer: "Memoire sur la Fabrication du Fer a Grains et de l'Acier Naturel aux Fours a Puddler", Annales des Mines, Vol. xv, 1859, pp. 147-167 and L. E. Gruner: "Notes Diverses concernant l'Acier Puddlé et l'Acier de Forge", Annales des Mines, Vol. xv, 1859, pp. 291-336. In the former paper it is commented that, without fear of contradiction, puddling for steel is one of the most important recent conquests in the field of iron and steel metallurgy.

52 L.Ansiaux and L. Masion: "Traité Pratique de la Fabrication du Fer et de l'Acier Puddlé", Liege, 1861. The puddling process for steel is on pp. 64-76. (A translation may be found in "Steelmaking Before Bessemer, Vol. 2, pp. 309–313.)

was now very small compared with formerly and that the process was disappearing.[53] Despite this, however, at the time of the Universal Exhibition at Paris in 1900, puddled steel was still being made at Assailly, Unieux and Montluçon and there was an impressive exhibit of agricultural items, including forks, shovels, plough fittings and cart tyres, all in puddled steel, put on by the house of Gouvy and Company of Dieulouard.[54] Styling himself "Maitre de Forges de la Maison Gouvy", as late as 1955 F. Gouvy, lamenting the demise of the process during the 1914–1918 War, referred to puddled steel in the following terms:[55]

> The craftsmanship and knowledge of the secrets of the process were an adequate substitute for the measuring instruments so much in use today, without any prejudice to the quality, which was probably higher than that of the modern processes. It was easier to guarantee the tools made with such steels.

There are no production statistics available, but it is clear that the use of the process in Britain expanded over the same period, since Dr. Percy, writing in 1864, gave his considered opinion[56] that: "puddled steel is now an article of great commercial importance".

It seems that one of the uses of puddled steel was in plates for shipbuilding. In a list of ships incorporating steel between 1860 and 1880, fourteen of the first twenty contained puddled steel; the suppliers were the Mersey Iron and Steel Company, the Weardale Iron Company and the two Sheffield steel firms Thomas Firth and Sons (who supplied angles and plates in 1861) and John Brown and Company, supplying plates from 1859 to 1861 as well as "Atlas Toughened Steel" from 1864 to 1866.[57] It is known that John Brown installed puddling furnaces at his Atlas Works in 1858[58] and that others were added over the next three years, making a total of 72.[59] Over the same period, at the neighbouring Cyclops Works, Charles Cammell was installing some 60 puddling furnaces. Tradition has it that the puddling furnaces at both works were installed for the provision of wrought iron armour plates but it is quite evident that both works produced puddled steel. One French report indicates that John Brown and Company made fine iron for armour plates

[53] S. Jordan: "The Iron and Steel Manufacture in France", J.I.S.I., 1889, Pt. II, pp. 20-22.

[54] H. Bauermann: "Iron and Steel at the Universal Exhibition, Paris 1900", J.I.S.I., 1900, Pt. 2, pp. 160-168.

[55] F. Gouvy: "La Maison Gouvy depuis sa Fondation jusqu'a Nos Jours", Le Fer a Travers les Ages, Nancy, 1956, pp. 351–352.

[56] Percy, op. cit., p. 791.

[57] Jeans, op. cit., pp. 716-717.

[58] J. Hunter (ed. A. Gatty): "The History and Topography of the Parish of Sheffield in the County of York", Sheffield, 1869, pp. 214-215.

[59] S. Griffiths (ed. W. K. V. Gale): "Guide to the Iron Trade of Great Britain, 1871", Newton Abbot, 1967, p. 279.

as well as puddled steel for springs and tyres.[60] Translation of a passage from another French report reads as follows:[61]

> From the time of the introduction of puddled steel in Sheffield may be dated the production of rails, tyres and large plates in steel, not before produced in this area. Also, following the introduction of this process, cast irons flow into Sheffield from all parts of the globe.

The same report gives an assessment of the Sheffield costs of producing puddled steel, using Swedish cast iron at £7.8.0 per ton, which indicates that for John Brown to sell this at £13–£14 per ton would give him a very reasonable profit. The records clearly show the availability of "John Brown Melting Base" in the 1860s.[62] This was purchased by the smaller crucible steelmakers in Sheffield as an economic and viable substitute for the conventional common grade blister steel used in crucible charges and it is reasonable to assume that this, and the "Atlas Toughened Steel", were products of the puddling furnaces. Thomas Firth and Sons specialised in guns and projectiles rather than armour plate but nevertheless they thought fit to set up a works at Whittington, near Chesterfield, in 1856:[63]

> At these works, the manufacture of Firths famous puddled steel was carried out . . . [it] had the reputation of being the finest made in the country and they supplied Wentworths [presumably a misprint for Whitworths], Vavasseurs and Armstrongs, beside the Government Works at Woolwich.

Puddled steel was probably also made by R. F. Mushet at his "Secret Steelworks" in the Forest of Dean; it was, on the other hand, quite definitely one of the products at Ebbw Vale, made by George Parry,[64] who made the very valid point that the puddling process was quite capable of removing phosphorus from cast iron. There are references to the making of puddled steel by John Gjers at Middlesbrough, using Linthorpe No. 3 iron, in 1868, and also by James Kitson at Monkbridge in 1876.[65] The "reminiscences" of

60 S. Jordan: "Revue de l'Industrie du Fer en 1867", Vol. 4, Paris 1871, p. 329.

61 Gruner and Lan, op. cit., p. 769, p. 796.

62 This is included in a number of the price lists issued by Daniel Doncaster in the 1860s (Doncaster Archives, Kelham Island Museum, Sheffield) and also occurs in the Brittain Accounts (Sheffield City Libraries, LD 266).

63 A. C. Marshall and H. Newbould: "The History of Firths'", Sheffield 1924, p. 22.

64 G. Parry: "On Puddled Steel", Proc. South Wales Institute of Engineers, Vol. iii, (1862-3), pp. 74–81.

65 Both these items of information come from private communications received by Dr. Percy, when he was collecting information for his projected revision of his 1864 treatise on Iron and Steel. The full collection of these "Percy Papers" is now held by the Institute of Metals.

a Staffordshire man on the manufacture of steel for railroad springs refer to the use of cast iron from Barrow, Forest of Dean, Tredegar or Blaenavon – all low phosphorus materials – and the fettling of the hearth with Cumberland ore; the date and location of such operations, however, are not quoted.[66]

The evidence presented proves conclusively that the production of puddled steel was a most important contribution to the engineering industry between about 1855, when it came into prominence in Europe, and about 1885 when it was relegated to a minor role following the introduction of the basic steelmaking processes. As will become clear later in this study, it had an advantage over the original Bessemer and Siemens processes, with their acid lined furnaces, since it was capable of producing relatively low phosphorus steel from phosphoric cast irons and, provided the application was not one involving critical stress patterns, it could be used with considerable advantage over wrought iron on account of its enhanced strength, making thinner and less weighty constructions possible. In addition, it was a perfectly suitable material for such applications as agricultural tools, vehicle springs, railway locomotive and carriage tyres and axles, marine shafts, lock and harbour installations and even boilers. Up to 1880, indeed, puddled steel could, in a sense, be considered as complementary to Bessemer steel, the one being used with phosphoric irons and the other with haematite irons. This, in turn, explains why the puddled steel process assumed much greater importance on the Continent, since low phosphorus ores were relatively scarce in both France and Germany. Although the making of steel in this way has now been almost forgotten, its value at the time should not be underestimated. A comment from 1880 expresses this well:[67]

> . . . the production of steel by the process of puddling has almost entirely been superseded by the more simple, reliable and economical processes of the pneumatic converter and the Open Hearth; but the puddling process has, nevertheless, continued to be one of the practical aspects of steel manufacture and one that may fairly claim a high degree of consideration.

Methods involving Decarburisation of Cast iron to produce Liquid Steel

This section obviously covers the two major successes in the search for bulk steel manufacture, the **pneumatic process**, associated with the name of Bessemer, and the **Open Hearth process** derived by Siemens. These two

[66] M. Millard: "Old Methods of Ironmaking", Proc. Staffordshire Iron and Steel Institute, Vol. xxvii (1911–1912), pp. 188–191.

[67] Jeans, op. cit., p. 518.

processes provided the foundations for steelmaking activities for the next hundred years and, accordingly, become the subject matter for the bulk of the present study. There were several other workers in the same field and a number of other processes which appeared promising and then faded and it is not proposed to go into detail on most of these. Some will find mention later but only one, relatively minor, process appears worthy of comment here, since it links the old procedures into the new and it persisted, in Sweden at any rate, for almost a hundred years.

The **Uchatius process**,[68] a modification of the crucible process, depended on the granulation of cast iron, achieved by pouring the stream of the molten metal into water, and the mixing of the granules with about 20% of their own weight of crushed and powdered high grade spathic iron ore, rich in manganese (or a good quality haematite ore to which oxide of manganese was added), together with about 5% of fireclay. The mixture was then charged into crucibles and the process carried out as normal for the melting of the charge; the oxides present reacted with the carbon in the cast iron, producing a steel. The yield was some 6% greater than the weight of cast iron charged. The method was of Austrian origin and the British patent[69] was soon taken up by the Newburn Steelworks on the Tyne, the manager there providing a most elegant description of what occurred in the crucible:[70]

> Each granule [of cast iron] being surrounded by the pulverised oxides, the decarbonisation takes place first on the outside of each granule and so progresses towards the centre as the heat increases, the oxygen in the ores combining with the carbon in the granules and passing off as carbonic acid gas; if, therefore, during the process the granules could be examined, it would be found that the outside of each is entirely deprived of its carbon, the next portion partially decarbonised and the centre not decarbonised at all; so that each granule would be composed of pure wrought iron, steel and cast iron. By increasing the heat the cast iron portion of the granule first becomes fluid and then falls by its own weight to the bottom of the crucible. At the same time, the earths mixed with the ores melt and rise to the top, forming a layer of scoria or dross floating on the surface of the melted iron. Each granule of melted metal has therefore, in falling, to pass through the rising scoria and it is in passing through that the combination of the impurities in the metal with the alkaline earths takes place, so that the decarbonised iron on reaching the bottom of the crucible is cleansed from all impurities. The heat continuing to increase melts the outside portions of the granules and the whole is reduced to one homogeneous fluid mass in the crucible, which is then ready for being poured into ingot moulds. . . . The oxides employed in this process are iron

68 The inventor, Franz Uchatius, was a Colonel in the Austrian Army.

69 F. Uchatius, British Patent 2189, 1 October 1855.

70 T. Spencer:"On the Manufacture of the Uchatius Cast Steel", Proc. Institute of Mechanical Engineers, August 1858, pp. 146–154.

> ores of the finest qualities, such as spathose or haematite, which are previously calcined and pulverised. The proportion of oxide to the granulated iron is according to the hardness of the steel required, say from 20% to 30%; the greater the quantity of oxide employed the greater the decarbonisation and consequently the softer will be the steel produced.

The writer of this paper claimed a rapid conversion of cast iron into steel by a direct method, which gave a uniform quality from a predetermined mixture of cast iron and ore at a cost that was less than half that of remelting blister steel for comparable quality. It is clear that the works at Newburn did provide material of most acceptable quality. Some twenty years later, however, the process had been abandoned there.[71]

In one organisation in Sweden, nevertheless, the process was eminently successful. The works at Wikmanshyttan was set up in 1859, designed specifically to operate the Uchatius process.[72] Local charcoal smelted cast iron was used, granulated as it came out from the furnace. The granules were then mixed with calcined and powdered pure Bispberg ore, charged into crucibles made from a Belgian clay, sometimes adding extra charcoal to the crucible, and the charges melted in coke fired crucible furnaces. The fuel was imported but it is recorded that only 2.2 tons of coke were needed per ton of steel produced, only some 60% to 70% of the coke consumption in Sheffield furnaces operating the standard crucible melting procedures with all metal charges. Specimens of steel made at Wikmanshyttan by the Uchatius process were exhibited at the International Exhibition of 1862; the excellence of these exhibits, particularly of the sword blades, won high praise.[73] At the Paris Exhibition of 1867 even more adulation was lavished on the Wikmanshyttan exhibits; an as-cast ingot showed hardly any grain structure in the fracture; the forged bars showed an excellent fine grain and their strength was the equal of the best British steel.[74] The comment was made that this works seemed to be the only one able to make a success of the Uchatius process, having found a reproducibility in the quality of the cast iron and a richness and purity in the iron ore which it would be difficult to match anywhere but in Sweden. These conditions continued to operate in favour of the operations at Wikmanshyttan. Production grew steadily, furnace additions being made in 1898, 1906 and 1912; a peak output of just over 2,000 tons in the year was achieved in 1915 and interesting photographs of the granulation of the cast

71 J. S. Jeans: "Notes on the Northern Industries", London, 1878, pp. 79–82.

72 C. Sahlin: "De Svenska Degelstålsverken", Med Hammare och Fackla, IV, 1932, pp. 79–91.

73 Percy, op. cit., p. 803.

74 S. Jordan: "Revue de l'Industrie du Fer en 1867", Vol. 4, Paris, 1871, pp. 314–315.

iron and the teeming of the metal, taken in 1929,[75] have survived. The original process appears to have been operated virtually unchanged for over eighty years, eventually being discontinued in 1941.

The stage is now set for a detailed consideration of the development and growth of the two major processes in this category of decarburisation of cast iron to produce liquid steel.

[75] These photographs, taken by F. Walldow in 1929, are reproduced in the article by Sahlin, op. cit.

3 Bessemer and the Development of the Pneumatic Process

Henry Bessemer (1813–1898) is probably best described as a professional inventor. He took out his first patent in 1838, at the age of twenty five, concerned with the manufacture of type for printing. Over the next 45 years there followed a further 116 patents, some 36 of which concerned the manufacture of iron and steel. The range of interests of the remainder is fascinating, including glass manufacture, metallic powers for paints, embossed velvet, sugar manufacture, projectiles and guns, steamships for the prevention of sea sickness and the manufacture of telescope lenses.[1]

The major source of information on the manner in which Bessemer became involved in iron and steelmaking and the problems of the development period is in his own writings and, in particular, the "Autobiography" which he wrote as reminiscences late in life.[2] The account which follows is, accordingly, largely abridged from his own reports.

It was his interest in projectiles which led to his experimentation on refining cast iron. In 1853 he had devised the idea of an elongated projectile in the outer surface of which a series of grooves would lead to a rotary motion, using a smooth bore gun. It seems that the British Government was not interested; by chance, however, whilst Bessemer was in Paris, his ideas became known to the Emperor Napoleon III, who invited him to arrange for trials at Vincennes, near Paris. Bessemer returned to London and continued his work, eventually taking over a series of his new projectiles to Vincennes, where the firing trials were eminently successful; nevertheless, the considered opinion was that, unless a stronger material could be found for the guns themselves, the

[1] H. Bessemer: "An Autobiography", London, 1905, pp. 328–332. This provides a complete list of all the Bessemer patents. A facsimile reprint of the Autobiography has recently been issued by the Institute of Metals (1989)and its improved availability is to be welcomed.

[2] The Autobiography was, in fact, incomplete when Bessemer died in 1898. An epilogue was provided by his son, relying on memories and "some memoranda and letters", his father's copious notes having been unfortunately destroyed.

heavier projectiles, which Bessemer's invention made possible, would be of little practical value. Thereupon Bessemer took upon himself the endeavour of providing such a stronger material. As should now be clear to the reader, the common engineering materials of the time were:

1 **Cast iron**: brittle and intractible in tension, but strong in compression, having good wear resistance and, above all, capable of being cast to shape;
2 **Wrought iron**: ductile and malleable, capable of being forge welded into relatively large masses but with little wear resistance;
3 **Crucible steel**: exhibiting a most useful combination of strength, ductility, wear resistance and capability of being forged to desired shape, but costing up to ten times as much as cast iron and five times as much as wrought iron.

At the time of Bessemer's French trials his gun barrel was of cast iron; other designs of guns at the time had cast iron gun tubes wrapped round with wrought iron bands; the day of the steel gun tube was some years away.[3] Bessemer, therefore, had in mind the provision of a material with better properties which might be obtained by the refining of cast iron.

Bessemer at this time had little knowledge of ferrous metallurgy. It seems he looked upon this as an advantage:[4]

> I find that persons wholly unconnected with any particular business have their minds so free and untrammelled to view things as they are, and as they would present themselves to an independent observer, that they are the men who eventually produce the greatest changes.

A later observer put it rather differently:[5]

> Bessemer, at one time an engraver of labels, was the inventor of many things, but he was not associated with the steel trade until he conceived the amazing idea that steel might be made from molten pig iron by blowing air through it ... an idea that would never have entered into a steel maker's head. The idea worked, but it was an amazing piece of indiscretion to assume that it would lead to anything but failure.

[3] One of the earliest trials of a steel gun tube was made in 1862 when a forging in crucible steel, made by Naylor, Vickers and Company at their Millsands Works in Sheffield, was converted by the Woolwich Arsenal into a twenty pounder rifled gun and submitted to extended tests. These included the firing of specially extended projectiles using double the normal charges. The only effect on the gun was a slight expansion of the powder and shot chambers, the bore being reported as being free of flaws at the conclusion of the trials. The gun was for many years an exhibit on the office corridor of the River Don Works; it has recently been transferred to the head office of British Steel in London.

[4] Autobiography, p. 136.

[5] H. Brearley: "Steelmakers", Sheffield, 1933, p. 121.

Bessemer returned to London, proceeded to read up what he could on the subject, visited various ironworks to try to gain some practical knowledge and eventually set up an experimental works in St. Pancras. With the usual foundry equipment of a cupola and a reverberatory hearth (or air furnace), Bessemer carried out experiment after experiment. During one melt in the reverberatory furnace, he noted two pieces of cast iron on the edge of the hearth which, apparently, had not melted; he therefore increased the air flow so as to increase the heat but after a further thirty minutes they were still there; on pulling them into the hearth he found they were only shells and, on further examination, found them to be virtually pure iron, completely decarburised by the action of the air. As he himself reported:[6]

> Thus a new direction was given to my thoughts and, after due deliberation, I became convinced that, if air could be brought into contact with a sufficiently extensive surface of molten crude iron, it would rapidly convert it to malleable iron.

Tests he carried out in crucibles of molten metal with air passing through a blowpipe into the metal convinced him that this was a possibility and his next step was to test the effect of blowing air through the bottom of a vessel into which molten metal from the cupola could be poured.[7]

> I had very little data to guide me, as the crucible converter was hidden from view in the furnace during the blow. I found, however, that slag was produced during the process and escaped through the holes in the lid. Owing to this, I determined on constructing a very simple form of cylindrical converter, about 4 ft. in height in the interior, which was sufficiently tall and capacious, as I believed, to prevent anything but a few sparks and heated gases from escaping through a central hole made in the flat top of the vessel for that purpose. [See Fig. 6] The converter had six horizontal tuyeres arranged around the lower part of it; these were connected by six adjustable branch pipes, deriving their supply of air from an annular rectangular chamber, extending around the converter, as shown.
>
> All this being arranged, and a blast of 10 to 15 lb. pressure turned on, about 7 cwt. of molten pig iron was run into the hopper provided on one side of the converter for that purpose. All went on quietly for about ten minutes; sparks such as are commonly seen when tapping the cupola, accompanied by hot gases, ascended through the opening on top of the converter, just as I supposed would be the case. But soon after a rapid change took place; in fact the silicon had been quietly consumed, and the oxygen, next uniting with the carbon, sent up an ever–increasing stream of sparks and a voluminous white flame. Then followed a succession of mild explosions, throwing molten slags and splashes of metal high up into the air, the apparatus becoming a veritable volcano in a state of active eruption. No one could approach the converter to turn off the blast and some low, flat, zinc–covered roofs close at hand were

[6] Autobiography, p. 142.
[7] Autobiography, pp. 143–144.

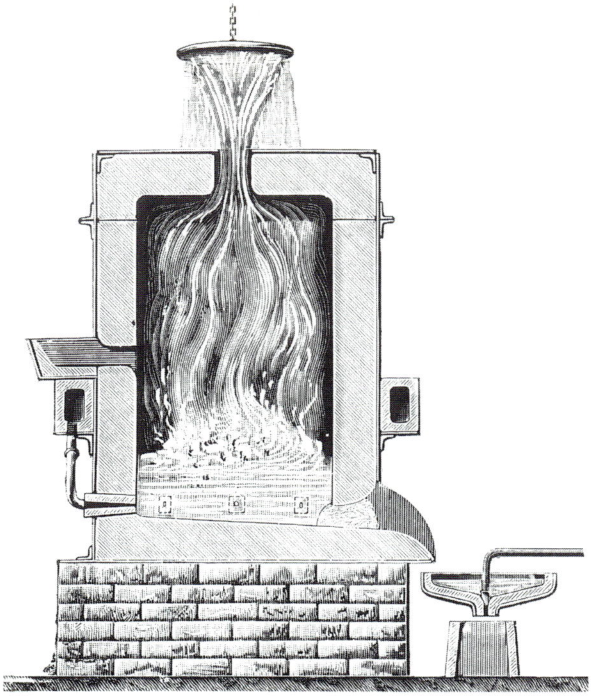

Fig. 6. The original Experimental Converter as used by Bessemer in his early trials in 1856. This drawing is reproduced from Fig. 39 in Bessemer's "Autobiography". The vertical converter was four feet in diameter and had six horizontal tuyeres arranged near its base. It held 7 cwt. of molten iron and the air blast was at a pressure of 10 to 15 lb./sq.in. The cast iron plate suspended above the upper opening was completely melted by the heat of the ensuing reaction.

> in danger of being set on fire by the shower of red-hot material falling on them. All this was a revelation to me, as I had in no way anticipated such violent results. However, in ten minutes more the eruption had ceased, the flame died down, and the process was complete. On tapping the converter into a shallow pan or ladle and forming the metal into an ingot, it was found to be wholly decarburised malleable iron.
>
> Such were the conditions under which the first charge of pig iron was converted in a vessel neither internally or externally heated by fire.

Bessemer goes on to relate that, in an attempt to cut down the emission of material, he suspended a cast iron checker plate on chains over the central opening at the top of the furnace for the next trial; the eruption, once started, heated up the plate and eventually melted it away completely. He thereupon redesigned his converter, providing one with an arched roof and a conical sloping floor [See Fig. 7] together with a casting station in which the metal could be run into an ingot mould and the ingot then hydraulically pushed out from the mould. This still operated on a 7 cwt. charge, providing a 10 in. square ingot. The whole apparatus was erected in the St. Pancras workshop and, early in August 1856, almost the first to witness its operation was George Rennie, one of the foremost British engineers, who was so impressed by what he saw that he immediately persuaded Bessemer to make the matter public, arranging for the presentation of a paper at the Chelten-

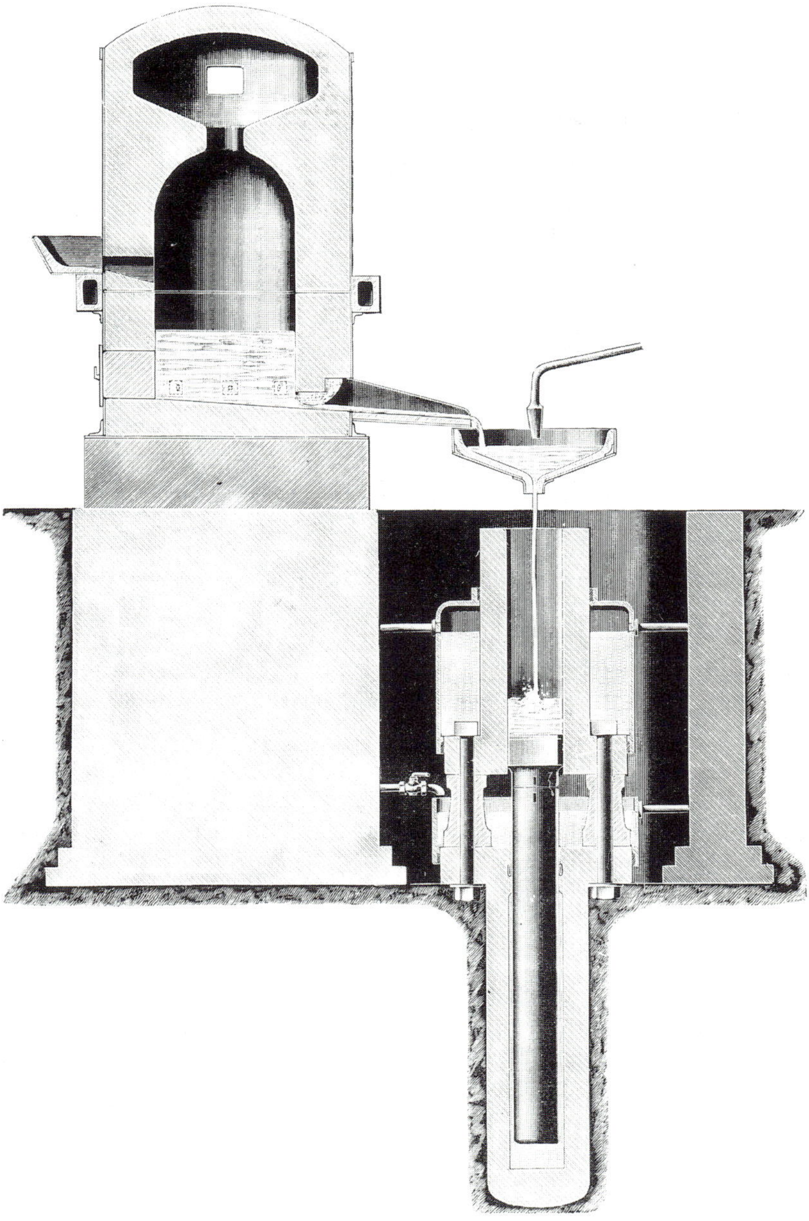

Fig. 7. Further design of Fixed Converter as modified by Bessemer. This is a reproduction of Fig. 42 from the "Autobiography". Based on the experience with the original vessel, Bessemer provided a more solid vessel with an upper chamber to contain the metal and slag particles which were previously ejected. He also provided a sloping floor to the chamber, a hopper into which the molten metal could be poured from the cupola and a spout through which the metal could be run after reaction. He also provided a stoppered tundish through which the metal could be run into an ingot mould, this being supported by a hydraulic ram, which could be used to lift the solid ingot above floor level.

ham Meeting of the British Association on 11th August 1856. Bessemer chose as his title "The Manufacture of Malleable Iron without Fuel". The sheer outrageousness of the title ensured him a large audience, most of whom, like the "Times" reporter, were extremely sceptical and initially took little notice, but eventually were convinced that here was an important contribution. The reporter approached Bessemer afterwards, apologised that he had not taken sufficient notes and asked if he might take Bessemer's paper with him. So it was that the text appeared in extenso in the The Times of 14th August 1856.[8]

As was his practice, Bessemer had covered himself well with patents; in consequence, he was approached by several ironmasters and the workshop at St. Pancras was kept busy demonstrating the process. Two ingots were shipped to the Dowlais Mill and rolled eminently satisfactorily into rails. Bessemer had a stormy interview with Thomas Brown of the Ebbw Vale Works who wanted to buy the full patent rights for £50,000, a sum which the inventor considered derisory; eventually he decided to divide the country into five regions and to issue licences to one major operator only in each area. Thus, before the end of August, the Dowlais Works, the Butterley Iron Company, Galloways of Manchester, the Govan Iron Works and a South Wales tinplate works had between them paid £27,000 in royalties.

Then disaster fell. First came the trials at Dowlais. Menelaus, the works manger, indicated he had seventeen furnaces in blast and offered to detail the burdens of each, whereupon Bessemer is reputed to have replied that it did not matter since his apparatus would work with any kind of iron. As there were the remains of a building adjacent to one particular furnace, which was making iron for puddling to rails, the converter was placed there. The metal was run in and it showed all the expected fireworks and cast quite satisfactorily into the mould. When it came to the working down of the ingot, however, it broke up and was useless. A repeat of the experiment led to precisely the same results. Bessemer is said to have departed from Dowlais with serious apprehensions as to the success of his invention.[9] Experiments at the Govan Works in Glasgow, Butterleys in Derbyshire and Galloways in Manchester were just as catastrophic; repeat trials at St. Pancras, however, were still satisfactory. Bessemer was quite exhausted with his efforts and at a loss to understand the reasons for the debacle. He wrote:[10]

[8] It is significant that, in view of the later adverse comments on field trials of the process, it was considered prudent by the British Association editorial committee simply to record the title of Bessemer's contribution without any text.

[9] W. T. Jeans: "Creators of the Age of Steel", 2nd. ed., London 1885, p. 47.

[10] Autobiography, p. 170.

> The transition from what appeared to be a crowning success to one of utter failure well nigh paralysed all my energies. Day by day reports of failures arrived; the cry was taken up by the press; every paper had its letters from correspondents and its leaders, denouncing the whole scheme as the dream of a wild enthusiast, such as no sensible man could for a moment have entertained. I well remember one paper, after rating me in pretty strong terms, spoke of my invention as 'a brilliant meteor that had flitted across the metallurgical horizon for a short space, only to die out in a train of sparks and then vanish into total darkness'.

Bessemer rearranged his finances so that his family would be protected and set out, if necessary, to use the sum total of his remaining resources in solving the problem. He called in the services of Dr. Henry to analyse all the raw materials used and on Edward Riley and John Percy to advise him; he also studied the information available from the analytical work done by Robert Hunt of the School of Mines. In the course of time, it appeared to be quite clear that the problem was connected with the element phosphorus. As it so happened, Bessemer had used, quite by chance, one of the very few low phosphorus cast irons which are available in Britain, that from Blaenavon, and in this way had used an ideal raw material. All other trials had been carried out on metal which was produced from the generality of British iron ores, which contain significant amounts of phosphorus. The metal produced from such irons in Bessemer's converter would also be high in phosphorus – this was considered to be the cause of the inability of the metal to stand up to the hammer or the rolling mill. Bessemer spent considerable effort in his workshop endeavouring to find some way of removing the phosphorus, but all to no avail. He had ordered some cast iron from Sweden; when this arrived, he was relieved to find that he could again, with this low phosphorus iron, produce workable material.

To remove phosphorus had proved impossible; therefore, a search for a low phosphorus indigenous cast iron should be made. Armed with the analyses of the iron ores, it appeared that the haematite ores of Lancashire and Cumberland would provide a suitable product. Samples of cast iron from Workington were, however, still high in phosphorus; Bessemer went there to examine the production and found that they were using, as flux to promote smooth running of the blast furnace, the red cinder which came back as ballast in the boats which took haematite ore to the Staffordshire furnaces. This was the source of the phosphorus: this cinder came from the Black Country puddling furnaces and contained much of the phosphorus from the local ores, transferred by the smelting operations in the blast furnace to the cast iron but, to a large degree, subsequently reoxidised during the puddling operation and, accordingly, held in the rich puddling slag. Bessemer persuaded the Workington management to substitute a local shale for the cinder. Analysis of the resulting material showed it to have the expected low phosphorus content; trials of the material in the converter showed it to be capable of yielding a satisfactory product. This type of low phosphorus cast iron, henceforth to be known as "Bessemer Pig", was destined to be used for the next twenty years as the major source of steel in this country.

Having provided himself with the solutions to his metallurgical problems he now found himself faced with a question of credibility.[11]

> I announced the fact of my complete success to the world and held in my hands the most undeniable proofs of the truth of my assertion, but no one would now believe it. They remembered, but too well, the great expectations that were excited two years previously by the first announcement of my invention at Cheltenham and were not again to be disturbed by the cry of 'Wolf!'. Thus it happened that, after the hard battle I had fought for so many years, I found myself as far as ever from the fruits of my labour, for not a single ironmaster or steel manufacturer in Great Britain could be induced to adopt the process.

Bessemer relates how he produced a quantity of steel at St. Pancras to the type of analysis used for workshop tools. This material was forged by a Sheffield steelmaker into bars of similar appearance to the standard crucible steel normally used for the purpose, being identified by a special punch mark by Bessemer or his partner, Longsdon, at the time of forging. The bars were then taken to Galloways, in Manchester, for general use in the engineering shops. In service over a two month period, the workmen found no difference from the usual tool steel, which cost some six times as much. Satisfied with this added confirmation, he then took the decision that, having twice been shunned by the Sheffield steelmakers, he would set up a works in their midst and force them to adopt his method by underselling them in their own market and at the same time making a handsome profit. This project was agreed between Bessemer, his partner, Robert Longsdon, his brother in law, William Allen, and the Manchester firm of Galloways:[12]

> Thus were established the first Bessemer Steel Works and, in less than twelve months from its commencement, we had built a dozen melting furnaces and erected the steam and tilt hammers, blast furnaces and converting apparatus, suitable for carrying on the new manufacture. This we commenced by bringing steel into the market at £10 to £15 below the quotations of other manufacturers.

The site chosen was on Carlisle Street, one of the two parallel roads which had been built on either side of the Sheffield–Rotherham railway and in the heart of the newly developed steelworks area.[13] The description of the works buildings is interesting and calls for later comment. For "blast furnaces" it

[11] Autobiography, p. 175.

[12] Autobiography, p. 176.

[13] The office block still stands on Carlisle Street in Sheffield. Until recently it had been used as the head office by Firth Brown Tools Limited. It is now empty but is listed as a building of historic interest. Originally it carried a date stone reading "1856" (although it was towards the end of 1858 when it was first occupied). This has now, unfortunately, been defaced. The works opposite eventually became part of Firth Brown Limited; although converted into machine shops, they still continued to be known as "Bessemer Department".

seems we should read "cupolas"; the significant point, however, is the erection of "melting furnaces", nothing less than the old established Huntsman type of crucible furnaces. The situation of Bessemer's works, however, proved to be of crucial importance, being separated only by a wall from those of John Brown and it was not long before the Sheffield steel-maker became curious about his new neighbour's activities. Having witnessed the blowing operations, John Brown accepted an ingot for trials, with the result that he agreed to become the first Sheffield licensee of Bessemer's patents. This was in 1859. It was not long before Bessemer's other neighbour, Charles Cammell, also took out a licence and Sheffield was to embark on a new trade, the making of steel rails.

******

Such is the story as derived from the writings of Bessemer himself. The development was, however, even more complicated than Bessemer was willing to admit.

The initial licencees were all ironmasters; the steelmakers, much to Bessemer's chagrin, kept aloof and it was this that eventually dictated his move to Sheffield. It must also be remembered that wrought iron had been made from phosphoric irons for a considerable period and that several of these wrought irons contained amounts of phosphorus up to 0.5% and were still forgeable at elevated temperatures; the presence of phosphorus made them **cold short** – brittle at low temperatures – rather than **hot short**, or lacking in forgeability at elevated temperatures. It was only when the carbon content was increased that the presence of phosphorus gave forgeability problems; for steel production, therefore, the availability of a raw material low in phosphorus was mandatory, as evidenced by the long tradition of the exclusive use of Swedish iron by the Sheffield crucible steelmakers.

The records of the early trials which proved so disastrous simply indicate that metal was blown down until the reaction ceased and that the metal, which Bessemer cared to refer to as "wholly decarburised malleable iron",[14] was then cast into ingots. There is no record at this stage of any addition being made to the metal after blowing and it is reasonable to believe that Bessemer was ignorant of the use of any corrective at this stage for what would now be termed **overoxidation** and was then looked upon as "rottenness", the iron being said to be "burnt". Whether the experiments at St. Pancras succeeded in producing sound or rotten material would seem to have been dependent on the exact moment at which the flow of air was turned off: if there was still a little carbon left, then good materials would be obtained; to continue the blow a little longer would produce over–oxidised material and it has to be observed that, even using the favourable low–phosphorus Blaenavon iron, Bessemer's brother in law, William D.

[14] Autobiography, p. 144.

Allen, who was in charge of much of the experimental work, reported many years later:[15]

> Sometimes they succeeded in making a piece of an ingot having all the characteristics of steel, but he was bound to say that their failures were more frequent than their successes.

It is reliably reported that, shortly after the Cheltenham meeting, Thomas Brown, manager of the Ebbw Vale Iron Company, visited Robert Mushet[16] at his home in the Forest of Dean, taking with him a copy of Bessemer's paper and some samples of the defective metal produced by Bessemer in one of the abortive trials. Mushet recognised that the badly cracked bar was "burnt"; he melted a portion in a crucible, added one sixteenth its weight of molten spiegeleisen and emptied the mixture into a small ingot mould. The ingot was smooth and had shrunk down in the head, reminiscent of a good cast steel ingot. Heated to a good red heat, it forged perfectly satisfactorily; Mushet himself forged the bar down into a chisel, hardened and tempered it and said that is was indeed "cast steel, worth 42s. per cwt." Mushet patented the use of his "triple compound" of iron-carbon-manganese – in reality, another name for spiegeleisen – only one month after the Cheltenham meeting;[17] unfortunately, he allowed himself to be persuaded to couple this with the use of a worthless steelmaking patent in the name of Martien;[18] this was held by the Ebbw Vale Company with whom Mushet was collaborating at the time. It is reported that Bessemer, having heard of Mushet's activities, journeyed to Coleford to meet him and solicited his help; this was at the time that he was gathering together his metallurgical team to investigate his problems. Mushet, however, felt he must retain his loyalties to Thomas Brown and Ebbw Vale and therefore regretted he could not assist Bessemer. This turned out to be a misplacement of trust, since Ebbw Vale eventually neglected to pay the stamp duty required to keep the patent in being and also omitted to tell Mushet of their neglect. The patent, thereby, became public property from early 1859 onwards. Mushet, in 1884, was to write:[19]

> I understood this process [the Bessemer process] as well in 1856 as I do now and had Sir Henry come to me before the Ebbw Vale Company got hold of me – and I wish he had – it would have saved him tens of thousands of pounds and unspeakable labour and anxiety, and he would never have left Dowlais with serious apprehensions as to the success of his process.

15 W.D. Allen, in his acceptance speech on being presented with the Bessemer Gold Medal, reported in J.I.S.I., 1890, Pt 1, p. 10.

16 F.M. Osborn: "The Story of the Mushets", London, 1952, pp. 139-140.

17 R. F. Mushet, British Patent 2219, 22 September 1856.

18 G. Martien, British Patent 2082, 15 September 1855.

19 R. F. Mushet in letter to "Iron", 1 February 1884.

There was, however, assistance coming to Bessemer from a completely different source. He had taken out a Swedish Patent, No. 35, in July 1856, covering the use of the pneumatic process and the converter, which had been taken over by the merchant firm of Hoare, Buxton and Company in London, who were well known in Sheffield steelmaking circles as suppliers of Swedish iron. They also imported timber from Sweden and one of their suppliers was Daniel Elfstrand and Company of Gefle. This particular firm had a blast furnace at Edsken and a forge at Hogbo and, since the activities of Bessemer had come to the notice of Goran Fredrick Goransson, general manager of Elfstrand, he came to England in May 1857 for a demonstration by Bessemer. Convinced of the validity of the patent, he agreed to purchase one fifth of the Swedish rights, paying a half instalment of £1,000 in June 1857, entitling him to produce up to 500 tons of steel per annum on payment of royalties of two shillings per ton. The production of steel without the use of fuel was a concept of considerable interest in a country where there was virtually no coal and the forest resources were being strained to their limit.

Accordingly, two furnaces and the necessary blowing equipment were ordered from Galloways and shipped to Gefle, whence they were transported, wih considerable difficulty, to Edsken, some forty miles away, to be erected alongside the blast furnace. The equipment arrived in August 1857. It seems that the furnaces were to a newly designed tilting pattern curently favoured by Bessemer and, by the beginning of November, they had been erected under the supervision of the British engineering firm. Initial trials, carried out in the presence of Mr. Leffler, a representative sent over by Bessemer, appeared promising and, within a month, tools produced from one of the ingots by Lindahl, an engineer in Gefle, were reported to be "equal to those from the best English cast steel", despite considerable manipulative difficulties with the furnace.[20]

By an unkind twist of fate, at this interesting stage of development, with world trade in the doldrums, Daniel Elfstrand and Company went bankrupt at the end of 1857. Fortunately for posterity, however, those placed in charge of the company's affairs considered that both Edsken and Hogbo should be kept in operation and paid the outstanding £1,000 due for the patent rights. Goransson thereupon applied for help to Jernkontoret, the Swedish Ironmasters' Association, sending samples from the initial trials as evidence. Their representatives witnessed the blowing operations at Edsken, proceeding to Hogbo to see the ingots forged. By the middle of January 1858, funds were provided for work to continue under the supervision of a Jernkontoret engineer. Goransson, having by now become disenchanted with the tilting converter, arranged to revert to the original fixed design.

[20] Per Carlberg: "Early Industrial Production of Bessemer Steel at Edsken", J.I.S.I., Vol. 189 (1958), pp. 201–204.

With a new converter installed and the blast furnace repaired, work recommenced in March 1858.

There were still problems and Goransson appears to have lost faith completely in Leffler's advice. He thereafter directed operations himself, making copious notes of each blow. It soon became clear that lack of heat at the end of the blow was the major problem, the slag not separating effectively from the metal; trials made using a ladle led to the formation of large skulls of solid metal. More or less as a final desperate effort, Goransson ignored all the advice given by Bessemer and Leffler with regard to air flow through the metal. Instead of concentrating on a controlled high pressure air supply, Goransson arranged twelve tuyeres, twice the original number, in the base of the converter, at the same time enlarging them from five eighths of an inch to seven eighths of an inch in diameter, thus allowing much more air to enter but at a lower pressure. In the words of Goransson:[21]

> ...the result was astonishing. The temperature of the fluid steel was very much raised, the slag came up to the top beautifully, the ingots turned out perfectly even in temper, free from slag and extremely malleable, more so than any made on the old process. The first charge with the converter so arranged was made on 18th July 1858 and **from that date the Bessemer method may be regarded as started.**

Goransson arranged for thirty tons of ingots made in this way to be sent to Sheffield in September 1858; it seems likely that these were six inches square. Half were sent to a forge in the town; Goransson implies that the Sheffield forgers did not feel it was in their interest to forward the Bessemer method, so they spoiled them by overheating them prior to forging. The remainder, however, were forged at Bessemer's own works and turned out to be eminently satisfactory when the product was fully tested in the making of knives, scissors, razors and tools. This was a real eye-opener for Bessemer. At that particular time, the procedure at Bessemer's own works was to blow the carbon out of the metal and then to granulate the product by pouring it into water, subsequently remelting the granules with cast iron and other additions in crucibles; this indeed was the reason for the inclusion of the twelve melting furnaces in the original scheme of things at the Sheffield works on Carlisle Street. It was in the month following Goransson's visit to Sheffield that Bessemer:[22]

> ...ordered from Sweden some pure charcoal pig-irons from which such excellent steel was made.

[21] E. F. Lange: "Bessemer, Goransson and Mushet: A contribution to Technical History", Manchester Memoirs, Vol l vii (1913), No. 17, p. 14. Here is reprinted the text of a letter written by Goransson on 6 November 1879 to Professor Akerman in Stockholm.

[22] Autobiography, p. 173.

and when it:[23]

> arrived at last, no time was lost in converting it into pure, soft, malleable iron and also into steel of various degrees of hardness"

It is significant that there is an entry in the diary of William D. Allen, Bessemer's partner at the Sheffield works, for 18th June 1859, "**First made steel direct**," just eleven months after Goransson's first successful blow.[24] It should also be noted that at this date, the free use of Mushet's patent would have permitted the use of "triple compound" or spiegeleisen as a corrective addition at the end of the blow. With sources of low phosphorus raw material now available, the Bessemer process may truly be stated to have started. Without the assistance of Mushet and of Goransson, together with the work done by Bessemer's own team, however, it must have remained a doubtful starter.

Goransson, was perhaps, in a rather unique position, in that the generally available cast iron in Sweden was, perforce, charcoal smelted from high grade ores; the blast furnace product was typically a white iron, with a low silicon content, a high manganese content and very low in both sulphur and phosphorus. What was presumably not appreciated by Bessemer was that Goransson had worked out that he could reasonably stop the blow within a few points of the desired carbon content, thus introducing the principle of "catching the carbon", still retaining a small proportion of both silicon and manganese. He thus avoided any over-oxidation of the iron and had no need for recourse to the addition of spiegeleisen, so typical of later British and American practice. Provided the temperature was sufficiently high, and his modifications to the tuyeres appear to have ensured this, his procedure had the potential for the production of cleaner steel; certainly Swedish Bessemer steel, in later years, had a very enviable reputation for quality.[25] A further innovation from Goransson was the introduction of a casting ladle to ensure as complete a separation of metal from slag as possible: this is certainly the earliest recorded use of a ladle in a steelworks. More significantly, however, Goransson deliberately set up his converter alongside the blast furnace so that liquid metal could be taken direct. This was important in a situation where fuel was at a

23 Autobiography, p. 174.

24 Lange, op. cit., p. 17.

25 From 1880 until 1940, it was the practice of both Doncaster's and Senior's to import Swedish Bessemer steel for their Sheffield market. This was originally imported in the form of bars; eventually, there was a very large trade in ingots, from 0.4 tons to 3 tons in weight, in grades from 0.1% carbon to 1.3% carbon. These ingots were forged or rolled in Sheffield and the bars used for toolmaking and other general purposes. Any Swedish Bessemer scrap arising was utilised as part of crucible furnace charges – it was good quality material with suitably low sulphur and phosphorus contents. Details of the trade may be traced in the ledgers from the Doncaster Archives, now located at the Kelham Island Industrial Museum in Sheffield.

premium.[26] As a postcript, indicating the importance of this work, it seems appropriate to quote the official report of the visit by the Iron and Steel Institute to Sweden in 1898:[27]

> The originator of the Sandvik Works is Mr. G. F. Goransson, a merchant of Gefle, who was the first, in 1858, to carry into commercial practice the method of producing steel direct from pig iron, which was the basis of Sir Henry Bessemer's patent of 1855 – a patent which at that time had not been practically worked in England.

It is, therefore, a most curious fact that, throughout his own discussion of the development of the pneumatic process, Bessemer appears to have studiously avoided any mention of the part played by Goransson. He does, on the other hand, quote the favourable impression he, himself, made on Prince Oscar of Sweden when he visited the works of Goransson to see the Bessemer process in operation; this, he reported,:[28]

> ...resulted in my being made an honorary member of the Iron Board of Sweden [Jernkontoret], in recognition of the value of my invention: a compliment which I shall ever highly esteem.

His reaction to Mushet was to go through all the previous patents covering the manufacture of iron and steel where the word "manganese" occurred and to list these as prior usage; there is, in fact, a whole chapter of his autobiography devoted to the question of "Manganese in Steelmaking".[29] Reading the argument today, one is struck with the confusion which existed with regard to the use of manganese as a metallic element and as oxide. It was Josiah Heath who, in 1839, had devised the idea of preparing what he called a carburet of manganese by reacting manganese oxide with carbonaceous matter and then adding this metallic matter to the crucible charge in order to arrive at a superior steel from lower grade materials.[30] Henry Unwin, who had been Heath's agent, soon began supplying to the Sheffield crucible steelmakers a mixture of manganese oxide and tar, as a cheaper replacement for Heath's carburet; the apparent economy appealed to them and it seems Unwin made a fortune from it. Heath challenged him in the courts, but died in poverty whilst the extended litigation was still in progress.

26 This innovation is normally ascribed to the works at Terrenoire, but this was several years after Goransson's work at Edsken.

27 This quotation is reprinted in "Ett Svenskt Jernwerk: Sandviken 1862–1937", p. 108. This volume was printed privately to mark the 75th anniversary of the establishment of the works at Sandviken by Goransson. The origin of the quotation appears to be a special issue of "The Ironmonger" to cover the visit of the Iron and Steel Institute to Sweden.

28 Autobiography, p. 278.

29 Autobiography, pp. 256–295.

30 J. M. Heath, British Patent 8021, 5 October 1839.

Brearley, a first class metallurgist and practical steelmaker, was later to make a significant comment on this:[31]

> In the train of the Heath persecution came superstition and avoidable loss, which lived on the back of the tool trade for half a century. Heath's real discovery was ignored in favour of the use of black oxide of manganese and the trade deluded itself into believing that the black oxide of manganese, known as physic', was potent for good, whereas it was not even the shadow of the substance of Heath's discovery. Up to 1900, hardly anyone dared to make crucible steel without adding an ounce or two of black oxide of manganese to the pot; and its main effect was this – along the wash line where the slag floated on the molten steel, the black oxide of manganese was drastically corrosive and frequently ate holes through the pot's side and wasted molten steel ran through the grate bars into the cellars. A heap of 'runnings', representing thousands of tons of wasted steel, would be a fit memorial to the greed of steel manufacturers who disputed Heath's claim and for his knowledge substituted their ignorance.

Reading the arguments put forward by Bessemer as to why he felt he owed no real debt to Mushet, one cannot but feel that he too, to a great extent, had substituted ignorance for knowledge in the same way and that Mushet had some validity in his claim that he had pointed the way, from his own metallurgical expertise, to the ultimate viability of the process initiated by Bessemer at Cheltenham, even to the extent of suggesting that it should be known as the Bessemer-Mushet process.[32] It is only proper to relate that, in 1866, when Mushet was ill and their home was threatened, Mushet's daughter, then a girl of only sixteen, came to see Bessemer and, as a result of her pleading, went home with a cheque sufficient to clear all their debts. The following year Bessemer, "for reasons which he did not care to give" thought fit to grant an annual pension to Mushet:[33]

> In the hope that an allowance to Mr. Mushet might have the effect of restraining these attacks on me, I offered to pay him £300 a year, aiming at abating an intolerable nuisance which I had no other means of preventing. While we were paying over £3,000 per annum in the form of income tax, the £300 was but a small additional tax on my resources, so I allowed it to drag on until Mr. Mushet's decease in 1891, having thus paid him over £7,000. So, naturally, ends this part of the history of my invention, so far as Mr. Mushet is concerned.

[31] Brearley, op. cit., pp. 21–22.

[32] In 1883, R. F. Mushet wrote his version of the development of the pneumatic process in a privately published pamphlet entitled "The Bessemer-Mushet Process or the Manufacture of Cheap Steel". The text is reprinted in Osborn, op. cit., pp. 136–154.

[33] Autobiography, pp. 294–295.

Bessemer, at the end of his term as President of the Iron and Steel Institute in 1873, arranged for funds to be available for the annual presentation of the "Bessemer Gold Medal" to a recipient considered to have made a significant contribution in the field of ferrous metallurgy. In 1876, when the President was William Menelaus of the Dowlais Iron Works, which had benefited as much as any from Mushet's addition of spiegeleisen, the award was made to Mushet. In his preamble to the award, Menelaus stated that it was needless to enquire very particularly what success attended Mr. Mushet's attempt to improve the old process, because they were all overshadowed by the beautiful invention of the spiegeleisen process, as applied to Mr. Bessemer's great invention, and it was on that ground that the Council resolved to pay Mr. Mushet the compliment that they did. He continued:[34]

> It was an invention which was worthy of being associated with the great invention of Mr. Bessemer. The two inventions would go down in history together, in fact the one was the complement of the other and no man would be better pleased than his friend Mr. Bessemer that the council had resolved to pay the compliment to Mr. Mushet. It had made the invention of Mr. Bessemer perfect and probably would be used as long as Bessemer metal was made.

Bessemer supported the award and concluded his remarks saying:

> There were reasons why I did not acknowledge the validity of Mr. Mushet's patent. I will not go into that subject. Mr. Mushet has never pressed the question and I may say that for many years past both I and Mr. Mushet have been perfectly in accord on the subject; we have buried the hatchet and I hope and believe we entertain the most mutual respect for each other. I am happy to know that Mr. Mushet has that medal presented to him and think he richly deserves it.

Henry Bessemer had retired from actual involvement in the business of steelmaking in 1872, his last patent referring to steel having been taken out three years earlier. He became interested in astronomy, having constructed a telescope for the observatory at his residence set in forty acres of ground at Denmark Hill, then a London suburb. He worked on the grinding and polishing of lenses and also on the provision of a furnace powered by solar heat. Obviously, he continued to take an interest in steelmaking matters, as reference to the discussions at the meetings of the Iron and Steel Institute will show. He was

[34] The report of the speeches with reference to the presentation of the Bessemer Gold Medal to R. F. Mushet is given in J.I.S.I., 1876, Pt. 1, pp. 7–8. Mushet himself was recovering from illness and could not, unfortunately, be present to receive the public recognition which he so richly deserved. The President of the Institute subsequently travelled to Tressilian in Cornwall, where Mushet was convalescing, to present the medal.

knighted in 1879. He presented a paper to the Iron and Steel Institute in 1886,[35] dealing with the development of the Bessemer converter itself. In 1896, just forty years after that fateful meeting in Cheltenham, he was made an honorary member of the American Society of Mechanical Engineers and asked to present a paper. Never having been a good sailor – indeed one of his later projects was the production of a ship which had a swinging saloon on a gyroscopic principle to insulate the passengers against sea sickness – and in view of his age, he forwarded a written presentation on the origins of the Bessemer Process.[36] In the discussion of this paper, Robert W. Hunt, who had been in charge of the pioneering operations in Bessemer steelmaking in the United States, made the following comment:

> I have great respect for Mr. Bessemer and it is with great hesitation that in his advanced age I would permit myself to say anything which would seem to take away an atom from his honor and glory, but I wish he were just a little more generous and a little more just.

Considering his reaction to the contributions made by Goransson and Mushet and, as will be noted later, to the American William Kelly, such a comment does not appear inappropriate. Notwithstanding this, however, the value and impact of the work of Bessemer for the future of the steel industry and, in consequence, for the growth of engineering and thus of the general economic growth in the latter half of the nineteenth century cannot ever be overestimated.

Sir Henry's wife died in 1897; they had been married for sixty years and he never recovered from his grief, dying the following year at the age of eighty-five.

[35] H. Bessemer: "On Some Early Forms of Bessemer Converter", J.I.S.I., 1886, Pt. 2, pp. 638–649.

[36] H. Bessemer: "Historical and Technical Sketch of the Origin of the Bessemer Process", Trans. Amer. Soc. Mech. Engs, Vol. 18 (1897), pp. 455–475; discussion pp. 480–491.

4 The Bessemer Process in Britain

The Bessemer Process in Britain may be considered to have developed to the commercial stage at the time that W. D. Allen reported in his diary that steel had been made by the direct process. This was operated in the tilting converter which had been developed at the Sheffield works set up by Bessemer and his partners, casting the metal into a ladle and thence into ingot moulds. It is a remarkable fact that this system, with some minor modifications, was to be standard practice, not only in Britain, but virtually world wide, for the next fifty years or more; in fact, the last converter to operate in Britain, which finished its productive life at Workington in 1974 and now stands outside the Kelham Island Industrial Museum in Sheffield, is recognisably of the same design.

The development of the converter itself from the first fixed cylindrical type at St. Pancras to the conventional vessel on trunnions is a matter of considerable interest, particularly as Bessemer himself provided an historical survey.[1] The inventor soon became aware of the advantages of a movable vessel, which could be blown in the upright position and could then be rotated to run the metal out. Indeed, he patented such a furnace as early as the end of 1855;[2] this was virtually spherical and mounted on trunnions, with a pouring lip and a charging spout/flame exit at opposite ends of a chord in the upper part of the vessel [Fig. 8]. The tuyere entered half way between them, at an angle of 45° to the surface of the metal in the blowing position, being inserted before blowing and removed at the end of the blow before tapping. Bessemer points out that this form was never constructed, the upright converter being preferred for the early experiments. The first tilting converter actually put into operation was the one installed originally in the Sheffield works and made by Galloways of Manchester.[3] This had a single tuyere blowing upwards, with a retention chamber on top of the vessel to trap

[1] H. Bessemer: "On Some Early Forms of Bessemer Converter", J.I.S.I., 1886, Pt. 2, pp. 638–649.

[2] H. Bessemer, British Patent 2768, 7 December 1855.

[3] Presumably this was the pattern of converter supplied by Galloways to G. F. Goransson at Edsken in 1857 and subsequently discarded by him in favour of the earlier fixed converter, but with twelve tuyeres set tangentially to the circumference, giving rotation to he metal in the converter.

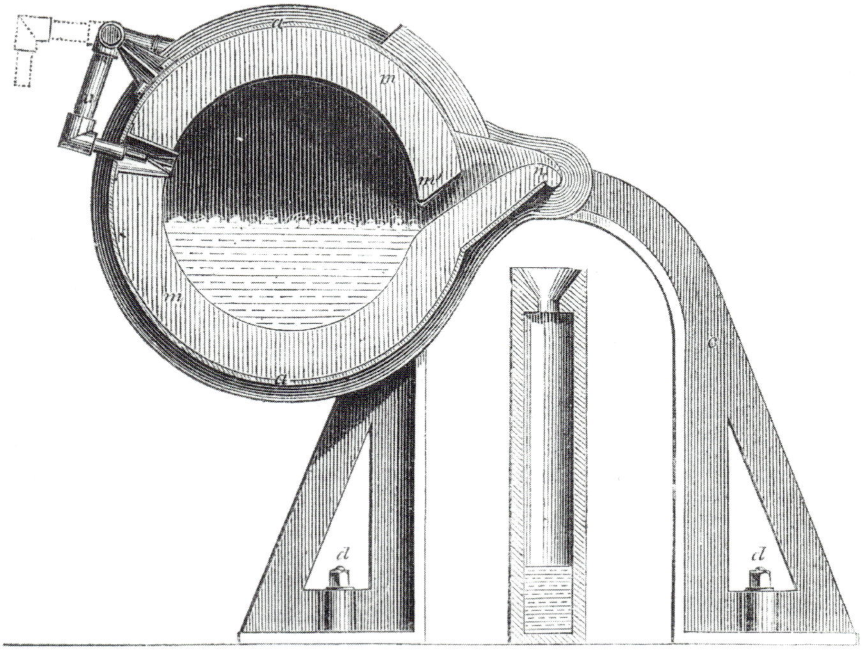

Fig. 8. Earliest form of rotative converter designed by Bessemer. This was a form of vessel which was never actually constructed but was patented at the end of 1855; Bessemer thought fit to record it as part of his presentation on the origin of the Bessemer Process both to the Iron and Steel Institute in 1886 and to the American Society of Mechanical Engineers in 1896; the illustration is reproduced from the report on the latter in *Engineering*, 11 December 1896, p. 747.

most of the splashed out material. This suffered from the defect of not raising the tuyere above the level of the metal during pouring; Bessemer states, however, that this was of little consequence since it was chiefly used for the production of a pure malleable iron from Swedish cast iron, the product then being poured into water to granulate it. A further modification came with the design of a side blown converter on trunnions which could be rotated so as to put the tuyeres out of contact with the metal on charging and tapping. This led logically to the bottom blown converter on trunnions, which is the form which was to be a common sight in steelworks throughout the world for nearly a century; the original form appears to have been installed in Bessemer's Sheffield works early in 1859.[4] It was, then, this form which was seen by John Brown and other Sheffield steelmakers when they looked into the works on Carlisle Street [Fig. 9].

The inventor's own description of the rationale of his process in its fully developed form seems appropriate at this stage. He reported the operations using the bottom blown tilting converter in his own Sheffield works in May 1859.[4] It seems probable that this had a capacity of 33 cwt of iron.[5] He later reported more fully in his description of the operations at John Brown's works in 1861 using a 4 ton converter:[6]

> This process has been in daily operation in Sheffield for the last two years. The apparatus by which it is effected is shown in Plates 30 to 33 [here reproduced as Figs. 10 to 13], which represent the arrangement at Messrs. John Brown and Co's Atlas Steel Works, Sheffield; Fig. 1 [Fig. 10] is a side elevation and Figs. 3 and 4 [Fig. 11] a front elevation and plan. The crude pig iron chiefly used in this process has been the hot-blast haematite pig smelted with coke, which is melted in a reverberatory furnace adjoining and is then run into the converting vessel in which its conversion into steel is to be effected. The converting vessel, A, is shown enlarged in section in Fig. 5 [Fig. 12], which represents its position in filling, the melted pig iron being run into it by the spout, B, direct from the furnace. It is made of stout boiler plate and lined with a powdered siliceous stone found in the neighbourhood of Sheffield below the coal and known as "ganister". The rapid destruction of the lining of the converting vessel was one of the greatest difficulties met with in the early stages of the invention: the excessive temperature generated in the vessel during the 30 minutes required to convert a single charge of iron into steel. The ganister now used is not only much cheaper than firebricks, costing only about 11s. [55p.] per ton in the powdered state, but it is also very durable. [A portion of lining which had withstood 96 consecutive conversions before its removal was exhibited at the meeting.] The converting vessel, A, is mounted on bearings which rest on stout iron standards, CC, Figs. 3 and 4 [Fig. 11] and by means of the gearing and handle, D, it may be

[4] H. Bessemer in a lecture presented to the Institute of Civil Engineers in London on 24th May 1859 entitled: "On the Manufacture of Malleable Iron and Steel" (Reported in "The Mechanics' Magazine", 17.6.59 and 24.6.59, pp. 397–398 and 405–407).

[5] S. Jordan: "Revue de l'Industrie du Fer en 1867", Vol. 4, Paris, 1871, p. 363.

[6] H. Bessemer: "On the Manufacture of Cast Steel and its Application to Constructive Purposes", Proc. Inst. Mech. Eng., 1861, pp. 133–157.

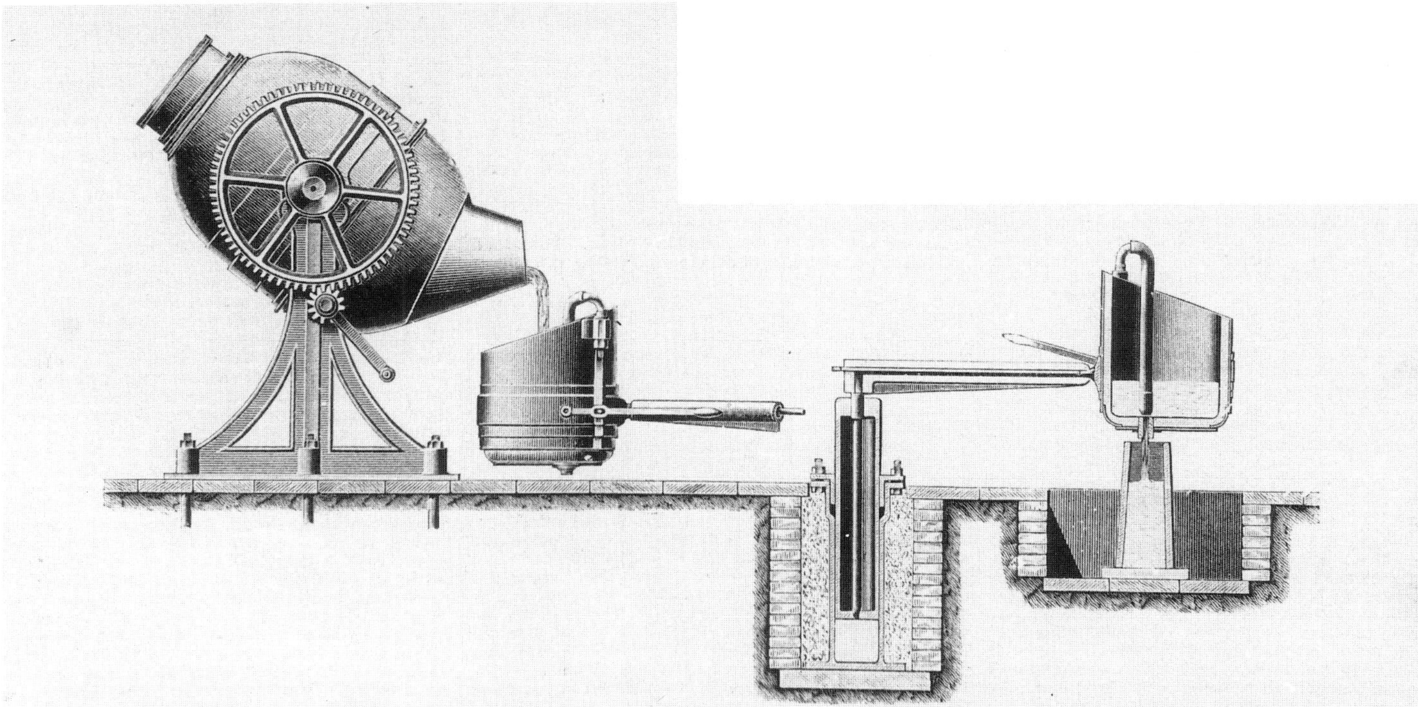

Fig. 9. An early form of the Standard Rotative Bessemer Converter Vessel. This is a reproduction of Fig. 45 from *Autobiography*. It is presumed that it represents the first commercial scale plant erected at Bessemer's own works in Sheffield around 1860, the type of plant which would be seen in operation there by John Brown, Charles Cammell and the other Sheffield steelmakers who were later to adopt the process in their own works.

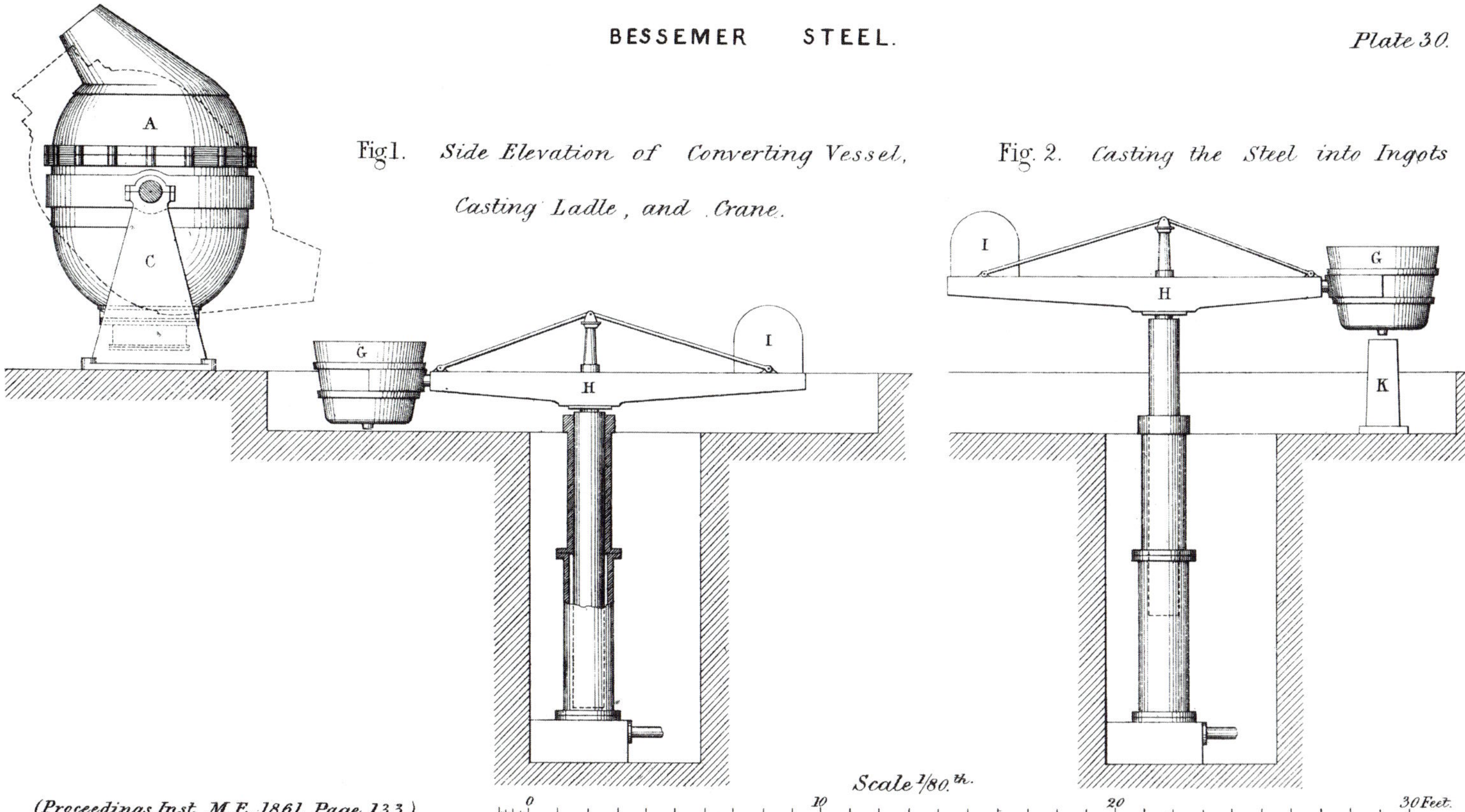

Figs. 10-13. Further details of the early Rotative Bessemer Converter. These illustrations come from the Proceedings of the Institute of Mechanical Engineers, 1861 and are the illustrations from a paper "On the Manufacture of Cast Steel and Its Application to Constructive Purposes" given by Bessemer.

turned into any required position. There is an opening at the top for filling and pouring out the metal; and at the bottom of the vessel are inserted seven fireclay tuyeres, Fig. 9 [Fig. 13], each having seven holes, as shown enlarged in the longitudinal section and plan, Figs. 10 and 11 [Fig. 13]. The blast from the engine is conveyed through one of the bearings, E, of the vessel, Fig.3 [Fig. 11], into the tuyere box, F, and enters the tuyeres at a pressure of about 14 lbs. per square inch, which is more than sufficient to prevent the fluid metal from entering the tuyeres.

Before commencing with the first charge of metal, the interior of the converting vessel is thoroughly heated by coke, with a blast through the tuyeres to urge the fire; when sufficiently heated it is turned upside down and all the unburnt coke falls out. The vessel is then turned into the position shown in Fig. 5 [Fig. 12] and the melted pig iron is run in from the furnace by the spout, B, the vessel being kept in such a position during the time it is being filled that the holes of the tuyeres are above the surface of the metal. When the proper charge of iron has been run in, the blast is turned on and the vessel quickly moved up to the position shown in Fig. 6 [Fig. 12]. The blast now rushes upwards into the fluid metal from each of the 49 holes of the tuyeres, producing a most violent agitation of the whole mass. The silicium always present in greater or less quantities in pig iron is first attacked and unites readily with the oxygen of the air, producing silicic acid; at the same time a small portion of the iron undergoes oxidation and hence a fluid silicate of iron is formed, a little carbon being simultaneously burnt off. The heat is thus gradually increased until nearly all the silicium is oxidised, which generally takes place in about 12 minutes from the commencement of the process. The carbon of the pig iron now begins to unite more freely with the oxygen of the air, producing at first a small flame, which rapidly increases, and in about 3 minutes from its first appearance a most intense combustion is going on: the metal rises higher and higher in the vessel, sometimes occupying more than double its former space, and in this frothy fluid state it presents an enormous surface to the action of the air, which unites rapidly with the carbon contained in the crude iron and produces a most intense combustion, the whole mass being in fact a perfect mixture of metal and fire. The carbon is now burnt off so rapidly as to produce a series of harmless explosions, throwing out the fluid slag in great quantities; while the combustion of the gases is so perfect that a voluminous white flame rushes from the mouth of the vessel, illuminating the whole building and indicating to the practised eye the precise condition of the metal inside. The blowing may thus be left off whenever the number of minutes from the commencement and the appearance of the flame indicate the required quality of metal. This is the mode preferred in the working of the process in Sweden. But at the works in Sheffield it is preferred to continue blowing the metal beyond this stage, until the flame suddenly drops, which it does just on the approach of the metal to the condition of malleable iron: a small measured quantity of charcoal pig iron containing a known proportion of carbon is then added, and thus steel is produced of any desired degree of carburation, the process having occupied about 28 minutes altogether from the commencement. The converting vessel is tipped forwards and the blast shut off for adding this small charge of pig iron, after which the blast is turned on again for a few seconds.

The vessel is then turned into the position shown in Fig. 7 [Fig.13] and the fluid steel run into the casting ladle, G, which is carried by the hydraulic crane, H, being counterbalanced by the weight, I, on the opposite end of the jib. When all the metal

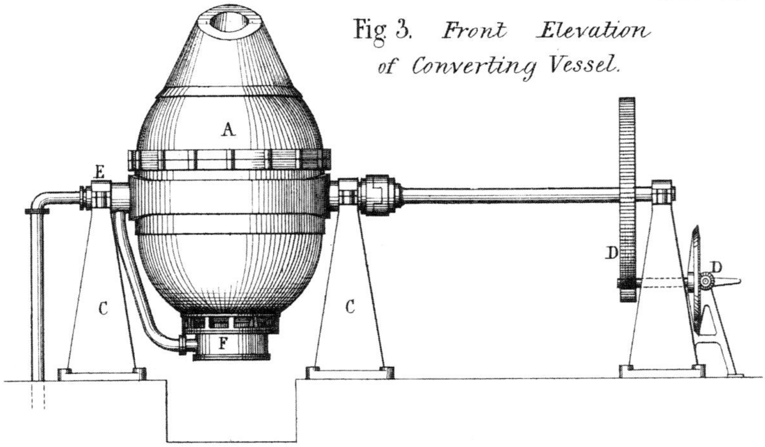

Fig 3. *Front Elevation of Converting Vessel.*

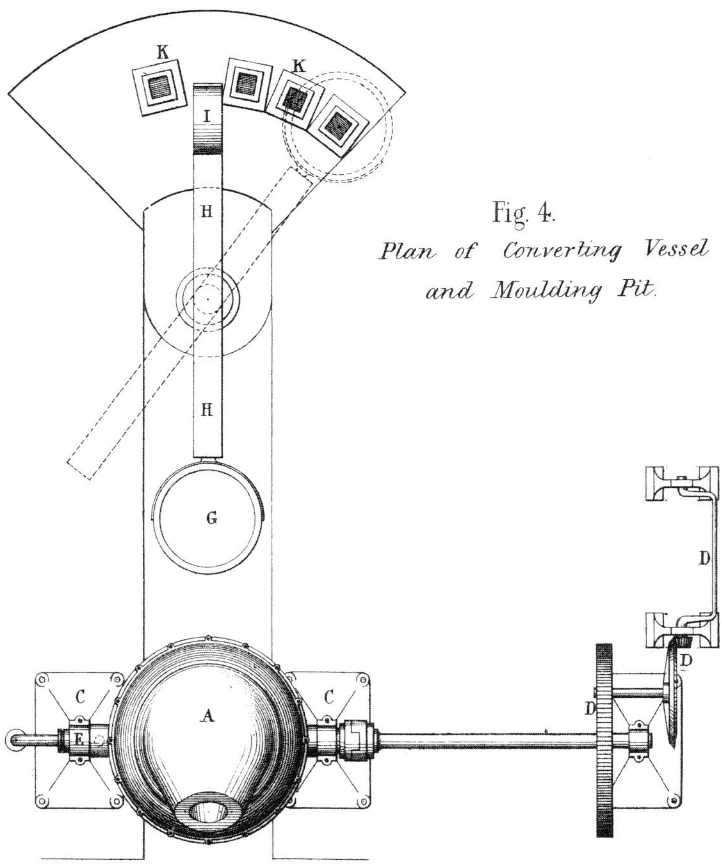

Fig. 4. *Plan of Converting Vessel and Moulding Pit.*

Scale 1/80th 0 5 10 15 20 *Feet*

(Proceedings Inst. M.E. 1861. Page 133.)

Fig. 11

BESSEMER STEEL.

Positions of Converting Vessel in filling and blowing.

Fig. 5. *Filling.*

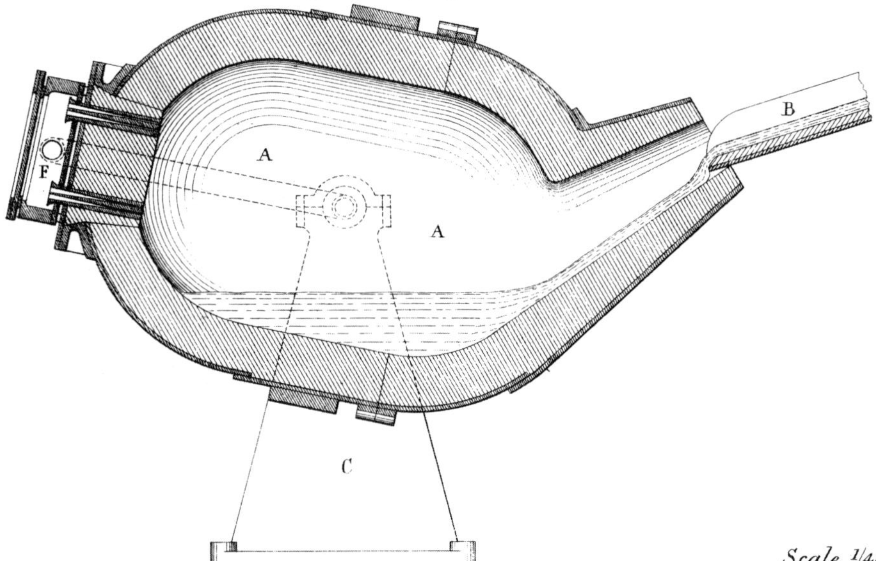

Fig. 6. *Blowing.*

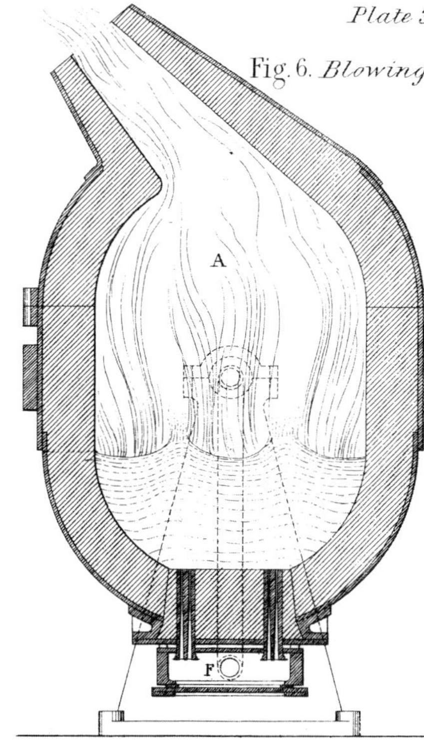

Scale 1/40th. 0 1 2 3 4 5 6 7 8 9 10 11 12 13 14 15 Feet.

(Proceedings Inst. M.E. 1861. Page 133.)

Fig. 12

> is poured out of the converting vessel, the crane is raised by water pressure and turned round, as shown in Fig. 2 [Fig. 10], for the purpose of running the steel into the ingot moulds, K. Instead of tilting the casting ladle for pouring into the moulds, it is made with a hole in the bottom fitted with a fireclay seating, L, Fig. 8 [Fig. 13], and closed by a conical plug of fireclay, M, forming a conical valve. The valve rod, N, is coated with loam and bent over at the top, and works in guides on the outside of the ladle, as shown in Fig. 7 [Fig. 13], with a handle, O, for opening and closing the valve. By thus tapping the metal from below, no scoria or other floating impurities are allowed to run into the mould and the stream of fluid steel is dropped straight down the centre of the mould right to the bottom, without coming in contact with the sides of the mould. The moulds are made of a slightly tapered form, as shown in Fig. 8 [Fig.13], so that as the ingot contracts in cooling it liberates itself from the mould completely on all sides; and the mould is removed by being lifted off the ingot when sufficiently set. The moulds are arranged in the moulding pit in an arc of the circle described by the casting ladle, as shown in the plan, Fig. 4 [Fig. 11]. By this process from 1 to 10 tons of crude iron may be converted into cast steel in 30 minutes, without employing any fuel except that required for melting the pig iron and for the preliminary heating of the converting vessel, the process being effected entirely without manipulation. The loss in weight of crude iron is from 14 to 18 per cent with English iron worked in small quantities; but the result of working with a purer iron in Sweden has been carefully noted for two consecutive weeks and the loss in weight of fluid iron tapped from the blast furnace was ascertained to be only 8.75 per cent. The largest sized apparatus at present erected is that in use at the Atlas Steel Works, Sheffield, as shown in the drawings already described, the converting vessel being capable of converting 4 tons at a time, which it converts into cast steel in 28 minutes. In consequence of the increased size of the converting vessel in this case, no metal is thrown out during conversion; and the loss of weight has fallen as low as 10 per cent, including the loss in melting the pig iron in the reverberatory furnace.

This description adequately describes the operation of the Bessemer process, whether it be in a 4-ton converter in 1861 in Sheffield or in a 30-ton converter in 1974 in Workington, the only essential difference being in the size of the converting vessel.

It does, however, yield just a little in effect as compared with Bessemer's earlier description[4] of the burning out of the carbon:

> . . . the increase in temperature which the metal undergoes, and which seems so disproportionate to the quantity of iron and carbon consumed, is doubtless owing to the favourable circumstances under which combustion takes place. There is no intercepting material to absorb the heat generated and to prevent it being taken up by the metal; for heat is evolved at thousands of points, distributed throughout the fluid, and when the metal boils, the whole mass rises far above its natural level, forming a sort of spongy froth, with an intensely vivid combustion going on, in every one of its numberless, ever-changing cavities. Thus, by the mere action of the blast, a temperature is obtained in the largest masses of metal, in ten or twelve minutes, that whole days of exposure in the most powerful furnaces would fail to produce. The changes in the colour and volume of the flame, and the kind of sparks thrown off, afford easy modes of judging the state of the metal,

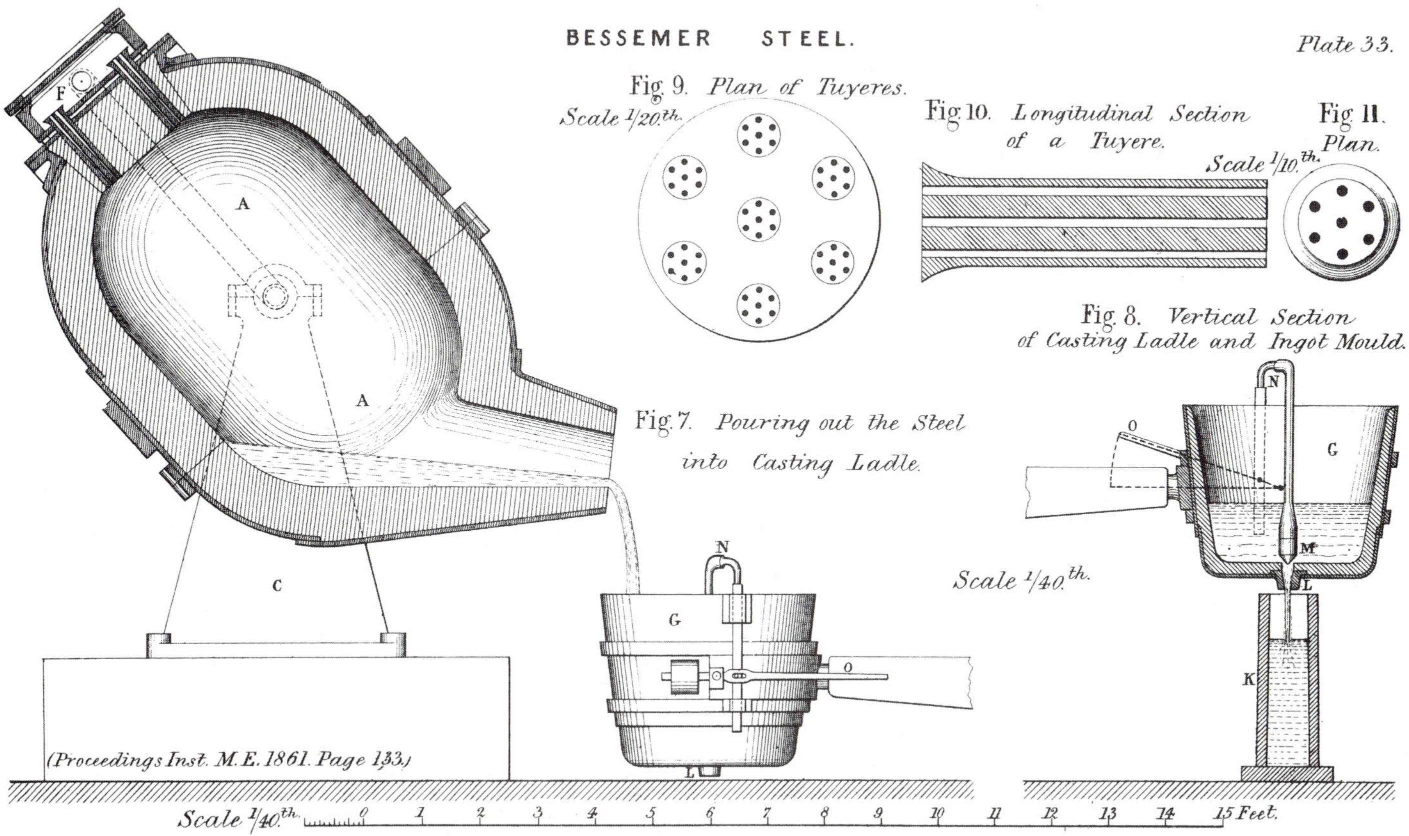
BESSEMER STEEL.
Plate 33.
Fig. 9. Plan of Tuyeres.
Scale 1/20th.
Fig. 10. Longitudinal Section of a Tuyere.
Fig. 11. Plan.
Scale 1/10th.
Fig. 8. Vertical Section of Casting Ladle and Ingot Mould.
Fig. 7. Pouring out the Steel into Casting Ladle.
Scale 1/40th.
F
A
A
C
G
N
O
L
M
K
(Proceedings Inst. M.E. 1861. Page 133.)
Scale 1/40th.
0 1 2 3 4 5 6 7 8 9 10 11 12 13 14 15 Feet.

Fig. 13

> since these are given off exteriorly and are not interfered with by the flame of the fuel. The sound which the metal produces in the suspended vessel affords, also, a good indication to the workman. Indeed, few processes appeal so strongly to the external senses.

One very interesting feature of both these descriptions, particularly in view of Bessemer's later writings in which he is at pains to make it clear that the use of manganese in steelmaking was something with which he had been familiar for many years, is that there is no mention of a final addition in the 1859 paper and only an addition of:

> a small measured quantity of charcoal pig iron containing a known proportion of carbon

in the 1861 paper. This is in contrast to the description given by a French eyewitness of operations in Bessemer's own works in 1862, to be quoted shortly, in which additions of from 5–8% of spiegeleisen are precisely quoted.

John Brown's entry into Bessemer steelmaking was a calculated one, based on his appreciation of the value of steel in the production of railway rails. Up to 1859, he had been proposing to use puddled steel for rail production and, indeed, had built up the number of puddling furnaces at Atlas Works to a total of 72.[7] Having seen Bessemer's converter in action, however, he changed his plans and, in 1859, he took out a licence from Bessemer, which he arranged so that the royalty would be only £1 per ton on steel for rails but £2 per ton on steel for other purposes.[8] Up to that date, rails were produced in wrought iron, but these wore out quickly, particularly where the traffic was heavy. The first ever steel rail was produced late in 1856. The scrap from an abortive Bessemer trial at Ebbw Vale was remelted by Mushet[9] in crucibles: 44 lb. of scrap were charged to each of sixteen crucibles and, when molten, 3 lb. of molten spiegeleisen were added to each crucible, the contents of the sixteen crucibles then being poured to give a single ingot. The ingot was sent to the Ebbw Vale Company and rolled to give a sound double-headed rail. The rail was forwarded to Derby Station early in 1857 and was laid down at a part of the track where, it was stated, the iron rails had to be renewed every six months or sooner. Mushet asked that it be returned to him when it was removed. It seems that he was reminded of this by reference to his old note book in June 1873 – sixteen years later! – and, on contacting the Midland Railway Company was informed that it had been taken up and scrapped only ten days earlier, an act which Mushet characteristically described as vandalism! There is also a "legend" which concerns another stretch of the Midland Railway, that which

[7] S. Griffiths (ed. W. K. V. Gale): "Guide to the Iron Trade of Great Britain, 1873", Newton Abbot, 1967, p. 279.

[8] A. J. Grant: "Steel and Ships", London, 1949, p. 19.

[9] F. M. Osborn: "The Story of the Mushets", London 1952, p. 42, p. 82.

ran between the two parts of John Brown's works.[10] John Devonshire Ellis, John Brown's partner, is said to have taken a gang of his own works platelayers, "one dark and stormy night", and surreptitiously removed one of the iron rails from this track, substituting a Bessemer steel rail of precisely the same section. He said nothing of this to the Midland Railway, but about six months later he was asked by one of their engineers to come to look at a rail which was showing no signs of wear, whereas the iron rails on either side had worn down as expected. He then confessed to the authorities what he had done and promptly received orders for the whole of the Midland Railway system. So the story goes; there is no doubt, however, that the extended life of steel rails soon became appreciated. They were, indeed, more expensive; figures quoted for 1864 indicate that iron rails from Cleveland sold at £7 per ton against £17.10.0 per ton for Bessemer steel rails.[11] On the other hand, apart from their increased life, the labour cost and traffic problems involved in rail replacement were greatly reduced and it soon became clear that there was a large market for this product. Bessemer's own works had supplied 500 tons of steel blooms to the London and North Western Railway Company's rail mill at Crewe in 1860[12] and steel rails rolled in this mill were exhibited at the London International Exhibition in 1862. Steel rails laid at Crewe Station in March 1863 were reported still in place eighteen years later.[13]

John Brown supplied his first rails in 1860; in 1862 there is a report that he was then operating three converters with a capacity of 4 to 5 tons each. Towards the end of 1861, Charles Cammell, whose Cyclops Works adjoined John Brown's Atlas Works, entered the same trade, with two 4 ton converters. During the same year, Bessemer plants, with 3 ton vessels, were also established by the Weardale Iron and Steel Company at Tudhoe in County Durham – this company supplied the first steel rails in the North East – and by Rowan and Company in Glasgow; it is not clear what was the intended product from the latter works. In 1862 Charles Cammell extended his Bessemer operations by taking over a works already built by Adamson at Penistone; Samuel Fox at Stocksbridge also took out a licence from Bessemer. In 1863 the Bolton Iron and Steel Company commenced the production of Bessemer steel rails, as did the Manchester Railway Steel and Plant Company at Newton Heath, Manchester. They were joined the following year by the Lancashire Steel Company at Gorton, whilst the London and North Western Railway Company installed Bessemer converters that year. It is said that 70 tons of Bessemer steel were being made

[10] Grant, op. cit., pp. 19–20.

[11] Royal Commission on the Depression of Trade, 2nd, report, 1886, p. 323.

[12] K. Warren: "The Sheffield Rail Trade, 1861–1930", Trans. Inst. British Geography, June 1964, p. 132.

[13] J. C. Carr and W. Taplin: "History of the British Steel Industry", Oxford, 1962, p. 27.

every week in Sheffield in 1863 and that by 1865 the three largest firms were between them providing over 400 tons per week.[14]

One of the most valuable sources of information on the progress of the Bessemer trade is an extensive report dealing with the iron and steel exhibits at the Universal Exhibition in Paris in 1867.[15] This also recalls that, at the similar Exhibition in London in 1862, Bessemer displayed a remarkable collection of items in Bessemer steel, all from material produced in Sheffield and including a broken section of a 14-inch octagonal ingot showing freedom from blowholes; an 18 inch ingot weighing 28 cwt (the 6,410th ingot made in his works), cannons and bullets, sabres, swords, breastplates, helmets, guns and pistols; axles, pistons and cylinders; a 40 foot rail; railway wheel tyres; steel wire, cutting tools, cutlery and metal pens made from Bessemer steel; and numerous examples of bend tests demonstrating the ductility of the steel, all showing close bends without any cracking.

This report is even more valuable, in that it records an eye witness account of some early operations in Bessemer's own works.[16] A translation gives the following information:

> In 1862 we followed several operations and collected information which proved to us that the method had been thoroughly worked out at that time and was now a practical proposition. Over a period of three weeks, 90 blows were carried out with the same ganister lining, originally about 4 to 5 inches thick; in the converter, which was about 4 feet in external diameter, were treated charges of from 22 to 33 cwt of iron, which formed a bath about 12 inches deep. This was blown through nine tuyeres, each with five holes, with air compressed to around 70 cm. mercury. To produce harder steel, Swedish irons were employed and, for the softer steels, haematite irons from Cleator. The operations were quite rapid and 4 to 8% of spiegeleisen were added for the final recarburisation.

The preamble to this section of the French report covers a general discussion of the development of the process. In this, criteria were laid down for the appropriate composition of the cast irons used.[17] These had to be sufficiently hot (that is, in the blowing) and it was made clear that the major provision of heat was from the combustion of the silicon in the iron, manganese giving some extra heat input, but the carbon contributing very little to the final metal temperature. It was pointed out, however, that an iron low in silicon and high in manganese did not produce a suitably fluid slag. It was suggested that a silicon content over 2.5% was also detrimental, since it was very likely to produce a steel in which all the silicon had not been

[14] Warren, op. cit., p. 135.
[15] Jordan, op. cit., pp. 342–395.
[16] Jordan, op. cit., p. 364.
[17] Jordan, op. cit., pp. 344–346.

removed and this would be brittle. Ideally, it was suggested the silicon content should be from 1.75–2.00%; in irons with high manganese there should be at least 1% silicon for every 2.75% manganese. Additionally, the iron should be sufficiently pure. In particular, this referred to absence of phosphorus. It was pointed out that sulphur was not so injurious, since some was eliminated during the blow, and what remained was neutralised in part by the manganese present in the final steel. The report also suggested that the presence of arsenic, aluminium and copper was undesirable.

By 1867 the addition of manganese was fully accepted but the report indicates that use of spiegeleisen was being questioned.[18] The addition of massive pieces to the metal in the converter could, in some cases, lead to inhomogeneity. The use of a remelted spiegeleisen, however, could result in serious losses of manganese from the metal. Some operators carefully remelted the spiegeleisen in closed crucibles and then granulated the metal by pouring into water; this retained the manganese in the metallic state, whilst the small physical size of the product ensured its ready solution in the liquid metal in the converter. There was, however, an increasing call for soft steels, in the main to replace wrought iron. The almost obligatory use of spiegeleisen was giving some problems in this connection; spiegeleisen usually contained between 4% and 5% carbon with around 8% to 10% manganese and the usual addition considered necessary to give fully deoxidised metal would give a final analysis of the order of 0.25–0.30% carbon. Searches were made, therefore, for materials with a higher manganese to carbon ratio and it was reported that the Phoenix Foundry in Glasgow had succeeded in producing an alloy with 25–30% manganese with 5% carbon. This matter was under constant survey and will be discussed more fully later.

A further interesting discussion centered around the means for identifying the instant at which the blow should be interrupted.[19] The general practice of observing the rapid drop in flame and cessation of the roaring noise and then immediately turning down the converter left something to be desired. Some works had developed the practice of dipping in an iron rod to sample the slag and to judge by the fluidity and the colour of the slag; at Hoerde in Westphalia, for instance, their practice gave rise to a grey blue, thin coating of slag on the rod when the conditions were correct, whilst a more yellow sample showed over oxidation. The most rational method so far employed, however, had proved to be the observation of the gases issuing from the converter by means of the spectroscope. This technique had been observed in operation at the Atlas Works of John Brown in Sheffield, being the application of work carried out by William Bragg, John Brown's

[18] Jordan, op. cit., pp. 352–353.
[19] Jordan, op. cit., pp. 354–357.

associate at the works.[20] This work had established that, during the early part of the blow, there were no distinguishable lines until the sodium line gradually established itself and then remained present throughout the cycle. Gradually, as the blow proceeded, numerous other lines appeared, some bright, some dark; those in the red region appeared to be related to manganese; those in the green region were characteristic of carbon. With the flame dying, at the moment when the reactions were complete, all the lines, except that of sodium, disappeared suddenly. On turning down the converter and adding the spiegeleisen, the carbon and manganese lines briefly reappeared and then died out. The method concentrated on the carbon lines in the green region and was found to be of universal application, whatever the size of converter or whatever type of cast iron was employed.

Professor H. E. Roscoe was a pioneer in the use of the spectroscope[21] and he later provided a most useful summary referring to work he carried out at the Atlas Works of John Brown and Company.[22] This included analyses taken at various times during the blow, although there is no indication of the time at which they were taken:

	Pig Iron	Sample 1	Sample 2	Sample 3	Final Steel
Carbon	3.57	3.04	1.64	0.19	0.37
Silicon	2.26	0.96	0.47	trace	trace
Manganese	0.41	trace	trace	trace	0.54
Sulphur	0.107	0.091	0.098	0.093	0.090
Phosphorus	0.073	0.070	0.070	0.070	0.059

20 Jordan reports that Professor Roscoe presented a lecture in Manchester dealing with the results obtained by applying the Spectroscope to Bessemer flames early in 1863. J. S. Jeans: "Steel: Its History, Manufacture, Properties and Uses", London, 1880, pp. 411–414, has an extended section on this topic and refers to a later contribution from Professor Roscoe. Other workers were M. W. Watts: "On the Spectrum of the Bessemer Flame", Phil. Mag., vol. xxxiv, pp. 318ff, J. M. Silliman: "On the Examination of the Bessemer Flame with Coloured Glasses and with the Spectroscope", Phil. Mag., Vol. xli, pp. 1ff and F. Rowan: "On the Examination of the Flame of the Bessemer Converter", Chem. News, Vol. xix, pp. 170ff. (All prior to 1871).

21 H. E. Roscoe: "On the Spectrum produced by the Flame evolved in the Manufacture of Steel by the Bessemer Process", Philosophical Magazine, vol. xxv, pp. 318ff.

22 H. E. Roscoe: "Spectrum Analysis", J.I.S.I., 1871, pp. 38–62

Then follows his table:

Time	Appearance to Eye	Appearance of Spectrum
First Stage		
0 to 4 min.	No flame seen	Faint continuous spectrum due to sparks of burning metal.
4 to 6 min.	Small pointed flame	Brighter with yellow sodium line flashing.
6 to 8 min.	Unsteady flame and explosions	Bright continuous spectrum with sodium line, permanent red lithium line and red and violet potassium lines.
Second Stage		
8 to 10 min.	Very bright and dense flame.	Bright carbon lines in red, green and blue together with all preceding spectrum.
10 to 14 min.	Flame less dense but bright	Bright green carbon lines more distinct.
Third Stage		
14 to 16 min.	Flame diminishing in size.	Bright green lines becoming faint.
16 to 18 min.	Flame drops, blow ends.	Bright green carbon lines suddenly disappear, leaving continuous spectrum.

Professor Roscoe went on to report that the method had been used successfully at the Atlas Works since 1863, was widely used in Styria and Saxony and that at Seraing in Belgium it had proved impossible to produce satisfactorily uniform steel until the spectroscope was applied.

The list of the British Bessemer installations reported in 1867 reads as follows:[23]

Henry Bessemer & Co, Sheffield	1 @ 0.5T; 2 @ 4T
John Brown & Co, Sheffield	2 @ 10–12T; 4 @ 4–6T
Charles Cammell & Co, Sheffield	2 @ 4T
Yorkshire Steel Company, Penistone	4 @ 7T
Thompson & Armstrong, Normanton	?
Samuel Fox & Co, Stocksbridge	2 @ 4T
Weardale Company, Tow Law	4 @ 3T
Rowan & Company, Glasgow	1 @ 3T

[23] Jordan, op. cit., pp. 370–371.

Workington Iron Company	?
Haematite Steel Company, Barrow	4 @ 5T; 6 @ 7T
Lancashire Steel Company, Gorton	2 @ 6T
Bolton Steel Works	2 @ 5T
Mersey Steelworks	2 @ 5T
Gibraltar Works, Manchester	4 @ 3T
L.N.W. Railway, Crewe	4 @ 5T
Lloyd, Foster & Company, Wednesbury	2 @ 3T
Dowlais Company	6 @ 5T
Ebbw Vale Company	1 @ 1.5T; 2 @ 5T
Bessemer Brothers, Greenwich	2 @ 5T

The comment is added that the installed capacity in 1867 was in excess of 300 tons, so that, counting only 4 operations daily for 300 days a year, this would lead to a Bessemer steel production in Britain of the order of 350,000 tons per annum.

Not only was John Brown and Company the pioneer organisation in operating the fully developed Bessemer Process in Britain; they were also looked upon as the leaders in understanding the capabilities of the process and, having operated three converters of 5 tons capacity, by 1869 had three Bessemer shops, two with two converters each of 5 to 6 tons and the third with two converters of 10 to 12 tons. They operated with cast irons from several sources, blending them as appropriate for the many and various grades of steel they produced, using, as has been pointed out, the spectroscope to assist in their control of the process. The principal products of the Atlas Works at this time were armour plates, rails, tyre rims, springs of all kinds, ship plate, vehicle axles, tool steels and general bars and billets. The weekly tonnage of Bessemer steel was of the order of 700 tons in 1867, with only four converters working; on full capacity it should have produced over 1000 tons per week.[24]

The largest Bessemer steelworks in Britain, however, was the Barrow Haematite Steel Company, with eleven blast furnaces operating on the high quality Furness ore. The steelworks building was over 700 feet long, with two bays, 85 feet and 105 feet wide. At one end of the narrower bay were four converters of 5 tons each, arranged in two pairs, back to back, behind each of which were four reverberatory furnaces for melting the iron. In the other bay were two groups of three 7 ton converters. The three furnaces in each group were capable of pouring their contents into the same ladle; they were all placed in such a position as to be capable of receiving the iron direct from the blast furnace. The steel from Barrow was well known for its softness and ductility; at the 1867 Exhibition were to be found examples of their plates, axles and rails, all illustrating the excellent quality of the metal.[25]

[24] Jordan, op. cit., pp. 366–368.
[25] Jordan, op. cit., pp. 368–369.

The works at Crewe, belonging to the London and North Western Railway Company, were singled out as being the most interesting of the British steelworks. They had four 5 ton converters which made, between them, some 12 to 14 blows per day, giving 325 tons per week. Most of this output was destined for railmaking. The ganister linings of the converters lasted for 90 to 120 blows – 450 to 600 tons per furnace before relining was required. Ingots for rails were reheated and rolled down in a cogging mill invented by Mr. Ramsbottom, the well known director of the works, before a further reheating for final rolling to rail section. The works at Crewe, with their various hammers, mills and workshops, were said to merit the attention of all engineers, being ample tribute to the genius of Ramsbottom, their creator.[26]

By 1871 the installed Bessemer capacity in Britain had risen by a third over the three years since the previous survey, to a total of 76 converters with a nominal total tonnage of 406, with a further 14 converters under construction.[27] The details of plant in being are as follows:

John Brown & Co	2 @ 15T; 2 @ 10T; 2 @ 7.5T
Charles Cammell: Sheffield	4 @ 5T
Penistone	4 @ 5T
Samuel Fox, Stocksbridge	2 @ 5T
Henry Bessemer & Co	2 @ 3.5T; 1 @ 1T; 2 @ 5T
Patent Shaft & Axle Co	2 @ 3.5T
Bolton Iron & Steel	4 @ 5T
Manchester Railway Steel & Plant	4 @ 3T
Bolckow, Vaughan & Co, Gorton	2 @ 5T
Whitworth & Co, Manchester	1 @ 4T
L.N.W. Railway, Crewe	4 @ 5T
Mersey Steel & Iron	4 @ 5T
Barrow Haematite Company	6 @ 5T; 12 @ 7T
Dowlais Works	6 @ 5T
Ebbw Vale Works	6 @ 5T
Weardale Iron and Steel	4 @ 3T

It should be noticed that the increase in capacity over this period was one of enlargement of existing facilities rather than the erection of new plants in fresh locations. In particular, the expansion at John Brown and Co and, even more striking, at Barrow, might be mentioned.

As regards new centres of production, however, these were now coming forward, as the list of plants under construction given below will indicate:

[26] Jordan, op. cit., p. 369–370.
[27] Collected Statistics, J.I.S.I., 1871, Pt. 2, p. xv.

Rowan & Co, Glasgow	2 @ 3T
West Cumberland Haematite Iron Co, Workington	4 @ 7.5T
Bolckow Vaughan, Middlesbrough	?
Carnforth Haematite Co	?
Hampton, Radcliffe & Co, the Ickles, Rotherham	2 @ 5T
Brown, Bayley and Dixon, Sheffield	2 @ 5T
Bessemer Steel and Ordnance, Greenwich	2 @ 4T
Owens Patent Wheel, Tyre & Axle Co	2 @ 5T

Production figures began to be collected in 1868 and for the period up to 1873 the ingot tons of Bessemer steel produced were recorded as follows:[28]

1868	1869	1870	1871	1872	1873
110,000	160,000	215,000	329,000	410,000	495,000

The general arrangements of the Bessemer Shop had followed those worked out in Bessemer's own Carlisle Street Works, with two converters placed opposite each other at the extremities of the casting pit. A typical "English Type" layout is shown in Fig. 14. John Brown, as usual the pioneer in such matters, altered the layout, bringing the converters together, placed side by side, an arrangement which allowed more room for the placing of the ingot moulds, removing the restrictions on volume casting placed by the circular pit.[29] The Atlas Works was visited by many interested parties[30] and there is no doubt that the later American Bessemer shops, with their improved so-called "American Type" layout [Fig. 15], owed not a little to what had been seen in Sheffield. John Brown also was the only steelmaker in Sheffield who saw fit to assure himself of an independent supply of iron by taking a joint share with Bolckow, Vaughan and Company in haematite ore mines in Spain and erecting two blast furnaces along Carlisle Street for the in-house production of iron for his Bessemer operations. There was also a smaller blast furnace for smelting ferromanganese from manganese ore.

28 Collected Statistics, J.I.S.I., 1875, p. 238.

29 Jeans, op. cit., p. 364.

30 In particular, William F. Durfee, who was to have considerable importance in establishing the Bessemer process in America is known to have visited Atlas Works. See J. McHugh: "Alexander Holley and the Makers of Steel", Baltimore, 1980, p. 144, This whole concept will be found under discussion in the section dealing with the Bessemer process in America. There is, however, a relevant paragraph in R. W. Hunt: "A History of the Bessemer Manufacture in America", Trans. Amer. Inst. Mining Eng., Vol. v (1876–7), pp. 214–215.

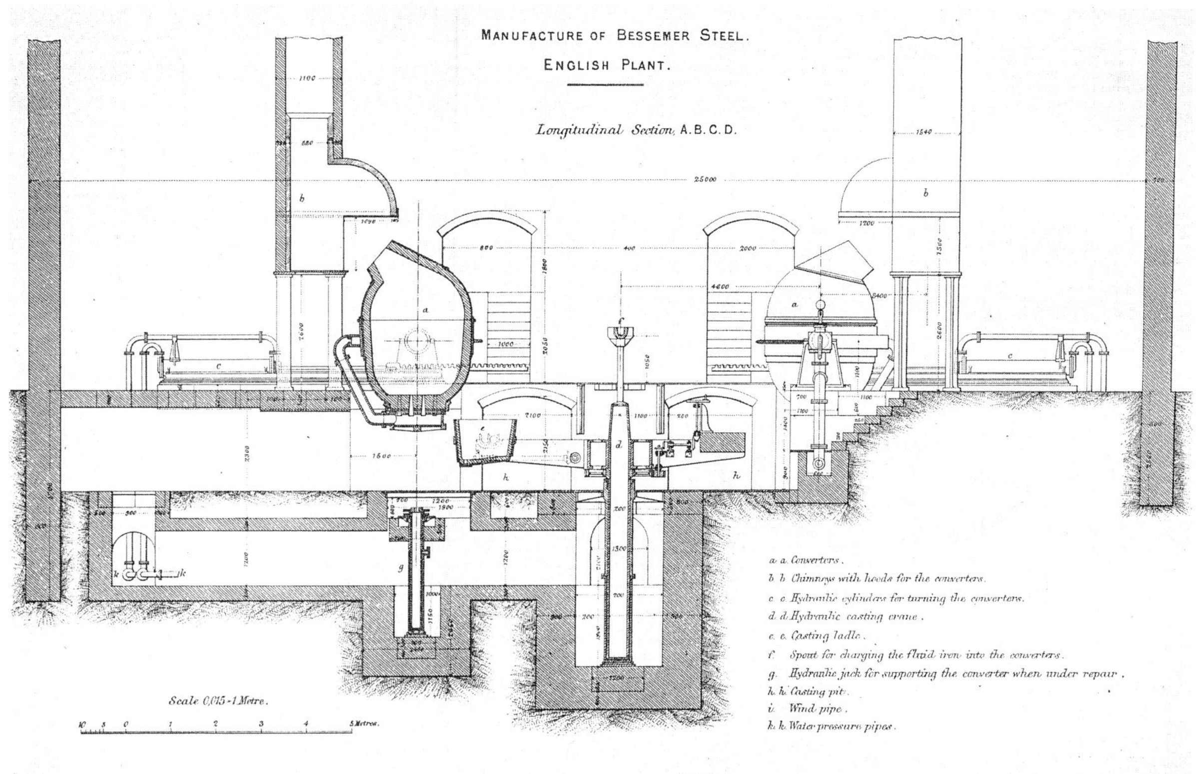

Fig. 14. Typical "English" Bessemer plant layout. This shows the typical layout of the early plants, with two vessels either side of a pit and a central rotating column carrying the ladles. The ingot moulds were set out along the two arcs of a circle away from the converter positions, so that they could be filled in turn from the ladle. The illustration is derived from J. S. Jeans: *Steel: Its History, Manufacture, Properties and Uses*, London 1880.

Fig. 15. Typical "American" Bessemer plant layout. This illustration is derived from an Encyclopaedia entitled *Chemistry: Theoretical, Practical and Analytical as Applied to the Arts and Manufactures.* This work was stated to be "by writers of eminence" and was published in 8 volumes by Wm. Mackenzie of London. The work is not dated but is clearly from the late 1870s. The drawing shows the open space in front of two vessels, set side by side and elevated above floor level. Casting was carried out at floor level and a sufficiency of crane power and rail transport ensured much higher outputs than in the earlier systems. This type of layout owed much to the engineering genius of Holley but seems to have germinated in the Atlas Works of John Brown in Sheffield.

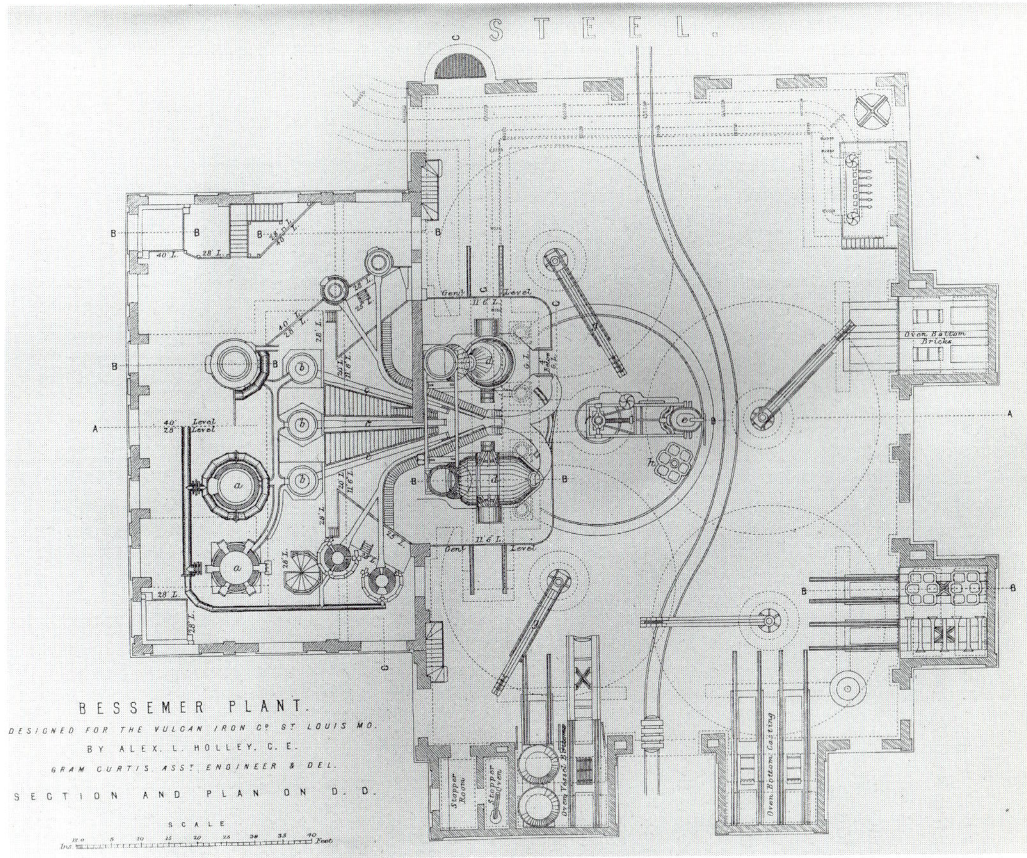

This was in 1872, the year in which the firm also purchased two collieries in the Rotherham area.[31]

It is not clear at what date the amendment of design of the converter took place to allow the base, which contained the tuyeres and wore much more rapidly than the lining, to become a replaceable unit; the base was fixed to the body by a number of cotter pins around the circumference and the blast pipe was made with a flange which would bolt on to the part fixed to the blast box and so to the tuyeres in the base [Fig. 16]. Certainly in the earlier days it was considered necessary to have a spare furnace on stand-by and this cut down the number of heats per day. The method of making the bottom good was to wait until the vessel was sufficiently cool for workmen to enter, to knock out the worn tuyeres, to place new ones in position and then to ram plastic ganister around the tuyeres from within. The whole had then to be fired until it was set. The vessel was normally cooled with water, to reduce the down time until the work could be done; this, obviously, did some damage to the lining and the bottom could not be back in service until the next day. The duplicate bottom was based on a scheme originally put forward by Bessemer in 1863,[32] although the earliest reference to it is from American practice reported in 1874.[33] A contemporary paper indicates that the plan to use what was technically referred to as a "dry bottom" was originally introduced by John Brown and Company in Sheffield and was by then used by most other makers in the district;[34] there is an inference that it was mainly to be found on larger vessels. Reports of operations at Workington and at Brown Bayley and Dixon's in Sheffield dating from early in 1875, two of the newer works which were in course of erection in 1871, clearly indicate that the movable bottom was standard practice at this time.[35] At Barrow in 1875, however, fixed bottoms were still in use. There, it was reported, the tuyeres lasted eight heats and the renewal of the bottom and the drying took 12 hours, during which time the other vessel of the pair would provide a further six heats; there was, therefore, no difficulty in keeping the output at

[31] Grant, op. cit., p. 32. In passing, there is a reference to the use of "John Brown Spiegeleisen", with 18–22% manganese, by Brown, Bayley and Dixon in 1874/5. (see footnote 35 below).

[32] H. Bessemer, British Patent 114, 13 January 1863.

[33] A. L. Holley: "Setting Bessemer Converter Bottoms", J.I.S.I., 1874, pp. 368–370. See also A. L. Holley: British Patent 1784, 22 June 1870.

[34] B. Walker: "Development of Latest Improvements in Appliances for the Manufacture of Bessemer Steel", J.I.S.I., 1874, p. 379.

[35] A. L. Holley: Private Publications: No. 4: West Cumberland Iron and Steel Works, Jan 1875, 38pp.; No. 6: Brown, Bayley and Dixon's Steel Works, May 1875, 18 pp.

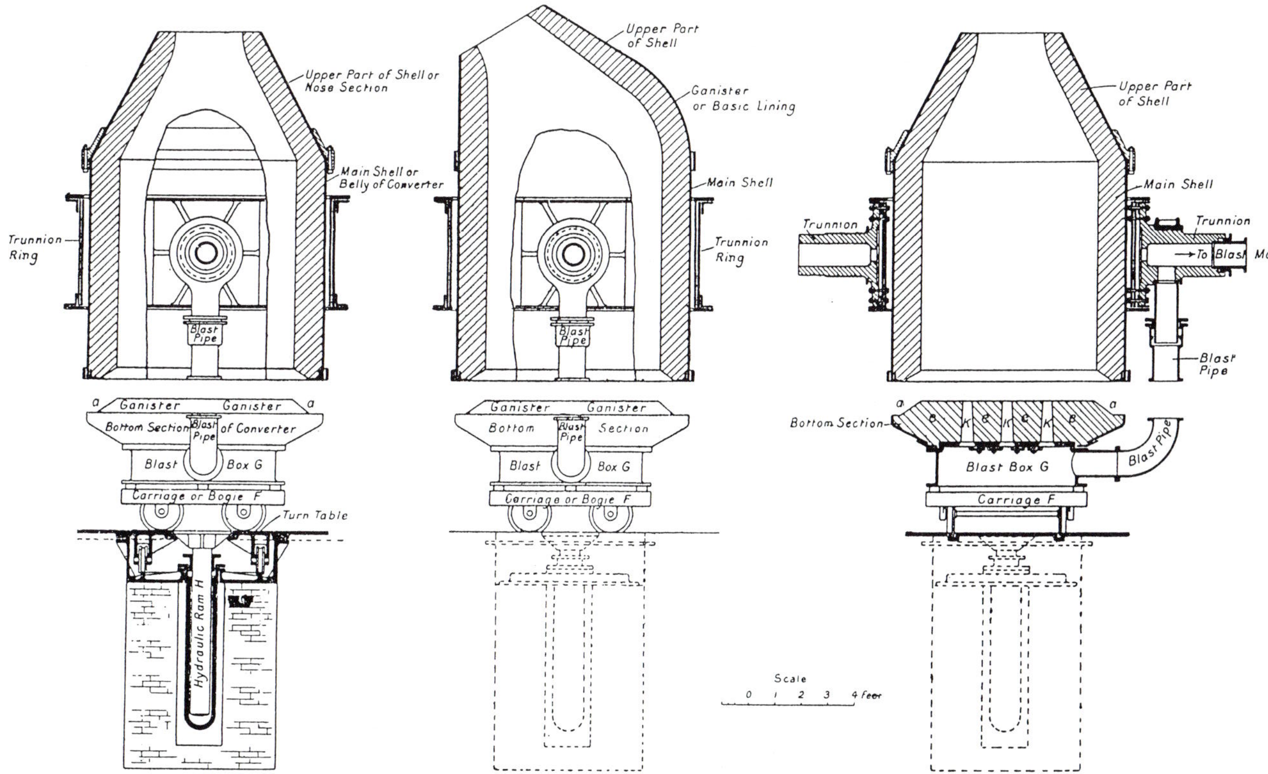

Fig. 16. A 16-Ton Converter Vessel showing the detachable bottom section. This is reproduced from Plate I of F. W. Harbord and J. W. Hall: *The Metallurgy of Steel*, 3rd. edition, London, 1907. The left hand section is from a Concentric Converter of the type favoured for basic operation by Windsor Richards; the central section is of a similar Eccentric Converter; the right hand section is taken at right angles to the others, indicating the tuyeres. Note the hydraulic ram arrangement for offering the bottom to the vessel prior to fixing with cotter pins.

this particular station with two furnaces.[36] This particular report has the interesting comment:

> I was surprised to find that old tuyere ends were not utilised in any better way than building them into walls and fences.

Details of ingot moulds from this period indicate that they were all used narrow end up.[37] Indeed this seems to have been the case well into the twentieth century.[38] There seems to have been some diversity of opinion as to whether a single piece mould or one split longitudinally was preferable. All the smaller moulds used in the crucible steel works were, of course, split moulds, held together with rings and wedges. With the advent of larger moulds, such as were needed for rolling into rails, the means of holding the two parts together became more critical and the provision of lugs and bolt holes rendered matters more complicated; some moulds were actually split on two opposite corners, with the faces planed so as to give a tight fit before bolting together. Such split moulds were used at Barrow in 1875 and a special design with a rotating clamp operated by a cam at Brown, Bayley and Dixon in 1875; at the same date, however, Workington were using solid moulds. There was always a tendency for split moulds to warp in service and the single piece mould gradually replaced them.

The method of casting altered with time. With more metal available, multiple ingot pouring looked attractive and in 1873 a method of uphill casting in clusters was proposed and seems to have been put into practice

[36] A. L. Holley: Private Publication: No. 7: Barrow Haematite Iron and Steel Works, July 1875, 31 pp. These "Private Publications" provide a most valuable source of information on steelmaking operations over the period 1874 to 1880. Each of them carries the following "Note" on the title page:

> "This is one of a series of papers which I am putting into print, not as a publication, but simply for the convenience of my clients. No copies will be distributed except to my clients. I have obtained the more important parts of the information these papers will contain on the understanding that no use whatever will be made of them except by my clients in their own business and I therefore communicate it on condition that it shall be considered as a confidential report."

The author acknowledges his gratitude to Miss Martha Goodway, Metallurgist at the Smithsonian Institution in Washington, for providing photocopies of these relatively unknown papers from the Library of Congress. Most intriguingly, most of the originals appear to have been the property of "Robt. Forsyth" since they carry his name handwritten on the outer covers – he had worked at the very earliest Bessemer plant at Troy and, in 1872, became Superintendent of the newly erected North Chicago Bessemer Works.

[37] Jeans: op. cit., pp. 391–393.

[38] F. W. Harbord and J. W. Hall: "The Metallurgy of Steel", 3rd. ed., London, 1907, pp. 30–37.

quite quickly.[39] The reports of 1875 indicated that all the three works were trying out uphill casting, with, it would seem, varying success. It was common practice, however, whether the metal was uphill cast or direct poured, to plate or cap off the ingot after casting to give as sound a top as possible. Brown, Bayley and Dixon inserted an iron plate with sand on top. At Barrow a cast iron plate was fixed into the top of each of the moulds used for bottom pouring so as to give a constant length of ingot, the gap being filled up with sand immediately after casting; a very similar device was used at Workington, there being provided, in this case, a central ventilation hole, 2.5 in. at the bottom and 0.5 in. at the top; the system was said, when carefully rammed with sand and clay, to work very well.

It is appropriate at this stage to return to the application of Bessemer steel to supply the rail trade. Barrow, Dowlais and the Mersey Steel and Iron Company had entered the trade in 1865 and Ebbw Vale in 1866 but through to 1870 Sheffield was still the centre of the trade, with John Brown and Charles Cammell the major suppliers, supplying vast quantities to the American railways. These two firms alone were exporting to America some three times the whole domestic American output in 1871 and in 1870–71 it is reported that Cammells added 800 men to their workforce, whilst Browns added 1000! Samuel Fox extended railmaking operations in 1871 and between the summer of that year and the end of 1872 there were five new Bessemer works and rail plants erected. At the beginning of 1872, Hampton, Radcliffe and Company built the Phoenix Bessemer Works at the Ickles, between Sheffield and Rotherham; this had two 5 ton converters and a weekly production capability of 600 tons of rails. A few months later, Brown, Bayley and Dixon set up a works in Attercliffe with a similar capacity, whilst Cammells, still feeling the need for extra output, set up a subsidiary company, Wilson, Cammell and Co at Dronfield, between Sheffield and Chesterfield, with four 6 ton converters. An illustration of operations at the Cyclops Works of Charles Cammell and Company can be seen in Fig. 17. By 1873, the Sheffield district had a Bessemer rail capability of some quarter of a million tons per annum.[40]

This situation illustrates one of those apparent paradoxes which operations in Sheffield regularly provided. The previous establishment of Sheffield as the centre of the crucible steel trade was another. In either case, the essential raw material had to be imported: Swedish iron for crucible steel

[39] B. D. Healey: "On a New Mode of Casting Ingots", J.I.S.I., 1873, Pt. 1, pp. 183–185. See also British Patent 1788, 16 May 1873. The German establishment at Hoerde, under their engineer, R. Pink, were also pioneers in what they referred to as "ascensional casting"; there are three patents taken out on their behalf by W. & J. Ireland for the "uphill casting of ingots in groups around a central git". These are British Patents No. 2548, 27 September 1871; No. 429, 10 February 1872; No. 263, 23 January 1873. There are references to "Pink's method" of casting in various of the Holley Private Publications.

[40] Warren, op. cit., pp. 135–139.

Fig. 17. Bessemer Converter at the Cyclops Works of Charles Cammell and Company, Sheffield, about 1870. This clearly shows the pit arrangement and the somewhat restricted working space typical of the early layout of the Bessemer facilities in Britain, as shown diagrammatically in Figure 14. This is reproduced by permission of Sheffield City Museums.

and Bessemer Pig for the rail trade. There were, of course, local advantages: good coal supplies, together with those of ganister and refractory clays; in addition, transport had been much improved since the opening of the railway in 1838. Nevertheless, Sheffield was relatively far from the coast and the trade was heavily dependent on exports. In addition, the trade was new to Sheffield and, to provide the rails, the appropriate rolling mills had to be built – these, obviously, would be of the best modern design and, when operated by workmen with innate skill and craftsmanship, the product would be of good standard. These intangible advantages were enhanced by the exceptionally poor opinion foreign customers had of the Welsh iron rails[41] and Ebbw Vale and Dowlais obviously had an uphill task; in addition, their concentration on their new steelmaking activities was, to some extent, blunted by the still considerable demand for the old iron rails. The giant works at Barrow, on the other hand, were unknown and had to build a reputation from scratch.

The trade, however, was not to last. The American demand fell away rapidly following the financial collapse in September 1873. This severely reduced the investment on new railway construction at a period when the American home production, which had been negligible up to 1871, was gaining momentum quickly. The following figures for rail imports (in thousands of tons) into the U.S.A. clearly show the disaster looming for the British rail producers, and for Sheffield in particular, since this had become the major centre for such production:[42]

1869	**1870**	**1871**	**1872**	**1873**	**1874**	**1875**	**1876**
280	356	505	474	231	97	17	0.2

The newer establishments which had been designed solely for rail production were particularly badly affected. The Pheonix Bessemer Steel Company failed in 1875 after just over two years production; it was, however, restructured under new management as Steel, Tozer and Hampton Limited and its product lines diversified. Brown, Bayley and Dixon made substantial losses in 1874 but managed to survive. Outside Sheffield, the Bessemer plant recently built by the Carnforth Haematite Iron Company was destined never to come into operation. John Brown and Company, who had a number of activities, as will have been noted from their wide ranging exhibits at the 1867 Exhibition, concentrated more and more on engineering

[41] L. E. Gruner and C. Lan: "Etat Present de la Metallurgie du Fer en Angleterre", , Paris, 1862, p. 561, refer to the Welsh product as "les rails les plus inferieux" – this needs no translation! A. S. Hewitt: "The Production of Iron and Steel in its Economic and Social Relations", Government Printing Office, Washington, 1868, pp. 10–11, comments: "it was humiliating to find that the vilest trash which could be dignified by the name of iron went universally by the name of the American rail."

[42] American Iron and Steel Association Reports.

materials and, from 1875 onwards, had little interest in rail production. Under these conditions, the three works run by Cammell remained as the principal sources of steel rails in the Sheffield area and all three carried on reasonable trading for the remainder of the 1870s. The reconstituted Phoenix Works also managed to find orders to keep them busy. The Sheffield trade was never to be the same again, however. The prices for rails fell with the lack of demand and there were other locations which began to have advantages over Sheffield. In the first place, the cost of transport on the pig iron, the cost of remelting it and the cost of transport of the finished rails to the ports increasingly became more significant. The turning point came in 1875, when Barrow started taking the liquid metal straight from the blast furnace, to be followed by Workington, Dowlais and Ebbw Vale – it is not clear whether John Brown ever did this from his own Sheffield blast furnaces. All these three works were, obviously, better situated than the Sheffield establishments for shipping the finished rails.

The effect of general diversification in steel production can be seen quite clearly from the fact that there was no slackening of the growth of Bessemer steel production over this period; the official figures[43] for annual ingot production, in thousands of tons, for the period immediately following what is usually considered the peak year of 1873 show no real slowing in momentum:

1873	**1874**	**1875**	**1876**	**1877**	**1878**	**1879**	**1880**
495	539	620	700	750	807	834	1,044

There is a further compilation of installed Bessemer converters available for the year 1878.[44] The details are as follows:

Henry Bessemer & Co	2 @ 3T; 2 @ 5T
Bolckow, Vaughan & Co	4 @ 8T
John Brown & Co	4 @ 7T; 2 @ 10T
Brown, Bayley & Dixon	4 @ 8T
Charles Cammell & Co, Sheffield	4 @ 4T
Penistone	2 @ 5T; 2 @ 7T
Wilson, Cammell & Co, Dronfield	4 @ 6T
Weardale Iron Co	4 @ 2.5T
Glasgow Bessemer Steel Co	2 @ 3T
Samuel Fox & Co	2 @ 5T
Bolton Iron & Steel	4 @ 6T
L.N.W. Railway, Crewe	2 @ 3T

43 British Iron and Steel Federation, Annual Statistcs.

44 Jeans, op. cit., p. 87.

Manchester, Sheffield & Lincolnshire Railway	2 @ 3.5T
Mersey Steel & Iron Co.	10 @ 5T
Manchester Steel & Railway Plant	4 @ 3T
Barrow Haematite Steel Company	16 @ 6T
Dowlais Iron Company	2 @ 7.5T; 2 @ 6T; 2 @ 5T
Ebbw Vale Company	4 @ 6T;2 @ 8T
West Cumberland Iron & Steel Co	2 @ 5T; 2 @ 8T
PhoenixBessemer Works	2 @ 7T; 2 @ 6T; 2 @ 2.5T
Carnforth Haematite Iron Co	2 @ 6T
Patent Shaft & Axletree	4 @ 3T
Moss Bay Haematite Iron and Steel Co	2 @ 7T
Rhymney Iron Co	3 @ 8T
Blaenavon Iron Company	2 @ 3T

A comparison of the information from the three available lists gives interesting information:

	1868	**1871**	**1878**
Numbers of Converters Installed	59	76	111
Total Installed Capacity, tons	297	406	615
Ingot Production, tons	110,000	325,000	807,000
Mean Converter Capacity, tons	5.03	5.34	5.54
Mean Annual Output per Installed Converter, tons	1,864	4,276	7,270
Mean Number of Blows per Annum per Converter	370	800	1312

The figures arrived at show clearly that, whilst the mean converter size hardly increased over the ten year period, the number of converters and, hence, the installed capacity almost doubled. More significantly, however, the production rose by a factor of over seven; the productivity in 1878, therefore, was around three and a half times what it had been in 1868 and the reduced amount of down-time brought about by the introduction of the removable bottom section must have had a marked effect on these figures.

A comparison of the 1871 and 1878 lists shows that 35% of the 1878 capacity, some 214 tons, was at new establishments; 87 tons was in the Sheffield area (Brown, Bayley and Dixon; Wilson, Cammell & Co; Phoenix Bessemer), 52 tons in Cumberland and Furness (Workington, Moss Bay and Carnforth), 32 tons in Cleveland (Bolckow, Vaughan) and 30 tons in South Wales (Rhymney and Blaenavon). Of the old established Sheffield firms, Cammells had increased their Penistone capacity by around 30%, Bessemer's own plant on Carlisle Street and Samuel Fox's installation at Stocksbridge had remained the same, whilst John Brown's capacity, following his withdrawal from the rail trade, had been reduced by about 30%. There were modest increases in capacity at both South Wales works of Dowlais and Ebbw Vale. The most marked increase in activity, however, was at the Mersey Steel and Iron Works, where a further six 5 ton converters were added to the previous four; William Clay, the manager there, had been an advocate of the production of puddled steel and it seems that he began to appreciate the value of the Bessemer process rather later than most of his contemporaries on account of this.

Statistics for Bessemer steel production for 1878–79–80 are available, together with some extra detail for 1879.[45]

District	**1878** Ingot tons	**Production for 1879** Operating Works	Fces	Ingot tons	Rail tons	**1880** Ingot tons
Sheffield	293,223	7	22	210,346	82,744	273,365
South Wales	203,552	4	15	252,573	189,503	308,233
Lancashire	139,463	4	14	152,130	73,870	167,850
Cumberland	90,990	2	7	127,163	103,969	140,869
N.E. Coast	74,007	2	6	85,299	69,255	144,000
Staffordshire	6,292	1	2	7,200	860	10,045
TOTAL	**807,527**	20	66	**834,511**	520,231	**1,044,362**

It should be noted that an annual production of Bessemer steel in Britain of one million tons was achieved by the year 1880. The figures for 1879, when there was a total availability of some 104 converters, 38 of these being idle, can be used to provide the following information:

District	**Ingot Tons per Converter**	**Rail Tons as % of Ingot Tons**
Sheffield	9,561	39.35%
South Wales	16,838	75.02%
Lancashire	10,866	48.56%
Cumberland	18,166	81.76%
N.E. Coast	14,217	81.19%
Staffordshire	3,600	11.94%
MEAN	**12,644**	**62.34%**

45 J.I.S.I., 1880, Pt. 1, pp. 241–244.

The fall in both production and productivity per furnace in Sheffield, relative to that in South Wales, is a marked feature of this period; at the same time, of the Sheffield production, only a relatively small proportion went into rails in 1879.

The situation by 1880 was well summarised by J. S. Jeans, Secretary of the Iron and Steel Institute.[46] Having pointed out that each erected converter, together with all its necessary appliances, represented an average capital outlay of some £20,000 and that, on this basis, there was already a capital of over two million pounds invested in the Bessemer steel trade in the United Kingdom, he continued:

> The extent to which the development of the Bessemer process has affected the iron trade, railways, shipbuilding and other industrial operations throughout the world can only be referred to here in the briefest possible manner. Allowing, as it now does, of steel of a high grade being manufactured with approximately one fourth of the fuel and one third of the labour of malleable iron, the Bessemer process has practically ruined the latter trade and has caused millions of capital to be depreciated to the verge of absolute loss. Steel, thanks to this process, is now being employed for almost every purpose for which iron was previously used.

The changes wrought by the introduction of Bessemer steel were, indeed, far ranging. Within less than a quarter of a century it had become possible to provide a stronger material than wrought iron in quantity at a cost only marginally higher. Moreover, it could be made in ingot form, in large masses if required, and could then be forged or rolled down to the desired shape without the necessity of forge welding together small masses of a hundred pounds or so as had been the case with wrought iron. It could be made in varying degrees of resilience or toughness, depending on the carbon content and the subsequent thermal and mechanical treatment it received. This was bulk steel, and now had to be distinguished from the more expensive steel from the crucible furnaces in Sheffield, which eventually came to be known as special steel. There was, obviously need for both. In 1873, when almost half a million tons of Bessemer steel were produced, the old Sheffield methods were still producing about a quarter of that amount – for edge tools, metal cutting tools and, surprisingly, for large forgings for ordnance and critical engineering parts. At this date, there were still almost three million tons of wrought iron made in Britain. This was, however, very much a peak and, when Jeans wrote the comments quoted above, the iron trade was in the doldrums; 1875 to 1879 were years of deepening depression; wrought iron output had probably fallen to little more than half what it had been in 1873 and the best estimate for the old Sheffield steel trade in 1879 would indicate that it, too, was running at only 60% of what it had been six years earlier. The significant feature, therefore, was that, even under these conditions, the output of Bessemer steel continued to rise.

[46] Jeans, op. cit., p.88.

5 The Bessemer Process in Europe, 1860–1880

Developments in Sweden

The earliest use of the Bessemer Process was, as previously mentioned, in Sweden, where the work carried out by G. F. Goransson at Edsken in 1858 was instrumental in bringing Bessemer's work to fruition.[1] Edsken was only one of the seven iron working properties run by Goransson; another was Hogbo, with an established blast furnace and forge for ironmaking, and it was here in 1860 that the work was transferred from Edsken, setting up two fixed converters to be fed from the blast furnace; the ingots produced were forged on the site. The results were sufficiently encouraging for a plant to be considered exclusively for the production of Bessemer steel and work was commenced on a site some five miles from Hogbo on the side of a lake in the spring of 1862. This was the foundation of the Sandvik Steel Company. A canal was dug to provide the steelworks with waterpower, a large blast furnace with an ore roasting kiln was erected, two 4 ton tilting Bessemer converters installed – the first of this type to be used in Sweden[2] – and a 15 ton steam hammer – built by Kirkstall Forge Limited and, at that time, the second largest in Europe – and two rolling mills installed. By the autumn of 1863 the plant was operational, the product found to be very satisfactory and soon being sold by the company's agents in Paris and St. Petersburg. At the beginning of 1865, however, the major investor in the company was forced into bankruptcy and the "Hogbo Bolag" was subsequently obliged to go into liquidation, work at Sandviken ceasing in March 1866. Goransson succeeded in buying back the works at Sandviken two months later, but the other forges and mills went into different hands. A new company, "Sandvikens Jernverks Aktiebolag", was formed the spring of 1868 and the famous Sandvik Steel works is still there.[3]

[1] K. F. Goransson: "The First Successful Bessemer Steel", Metal Progress, Jan 1956, pp. 85–87; the author was the grandson of the owner of Edsken. See also Chapter 3.

[2] A furnace from Sandvik Works of this capacity is preserved as an exhibit at Hogbo Bruk, the site of the old forge and now a leisure area.

[3] These details come from "Ett Svenskt Jernwerk: Sandviken, 1862-1937", published privately by the Sandvik Company to mark the 75th Anniversary of the founding of the works by G. F. Goransson. The comments on the procedures at Sandviken which follow are derived from O. Kollberg: "Berattelse for an 1864" [Annual Report for 1864], Jernkontorets Annaler, 1865, pp. 307-318.

It appears that the pig iron smelted in the newly erected blast furnaces at Sandviken — said to be the largest in Sweden at that date — had less than 0.010% sulphur and less than 0.015% phosphorus but, untypically, was low in manganese (0.2—0.7%). Accordingly, although the practice of catching the carbon was still used, it was found necessary to add spiegeleisen to the blown metal before casting. The converters were reported as being lined with "ganisterstone", imported from England. During 1864 there were 672 Bessemer blows at Sandviken, producing 43,050 centners (1,822 metric tons) of ingots; the ingot yield was 82.93%, with 5.53% scrap and 11.54% loss.

The continued use of the fixed converter was something which was almost unique to Sweden. Goransson at Edsken had originally been provided with the earliest form of tilting converter by Galloways of Manchester and, after finding it difficult to operate and the product mainly unsatisfactory, he reverted to the fixed type which Bessemer had used in his original work. This was set up according to Bessemer's instructions but again was found to be unsatisfactory until the tuyere arrangements were modified to allow more air at a lower pressure to be blown through the metal; at the same time, these circumferential tuyeres were placed at a slight angle so as to give a rotary movement to the liquid metal. In this way, Goransson established a pattern for all early Swedish Bessemer practice.[4] An artist's impression of the original furnace in operation can be seen in Fig. 18.

There were eight works with fixed converters in Sweden in 1868, as well as the two tilting converters at Sandvik:[5]

Savanaes	2 @ 2T	Kloster	2 @ 2T
Hogbo	2 @ 1.5T	Soderfors	2 @ 2T
Backa	2 @ 2T	Carlsdahl	2 @ 2T
Siljansfors	2 @ 2T	Fagersta	2 @ 2T

In 1864, Sweden produced 3,179 tons of Bessemer steel, 1,839 tons coming from Sandvik. The figures for the following year show Sandvik output was 2,938 tons from a total of over 6,000 for the output of the five works then in operation.

Details of one of these plants, that at Backa in Dalecarlia, have survived

[4] A small early test ingot from Edsken is still preserved at Sandvik Works. The head end shows several blowholes; the other end, which has been forged, appears quite sound. The analysis is quoted as 1.05% C, 0.01% Si, 0.02% Mn, 0.004% S, 0.021% P.

[5] S. Jordan: "Revue de l'Industrie du Fer en 1867", Vol. 4, Paris, 1871, p. 371. The nominal capacities of the converters stated by Jordan seem rather arbitrary. It is known from other sources that the charges at Backa were around 1.3 tons and those at Fagersta around 1.7T; both are stated here to be 2T.

Fig. 18. The Bessemer Blow at Edsken in Sweden in 1858. This is taken from a water colour by Carl Cantzler and is reproduced by permission of AB Sandvik Steel. It illustrates the use of the vertical converter to which Goransson reverted having found the rotative converter supplied from England to be unsatisfactory.

and a reproduction of the drawings can be seen in Fig. 19.[6] The author was the chief engineer of the works and he pointed out that the reluctance to use the tilting furnace was mainly a want of acquaintance with it; there was, however, some difficulty in lining a spheroidal vessel with refractory material, there being few suitable clays and no substitute for ganister to be found in Sweden. The fixed converter had, moreover, been found to give very satisfactory results in Sweden. Its only drawback had been the necessity of emptying the furnace as quickly as possible at the end of the refining, since the blast could not be shut off with liquid metal still in the furnace; this was only a problem in making the softer steels, where blowholes could not always be avoided.

The blast reservoir of the Edsken furnace had been external with a blast pipe which was fixed to the tuyere arrangement by means of a flange and nut. This had been modified, rather detrimentally it would seem, in the installation at Savenaes, where large cast iron blocks, moved by wedges, were set up in the ring shaped blast reservoir; each block had holes through which the tuyeres could be fixed. This having proved unreliable, the installation at Siljansfors was redesigned by the chief engineer, Stefanson, and the furnace at Backa [illustrated as Figs. 1 and 2 of the plate reproduced in our Fig. 19] was a minor modification of the Stefanson design to ensure that the leakage of the blast through the brickwork was as small as possible. This allowed easy replacement of individual tuyeres as necessary, the simple detachment of the base for renewal of all tuyeres and associated refractories and, above all, the prevention of any accidental leakage of metal should a tuyere fail. It was pointed out that the base of the converter required relining after about 25 blows but that the main furnace lining would be good for 150 to 200 blows; comments were made, however, that the Swedish bricks from Hoganas were inferior to the English "Newcastle Bricks"[7] which had been used at Edsken, as well as being more expensive.

Comments on the working of the adjacent charcoal fired blast furnace lead to the conclusion that the ore was well roasted before charging, to remove any traces of sulphur, and that the burden was adjusted to provide a mottled iron, high in carbon (around 4.5%) and moderately low in silicon

[6] L. E. Boman: "Jernfarskning efter Bessemers Metod", Jernkontorets Annaler, 1863, Pt. 3, pp. 184–192; Pt. 4, pp. 193–227. This obviously was considered an important contribution. It was translated into German as "Das Bessemern in Schweden in Seiner jetzigen Praxis" and published in Leipsig in 1864. From this an English translation was made and published in "The Engineer" (March 10 1865, p. 151; March 17, pp. 169–170; March 24, p. 183; March 31, pp. 198–199). It is obvious also that it was known in France, since when J. S. Jeans reported the details in "Steel: Its History, Manufacture, Properties and Uses", London, 1880, pp. 288–291, he acknowledged Professor Jordan's "Album du Cours de Metallurgie", Paris, 1875, as the source of the drawings of the plant, this being a direct copy of the Swedish original. It is from Jeans that Fig. 19 is reproduced here.

[7] One batch of the English firebricks used at Edsken was marked "EN" and the other "R. Dickinson, Consett".

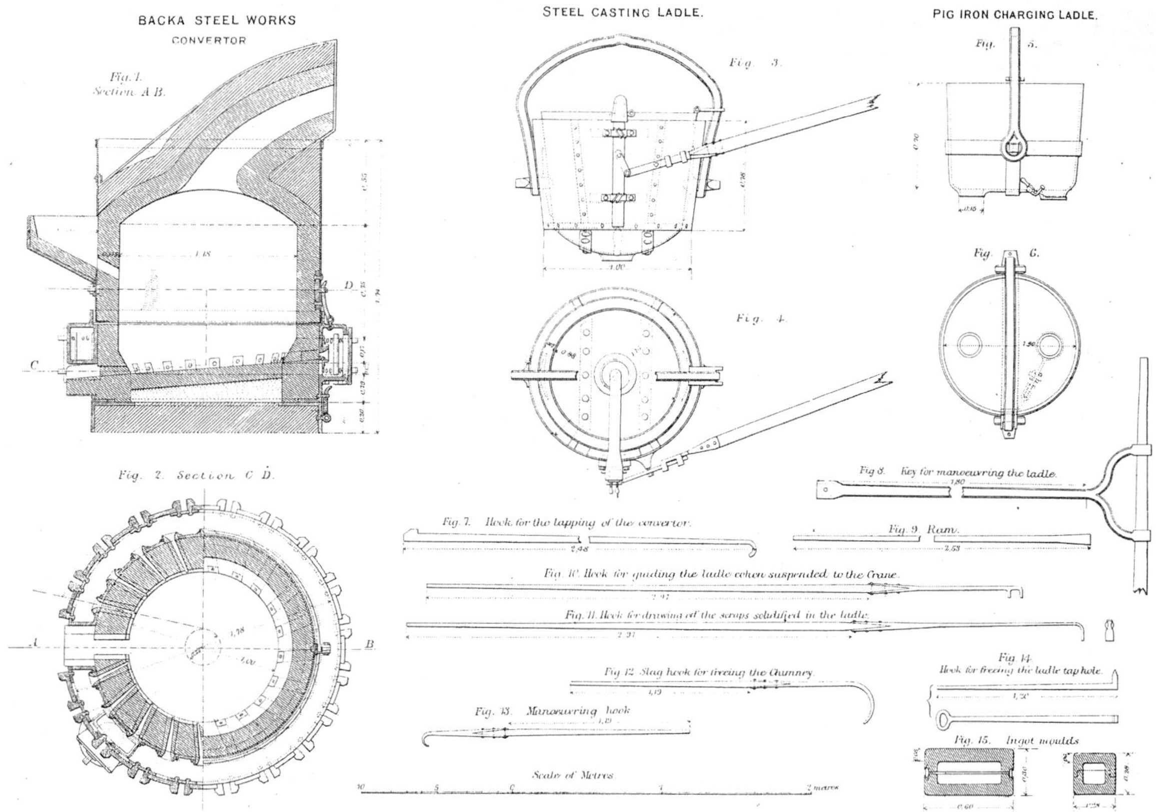

Fig. 19. Converter and ancillary equipment used at Backa Steel Works, Sweden. The illustration is taken from Plate 4 of J. S. Jeans: *Steel*... but has its origin in a paper by Boman in Jernkontorets Annaler, 1863. It shows the type of vertical converter which was in general use in Sweden in the 1860s and the converter is an obvious development from the Bessemer design shown in Fig. 7.

(1.0–1.5%). The local ores used were relatively low in manganese but also low in phosphorus. The floor level for the Bessemer converters was arranged to be some 12 feet below the blast furnace taphole, with the casting pit some six feet below the Bessemer floor level.

To operate the converter, the vessel would first be filled with charcoal with some burning coals put in near the bottom and a gentle blast blown through. When sufficiently hot, the remaining charcoal would be removed by a full blast. The taphole of the furnace was then closed by ramming in clay plugs until flush with the outside, when an iron plate would be placed in position over them. Blast furnace metal, tapped into the charging ladle as required and being carefully weighed, was taken over to the converter and, with the blast on, quickly run into the vessel. At first the flame was negligible, of a dirty yellow colour and with few sparks; a hissing sound could be heard and, occasionally, a beating noise. During this period the slag was being formed from the oxidation of the silicon.

Then came the "boiling" period. Its arrival was heralded by the appearance of dark blue stripes in the flame; a rumbling noise was heard, its loudness increasing with the liveliness and heat of the working of the furnace, sparks flying out plentifully and noisily. The flame became long, blue and with a dazzling white fringe. It was necessary at this stage to have a regulator valve to reduce the force of the blast somewhat, so as to lessen the amount of metal ejected with the flame, again increasing the blast as the ejection lessened. It was pointed out that the ejection of metal could be lessened by increasing the height of the furnace: extra courses of brick were subsequently incorporated between the original furnace top and the upper dome. When the boil slackened, the smoke and woolly sparks disappeared and the flame became clear with sparks of iron streaming out like an unbroken tail of light. The flame then shortened, drawing back into the throat of the furnace; at the same time, the rumbling ceased and the hissing noise returned. At this stage, the contents were fully decarburised.

In general, the aim was to stop the blow before this stage, so retaining an appropriate amount of carbon in the steel. It was pointed out that this was the most difficult part of the process and only by extended and careful working could proficiency be obtained. With practice, it was claimed, by consistent running of the blast furnace and the converter, the skilled operator, over a period of several working weeks, could keep the overall range of carbon contents to well within a quarter of one percent.

When the metal was ready to be tapped, the iron plate was removed from the tap hole and, at the signal from the steelmaker, the plug was knocked out and the metal run into the preheated ladle; it was pointed out that the tap hole should be large – about four inches square – so as to clear the metal quickly and prevent further action from the blast which must still be kept on. It is significant that there is no mention of any addition of such material as spiegeleisen to the finished metal. It is, however, mentioned in the general-discussion which follows that it had been noted that Mr. Bessemer, in making soft iron, had recommended the addition of a small amount of liquid pig iron to the casting ladle in an attempt to prevent the formation of honeycomb structure in the ingot. The metal was then run into ingot moulds by operating the ladle stopper. As regards moulds, those shown in the diagram are obviously two part ones, with a longitudinal joint; it will also be seen that both square ones, for the production of billets and bars, and rectangular ones, for the ultimate production of plates and sheets, were available. The comment is made in the report that they would have preferred to use single piece moulds, as in England, and so avoid the necessity of chipping off from the ingot the fash left from the seam in the moulds.

A further number of interesting points are made in this extensive article. In the first instance, it is stated that Bessemer steel in Sweden was classified according to a scheme originally devised by the smith at Edsken according to its hardening characteristics; No. 1 was the highest carbon content which could be forged and, therefore, gave the highest hardness on quenching; No. 5 was virtually carbon free, incapable of hardening on quenching and being the equivalent of wrought iron.

In 1862, Professor Eggertz, at the Mining School at Falun, had devised a rapid method of assessing the carbon content of steel. 0.1 grams of steel drillings were dissolved in dilute nitric acid (d. 1.20) on a water bath at 80°C. The solution was diluted with water in a graduated burette until it had the same colour as a solution of a steel of known carbon content contained in a tube of the same diameter and colour as the burette; the solutions were held, side by side, in front of a piece of thin filter paper against the light, the ratio of volumes multiplied by the known carbon content giving the carbon content of the steel. It was recommended that there should be three standard steels, representing the range of steels made at the works, for example, 0.4%, 0.8% and 1.2%, since the softer steels gave a rather more green solution and the hardest steel more of a red colour, compared with the brown colour of the intermediate one. This constitutes one of the earliest examples of steelworks analysis.[8]

Using Eggertz's method, there had been instituted at Siljansfors an extended scale; No. 1 was 2% carbon; No. 1.5 1.75% carbon; No. 2 1.5% carbon; and so on down to No. 4.5 which was 0.25% carbon, with No. 5 as carbon-free or wrought iron equivalent.[9] It was standard practice to

[8] It is of interest to note that the same method was still in use in some Sheffield laboratories when the author commenced his steelworks experience as an analyst in 1936.

[9] The Sheffield crucible steelmaker used fracture tests to assess the degree of carburisation of the steel and established a series of what were known as "tempers". Generally No. 1 was the softest, around 0.5% C, and No. 7 the hardest, around 1.5% carbon; in some cases, half tempers were used. Seebohm and Dieckstahl, very reputable steelmakers who came farily late into the Sheffield trade (they set up their operations in 1865) used a system very similar to that of the Swedish Bessemer steelmakers. Their "No. 0" was nominally 1.50%; "No. 0½", 1.40–1.45% C; "No. 1", 1.35% C; and so on down to "No. 6", around 0.90% C.

stamp each ingot with a cast number and, when known, the carbon content. An intriguing and practical suggestion was that, if Roman numerals were used, this would cut down on the complexity of stamps required, since a number of straight marks and half round marks would cover all requirements!

The working of the Stefanson furnace at Backa was summarised by Boman as follows:

Average charge of pig iron	1335 kg.	
Ingots produced	1020kg.	= 76.4%
Steel scrap made	65 kg.	= 4.9%
Total steel made	1085 kg.	= 81.3%
Metal ejected	80 kg.	= 6.0%
Loss by oxidation	170 kg.	= 12.7%
Slagging period		2 minutes
Boiling period		5.5 minutes
Fining period		2 minutes
Total time of blow		9.5 minutes

At the time of the 1867 Exhibition in Paris, the works at Sandvik were at a standstill and it may be assumed that all the exhibits of Bessemer steel from Sweden were from fixed furnaces. The works at Siljansfors, Kloster and Carlsdahl all had small exhibits; by far the most extensive, however, was that from Fagersta.[10] Here there were two converters of 1,740 kg. capacity. The exhibits covered a series showing the as-cast ingot, a forged bar and a final slag covering thirteen grades of steel from 0.10 to 1.30% carbon. The grain became more and more fine as the carbon content rose, the mildest steel showing an as-cast grain which was open and somewhat porous. There also was a drill in 1.5% carbon steel which had been used to drill a plate of 1.0% carbon steel which had been hardened prior to drilling and the drill was in quite good condition. The finished products from Fagersta were remarkably good, including railway axles with 0.3–0.4% carbon, locomotive and wagon tyres with 0.5% carbon, rifle barrels (both welded and solid drawn), solid drawn tubes, a very bright and very clear steel mirror, both fine and thick wire, straight and circular saw blades, sword blades, straight and curved sabres, knives, scissors, razors and files – all bearing the Fagersta mark of a flower with four petals.

Charles Aspelin, the proprietor of Fagersta, provided a list of the usages of the various grades of Bessemer steel as follows:

[10] Jordan, op. cit., pp. 372–376.

1.00–1.30% C	Files and Graving Tools
1.10%C	Turning Tools
1.00%–1.10% C	Mining Drills
1.00% C	Razors
0.80% C	Saws
0.60–0.80% C	Sword Blades
0.60–0.70% C	Hammers
0.60% C	Springs
0.50% C	Bayonets
0.30–0.50% C	Bored Rifle Barrels
0.10–0.30% C	Axles, Tyres, Sheets and Welded Rifle Barrels

This must be the first categorisation of steels suited for various purposes based on their composition and it is of some interest to compare it with a list put out in Sheffield of applications for imported Swedish Bessemer steel about the end of the century:[11]

1.10–1.25% C	Turning and Planing Tools, Twist Drills, Mill Files
1.00–1.05% C	Saws, Machine Knives, Reamers, Granite Chisels
0.90–0.95% C	Taps, Pneumatic Chisels, Picks, Gouges, Valve Springs
0.80–0.85% C	Punches, Cold Chisels, Masons Tools, Punches, Miners Drills, Spiral Springs, Wood Chisels
0.70–0.75% C	Press Tools, Shear Blades, Blacksmiths Tools, Coal Cutters, Piston Rings
0.60–0.65% C	Hammers, Stone Working Tools, Vice Jaws, Axes, Swaging Dies

The growth of Bessemer steel production in Sweden over the period 1869 to 1873 can be gauged by the following figures:[12]

[11] Information derived from the private collection of Denis A. Senior. Esq., relating to the trade in Bessemer Steel carried out by Senior and Co in Sheffield; a similar trade was run by Daniel Doncaster and Company from about 1880 to 1930.

[12] The figures for the total Swedish Bessemer output are from Jeans, op. cit., p. 279, converted to metric tons for ease of comparison with the Sandvik figures from "Ett Svenskt Jernverk", p. 398.

Year	Number of Bessemer Steelworks Operating	Ingots: Metric Tons Whole of Sweden	Sandvik
1869	4	5,406	3,128
1870	9	6,746	3,150
1871	6	4,608	4,152
1872	10	12,652	4,092
1873	16	15,946	4,415

A survey for 1870 actually quotes production at only 7 works as follows:[13]

Sandviken	76,061	centners	= 3,221	metric tons
Lenna	1,036		= 4	
Siljansfors	4,616		= 195	
Backa	23,952		= 1,014	
Fredshammar	9,615		= 407	
Vestanfors	40,349		= 1,709	
Carlsdal	413		= 17	
TOTAL	**156,054**	**centners**	**= 6,607**	**metric tons**

It was also reported that new Bessemer works were under construction at Abackshyttan, Svartnaes, Iggesund and Forsbacka.

Production from the various Swedish Bessemer plants can be quoted for both 1873 and 1874 as follows:[13]

[13] The 1873 figures are from J.I.S.I., 1875, p. 315; the 1874 figures from Jeans, op. cit., p. 279. Both sets are quoted in Swedish centners (23.62 Swedish centners = 1 metric ton).

Works	Province	Output 1873		Output 1874	
		Swedish Centners	Metric Tons	Swedish Centners	Metric Tons
Iggesund	Gefleborg	23,631	1,000	39,596	1,676
Forsbacka	do.	63,488	2,687	66,729	2,826
Sandviken	do.	109,253	4,625	113,024	4,785
Lenna	Stockholm	10,943	463	5,437	230
Siljansfors	Kopparberg	2,341	99	645	27
Fredshammar	do.	2,728	115	546	23
Backa	do.	6,994	296	5,707	243
Svartnaes	do.	22,144	938	34,386	1,456
Abackshyttan	do.	47,143	1,996	67,561	2,860
Ulfshyttan	do.	7,643	324	21,248	899
Langshyttan	do.	14,455	616	30,837	1,305
Vestanfors	Vestmanland	57,677	2,442	52,037	2,202
Bangbro	Orebro			19,251	815
Carlsdal	do.	381	16	165	7
Bjorneborg	Vermland			32,891	1,392
Borgvik	do.			1,717	73
Langbanshyttan	do.			2,740	116
Gustavsfors	do.			6,624	280
	TOTAL	**368,381**	**15,615**	**501,149**	**21,215**

The establishment at Vestanfors was the scene of some of the earliest work on the chemistry of the Bessemer Process; it was there in 1874 that Dr. Tamm investigated the composition of the gas emitted from the 3 ton tilting type of Bessemer converter during the blow.[14] Here the pig iron was rich in manganese and the blow was halted with carbon and manganese still present. Metal analyses relating to one particular blow are quoted as follows:

	C	Si	Mn	S	P
Pig iron	4.10	1.39	4.67	trace	0.015
Bessemer Ingot	0.20	0.03	0.17	trace	0.020

[14] M. A. Tamm, Jernkontorets Annaler, 1875, Pt. 5, pp. 257-273, abstracted in J.I.S.I., 1875, p. 669, under the title "Chemical Composition of the Gases Emitted from the Bessemer Converter during the Blow"

The corresponding final Bessemer slag, together with a sample of the condensed fume near the end of the blow gave the following analyses:

	SiO_2	Al_2O_3	FeO	MnO	MgO	CaO	P_2O_5
Slag	42.85	3.94	15.62	36.83	trace	0.70	0.015
Fume	17.92	2.08	15.55	61.61	0.25	0.60	—

Gas analyses were carried out at four different periods during the blow, giving the following results:

Minutes from		Gas Analysis, % by Volume			
Start of Blow	O_2	CO_2	CO	H_2	N_2
3–5	4.76	9.13	—	—	86.11
9–10	1.70	6.00	17.56	0.91	73.84
21-23	0.97	4.86	19.32	1.12	73.74
26–27	0.55	1.85	14.31	1.70	81.59

Dr. Tamm was involved in further work on slag and metal composition, covering five Swedish works which were using blast furnace products which differed widely in the amount of manganese present.[15] In the following table, the slag analysis quoted against the pig-iron is that of the typical blast furnace slag in normal production; the time quoted is that period which had elapsed after the commencement of the boil at which the sample was taken. The times to commencement of the boil for the various works were:

Langshyttan	2m.45s
Sandviken	2m.30s
Nykroppa	1m.30s
Bangbro	2m.45s
Vestanfors	2m.30s

[15] R. Akerman: "The Bessemer Process as Conducted in Sweden", Trans. Amer. Inst. Mining Eng., Vol. XXII (1893), pp. 276–278. He reports work by M. A. Tamm, Jernkontorets Annaler, 1878, pp. 444ff and 1880, pp. 462ff, together with a contribution from C. G. Dalerius, Jern. Ann., 1881, pp. 314ff.

Sample	Time m. s.	Metal C	Si	Mn	Slag FeO	MnO	MgO	CaO	Al_2O_3	SiO_2
Langshyttan										
Pig		3.94	1.14	**0.64**	0.29	2.24	10.26	36.85	5.14	45.16
Bessemer	2.15	4.20	0.04	0.12	34.72	13.95	0.24	2.60	0.78	48.76
	4.30	1.10	0.03	0.12	21.08	15.48	0.30	3.25	0.98	59.82
	5.30	0.05	0.01	0.06	35.82	12.29	0.21	2.35	0.72	48.48
Sandviken										
Pig		4.49	1.08	**0.83**	0:67	5.77	10.70	30.98	4.32	46.56
Bessemer	3.00	3.87	0.03	0.11	20.24	23.90	tr.	0.44	1.84	53.44
	5.00	1.30	0.03	0.09	17.04	22.80	tr.	0.46	1.94	57.80
	5.45	0.33	0.02	0.07	18.48	22.23	tr.	0.36	1.58	55.76
Nykroppa										
Pig		4.35	0.88	**1.05**	0.77	2.14	22.18	21.79	5.83	47.16
Bessemer	2.30	4.10	0.10	0.15	13.50	29.76	0.23	0.42	2.28	53.26
	5.30	1.00	0.05	0.15	9.54	23.70	0.28	0.60	3.90	62.34
	6.30	0.08	0.04	0.08	30.60	21.39	0.21	0.38	2.14	44.52
Bangbro										
Pig		4.00	1.02	**1.83**	0.54	5.67	8.47	30.71	5.80	47.70
Bessemer	3.00	4.30	0.03	0.22	14.20	26.31	0.22	0.62	2.86	55.26
	4.45	0.90	0.03	0.12	18.52	31.01	0.14	0.38	2.70	47.20
	5.45	0.10	0.03	0.09	31.19	25.43	0.11	0.32	2.24	40.50
Vestanfors										
Pig		4.22	1.06	**5.12**	0.23	7.44	13.08	30.60	3.90	42.76
Bessemer	4.15	4.20	0.43	3.26	4.20	46.38	0.54	1.26	3.08	45.87
	8.35	1.30	0.12	0.85	6.24	52.26	0.29	0.70	2.49	39.07
	9.20	0.55	0.07	0.43	9.45	48.92	0.46	1.00	2.94	37.63

These results indicate several matters of interest. In the first instance, the reduction in iron content in the Bessemer slag, particularly at Langshyttan, but also at Sandviken and Nykroppa, from early in the blow reveals that the slag itself plays an important part in the oxidation of the carbon – it is not simply a direct action of the oxygen in the blast. Furthermore, with increasing manganese in the blast furnace product, the major oxidising constituent in the slag becomes manganese oxide and not iron oxide. Particularly at Vestanfors, with the very high manganese content in the charge, this has the effect of slowing down the oxidation and, therefore, leaving a greater control of the process when the carbon content is to be "caught" – that is, the blow is to be stopped short of full oxidation, a procedure which goes back to the origin of the Bessemer process in Sweden at Edsken and something which was more difficult without the aid of manganese in the metal. Here at Vestanfors, and at other plants operating on a high manganese blast furnace metal, the cooling effect of the extended blow was offset by the extra heat from the combustion of the manganese; in addition, a further beneficial effect was the extra fluidity of the slag, and hence the ease of separation from the metal, conferred by the presence of the high manganese oxide content in the slag. In this context, Professor Akerman also quotes a series of figures provided by J. A. Brinell on the results of this procedure at Vestanfors in making Bessemer steel of differing carbon content:[16]

Carbon	1.30%	1.10%	0.90%	0.70%	0.50%	0.30%	0.20%	0.15%
Silicon	0.06%	0.05%	0.045%	0.045%	0.04%	0.03%	0.02%	0.015%
Manganese	0.60%	0.55%	0.50%	0.40%	0.30%	0.20%	0.15%	0.12%

The paper by Professor Akerman, however, also deals with the most important matter of the operation of the blast furnace to produce a metal suited to the needs of the Swedish Bessemer steelmakers.[17] Remembering that these furnaces were charcoal fired and, in general, would run cooler than the more conventional Western European coke fired blast furnace, the normal product – that used for the subsequent refining down to bar iron – was quite low in silicon, normally between 0.2–0.4%; such a metal was unsuited to Bessemer work as it would "blow cold", that is, there would not be sufficient heat generated by the combustion of the silicon to provide a fluid melt; similarly, however, with the limited blowing capability of the smaller Swedish converters, an excessively high silicon content would so prolong the blow as to give an equally cold blow. It was soon found that a compromise with around 0.9–1.1% silicon was right for this particular practice; this is to be compared with standard British practice in which a much higher silicon – at least double the amount common in Sweden – was the norm.

From the very beginning, it has to be remembered that Swedish practice, in contrast to that elsewhere, was to take the metal straight from the blast furnace. The uniformity of this metal was, therefore, of the utmost importance and the proper management of the blast furnace was, therefore, an integral part of the whole operation. As a result of a considerable amount of investigatory work, the outcome was the running of the blast furnace with a moderately basic slag, with charges made of carefully blended ore supplies and with a charcoal to ore ratio somewhat higher than that normally used for forge iron production. This led to reduction of the appropriate amount of silicon, the recovery of the maximum amount of manganese, the maximum removal of such small amount of sulphur as was present – and it is pointed out that some of the ores richest in manganese did contain iron pyrites, not all the sulphur being removed by the prior roasting – and the provision of as high a carbon content – from 4.0–4.5% – as was practicable.

In 1876 it was pointed out that only three of the oldest and smallest Bessemer plants in Sweden still operated the fixed converters. The universal practice, however, was still to take the metal direct from the blast furnace; some trials with metal remelted in cupolas had proved unsatisfactory and

16 Akerman, op. cit., p. 280.

17 Akerman, op. cit., pp. 268–271.

the procedure had been discontinued. Only one works had steam blowing equipment at this date; the remainder operated with water power. Converter sizes varied between 2.5 and 5 tons. It appears, however, that the practice of adding spiegeleisen in the ladle was becoming more prevalent.[18]

There was a further Universal Exhibition in 1876, held in Philadelphia, and the Swedish makers of Bessemer steel were there with interesting examples of their product.[19] Sandvik exhibited a 5.5 ton ingot of 0.6% carbon steel and planed forgings from similar ingots which showed not the slightest discoverable imperfections. There were railway tyres and also welded tools with up to 1.1% carbon; there was cutlery made from Bessemer steel; the cutlers of Eskilstuna were reported to have used Sandvik Bessemer steel for the past fifteen years.

From Fagersta came a series of 9-inch square ingots in a range of carbon contents from 0.15% to 1.00% carbon, each ingot fractured in half and the other half forged to blooms which were also exhibited. The comment was that all were uniformly sound. One very interesting exhibit was a series of small "test ingots" taken at various stages through a Bessemer blow of a typical blast furnace metal with 4.5% manganese; these had been broken across to show the fractures. In addition, slag samples corresponding to them were also shown. They were taken as follows:

3 minutes	after start of blow	4.00% C
8	do.	3.00% C
11	do.	1.40% C
13	do.	0.20% C

The metal fractures became steadily more open in texture through the series; the slags were porous and swollen in the early stages, the third was vitreous and the fourth stony and opaque. The steels offered by Fagersta were listed as follows:

[18] A. L. Holley and Lenox Smith, Engineering, 8 December 1876, pp. 477–478.
[19] Report on the Progress of Iron and Steel, J.I.S.I., 1876, p. 583.

	C	Si	Mn	S	P
Soft steel for plates	0.085	0.008	0.10	trace	0.025
Steel for gun barrels	0.25	0.036	0.23	trace	0.022
Soft steel for tools, saws	0.70	0.032	0.26	trace	0.023
Hard steel for chisels, etc	1.05	0.067	0.36	trace	0.023

The works at Motala showed ingots in Bessemer steel (as well as Open Hearth steel) in a range of analyses from 0.10 to 1.00% carbon.

Despite the early start made to Bessemer steelmaking in Sweden and the high quality of the material produced – due not only to the high quality of the Swedish iron ores but also to the valuable investigatory work carried out by a number of Swedish metallurgists – the output remained relatively small and by 1880 Sweden was at the bottom of the league of European Bessemer steelmakers as regards output. No complete series of production figures is readily available for the years immediately prior to 1880. The Swedish output of Bessemer steel for 1875 is quoted as 455,699 centners;[20] this equates to 19,769 metric tons; this figure had decreased to 368,831 centners (16,000 metric tons) by 1878. In 1879 there was a rise to 487,747 centners (21,159 metric tons); virtually 80% of this total come from the five chief producers, as under:[21]

Sandviken	133,472 centners	= 5,790 metric tons
Vestanfors	67,582	2,932
Bangbro	64,603	2,803
Langshyttan	61,898	2,685
Forsbacka	53,939	2,340

In 1880 the total had again risen, this time to 29,908 metric tons;[22] the main producers now included Iggesund, these six producers accounting for 85% of the total output:

Sandviken	8,840 metric tons
Bangbro	5,300
Vestanfors	3,547
Nykroppa	3,546
Iggesund	2,547
Langshyttan	1,742

20 J.I.S.I., 1877, Pt. 2, p. 252.
21 J.I.S.I., 1880, Pt. 1, p. 788.
22 J.I.S.I., 1882, Pt. 1, p. 410.

Although it falls just outside our period, there is a very interesting census of total Swedish steel production for the year 1882, which has considerable value in putting the situation in perspective.[23] For a total steel production of 1,463,120 Swedish centners = 63,472 metric tons, the sub-divisions were:

Shear steel	6 metric tons
Crucible steel	220
Puddled steel	245
Blister steel	734
Open Hearth steel	13,602
Bessemer steel	48,665

Developments in France

As it turned out, the first Bessemer steelmaking in France was carried out by the grandson of the Sheffield trained Birmingham man who had initiated the manufacture of crucible steel in France after the Napoleonic Wars. James Jackson et Fils had a well deserved reputation as steelmakers and William Jackson became a pioneer of the Bessemer process. He had, it seems, contemplated the production of puddled steel as a cheaper alternative to their crucible steel production. Having witnessed the Bessemer converter in action in London, however, he changed his plans and took a proportion of the French patent rights from Bessemer.[24] He soon installed a small tilting converter at one of their works at Saint Seurin sur l'Isle, near Bordeaux, where they had blast furnaces. Reports vary as to the size of this converter and the date of its installation; one authority gives a three-quarter ton vessel installed as early as 1856[25] whilst another account indicates that a furnace twice that size was installed two years later.[26] Both agree, however, that a further two converters, these being of 3 ton capacity each, were commissioned in 1859–60. For some time this was the only French steelworks using the new process. In 1862 it was decided to abandon the St. Seurin works and the converters were transferred to the site of a previous iron forge at Imphy on the Loire. Renamed "Societé d'Imphy-Saint-Seurin", it remained the major producer of Bessemer steel in France until 1867, at which date it had one 4 ton and two 5 ton converters, between them capable of producing up to 1500 tons per month. For the melting of the cast iron there were four

[23] J.I.S.I., 1883, Pt. 1, p. 315.
[24] H. Bessemer: "An Autobiography", London, 1905, p. 224.
[25] J. S. Jeans:"Steel: Its History, Manufacture, Properties and Uses", London, 1880, p. 210.
[26] Jordan, op. cit., p. 357.

reverberatory hearths and two cupolas, together with small hearths for the spiegeleisen. A converter could make 80 blows before relining. Each blow provided at least 11 ingots for railmaking, the ingots being circular in section, about 9 inches in diameter and 40 inches long. The cast irons used came from Givors, Saint Louis and Montluçon; the proportion of spiegeleisen used varied from 12–13% for rail ingots down to 8% for ingots for springs. At the 1867 Paris Exhibition, Imphy showed a fine collection of items in Bessemer steel; a wide selection of railway rails, points and crossings for all the main French railways, guns, shells and axles, in addition to an as-cast ingot weighing over 7 tons and remarkable for its smooth skin.[27]

At the same date that the furnaces were moved to Imphy, Petin, Gaudet et Cie were erecting Bessemer facilities at their works at Assailly. By 1867, they had overtaken the output of Imphy, having available two 6–7 ton converters and one of 9–10 tons capacity. At their Givors works, they installed two further 5–6 ton converters, alongside their blast furnaces. In these two works they refined coke smelted iron from the ores of Sardinia, Africa and Spain. At Assailly, the three converters, served by four hearths for the grey iron and three hearths for spiegeleisen, could make 40 to 50 blows per week; losses in the vessel amounted to 9–10%, with a further 5% loss as casting scrap and ladle skulls. The 7 ton converters had 9 tuyeres with 9 holes each; the 10 ton vessel had 10 similar tuyeres. The spectroscope was not used, control being by careful observation of four pressure gauges strategically placed at positions between the blast box and the entry to the tuyeres. In 1867, they exhibited an ingot 40 inches square weighing 25 tons; a parting cut on the lathe had been discontinued near the centre and the pieces fractured apart to exhibit the uniformity and absence of blowholes; it showed the large grain size common to soft steels and the surface was a little rough, as usual with such materials. They also showed naval and artillery guns, shells, rifle barrels, heavy plates and several forms of railway rails.[28]

The Compagnie des Fonderies et Forges de Terrenoire, Lavoulte et Bessèges commenced their Bessemer operations in 1863 and differed from the others in taking the metal direct from the blast furnaces; they were destined to become the major producers of Bessemer steel in France.[29] In 1867 they had four converters at Terrenoire and two at Bessèges, all of 5 tons capacity each. They had a joint capability of 30 blows every twenty four hours, giving up to 36,000 tons per annum. Using the spectroscope to determine the end of the blow, with their direct use of metal the process went extremely well and the metal was hot and fluid. It could be allowed five minutes standing period in the converter

[27] Jordan, op. cit., p. 359.
[28] Jordan, op. cit., pp. 357–358.
[29] Jordan, op. cit., pp. 360–361.

after the addition of the spiegeleisen as well as a few further minutes rest in the ladle. This led to very good ingots and, despite the lapse of time, casting scrap and ladle skulls did not amount to more than 2–3%. The products exhibited at Paris in 1867 were divided into three classes:

1 "Acier dur" (Hard Steel), for resistance to wear; these included rails, springs and axles.
2 "Acier doux" (Soft Steel), for applications requiring strength but a degree of flexibility: they exhibited a double T section, angles and plates, the latter with specimens which had been bent, flattened and then re-opened without rupture.
3 "Acier extra-doux" (Extra Soft Steel), for applications requiring flexibility and resistance to shock: boiler plate, rifle barrels, machine parts and various rolled bars.

The French production of Bessemer steel, in metric tons per annum, for the years 1863 to 1869 is quoted as follows:

Works	1863	1864	1865	1866	1867	1868	1869
Imphy-Saint Seurin	838	2,956	3,751	4,858	6,518	10,373	10,131
Petin, Gaudet et Cie.	977	2,897	3,059	3,851	7,238	8,777	9,881
Terrenoire	41	443	1,754	1,537	5,256	20,566	29,514
De Dietrich et Cie		271	841	485	764	1,540	1,261
Menans et Cie		183	346				
Chatillon Commentry				59	117	1,345	1,613
TOTAL	**1,856**	**6,750**	**9,751**	**10,790**	**19,893**	**42,601**	**52,400**

At the end of 1869 the following converters were in operation:

Imphy	1 @ 4T; 2 @ 5T
Assailly	2 @ 7T; 1 @ 10T
Terrenoire	4 @ 5T
Bessèges	2 @ 5T
Mouterhausen	2 @ 4T
Montluçon	2 @ 4T

It was commented that these 16 converters, totalling an installed capacity of 75 tons, were theoretically capable of producing 300 tons every twenty four hours or around 90,000 tons per annum. There was also the comment that both Le Creusot and Chatillon-Commentry were putting in hand large scale Bessemer production to come into service in 1870.[30]

[30] Jordan, op. cit., pp. 360–361.

To put this growth in the production of Bessemer steel in proportion, it is useful to have details of the overall steel production in France over the years 1863–1870, produced from the official French Government sources:[31]

Year	Bessemer	Crucible	Puddled & Forged	Total
1863	1,856	9,482	35,823	47,161
1864	6,750	9,410	21,442	37,602
1865	9,751	10,030	21,778	41,559
1866	10,790	5,209	24,575	40,574
1867	19,893	6,119	21,585	47,584
1868	43,601	6,062	—	48,261
1869	52,400	9,353	8,514	60,267
1870	90,000	7,610	12,621	110,231

There is something of a problem in tracing the growth of the Bessemer production beyond this date, since the official French statistics quote the combined ingot tons made by the Bessemer and the Siemens Open Hearth processes – the latter being referred to as "Martin Steel". The French, as will be seen later, were pioneers in the use of the Open Hearth furnace and several plants included both types of furnace; in general, the Bessemer converter was used to produce steel from the blast furnace product whilst the Open Hearth furnace was used to deal with the increasing amounts of scrap which became available. For the weight of steel products, rather than ingots, in France in 1872 and 1873, the following figures are available, expressed in metric tons:[32]

	Bessemer			**Martin, Crucible & Puddled**			**Grand Total**
	Rails	Bar & Plate	Total	Rails	Bar & Plate	Total	
1872	**66,904**	**20,797**	**87,801**	**15,317**	**35,791**	**51,289**	**139,090**
1873							
Bassin de Paris					756	756	756
Centre	37,895	3,839	41,734	5,281	2,996	8,277	50,011
Gard					925	925	925
Loire	41,310	20,817	61,497	17,595	36,254	53,849	115,346
Sud Ouest					630	630	630
Ardennes					4	4	4
TOTAL	**79,205**	**24,026**	**103,231**	**22,876**	**41,565**	**64,441**	**167,672**

31 Jeans, op. cit., p. 231. There is something of a problem in accepting these figures at their face value , since there is no indication of any Siemens Martin production. It is possible that was included in the category "puddled and forged", as it quite clearly is in the following table (Footnote 32) but this requires some explanation for the years 1868 to 1870.

32 J.I.S.I., 1874, p. 187.

There is a list of Bessemer Converters in operation in France in 1875.[33] It does not, however, give any indication as to their size:

Compagnie de Chatillon et Commentry	2
Societé de Commentry Fourchambault	3
Societé de Denain et Anzin	2
Compagnie des Hauts Fourneaux, Forges et Acieries de la Marine et des Chemins de Fer (previously Petin, Gaudet et Cie)	3
Compagnie des Forges, Fonderies et Acieries de Saint Etienne	2
M.M. Schneider et Cie (Le Creusot)	4
Compagnie de Terrenoire, la Voulte et Bessèges	8
TOTAL	24

Some further clarification, but also some confusion, can be found elsewhere; Holley visited the works at Le Creusot, at Terrenoire and at St. Chamond in 1875.[34]

He pointed out that his visit to Le Creusot was an unusual and personal favour; for the past three years, Henri Schneider had refused admission to the steel department to all but his own friends, largely on account of the newer Open Hearth furnaces there. As far as the Bessemer plant at Le Creusot was concerned, against the above figure of four converters, Holley describes six converters arranged in three pairs and, moreover, indicates that the design had been copied substantially from the John Brown plant in Sheffield.[35] The converters are described as nominally 10 ton capacity but normally operating on eight ton charges. The tuyere boxes were not removable, but the whole lower sections of the vessels were detachable just below the trunnions when the tuyeres were worn, and another lower section, with tuyeres previously set and dried, could be attached. This was rather a more arduous operation than that in general use on the more recent design with a renewable bottom tuyere section and took rather more time but was, it seems, only necessary once every twenty heats, the tuyeres and refractory linings all being in native materials, the result of much evidently successful experimentation. The converters were directly fed with liquid metal from the blast furnaces; this metal contained around 3% manganese. Nevertheless, it was blown down to remove all the carbon. Blowing lasted about 15 to

[33] J.I.S.I., 1875, p. 194.

[34] A. L. Holley: Private Publication No. 8: Steel Manufacture in France and Belgium, May 1876, 83 pp.

[35] It was noted by Jordan, op. cit., p. 362, that the plant at Le Creusot was under construction in 1870.

17 minutes, finishing with a thick brownish smoke, making the use of the spectroscope vital for determining the exact end point of the blow. The blow was reasonably hot; if too hot, a certain amount of scrap would be added after blowing for cooling purposes. The metal destined for railmaking was then recarburised with spiegeleisen, that from St. Louis, near Marseilles containing as much as 12% manganese. Milder steel, for beams, boiler plates and "ship iron", was made with a 2% addition of a ferromanganese with 50% manganese. Typical analyses of the two types of material are quoted as follows:

	Rail Steel	Boiler Plate
Carbon	0.345	0.17
Silicon	0.055	0.02
Manganese	0.60	0.55
Sulphur	0.034	0.015
Phosphorus	0.070	0.077

The overall loss was said to be around 10–11%, lower than in most Bessemer plants. Around 75% of the ingots made went into rail production, the ingots for this purpose being either 9 inch or 11 inch square, all of them top poured. Holley made the comment that the Bessemer plant at Le Creusot was among the finest he had visited abroad.

At Terrenoire, as at Le Creusot, the Bessemer plant constituted the older part of the melting equipment, since it had recently been supplemented by an impressive range of Open Hearth furnaces. Indeed, the Bessemer plant dated back to around 1863 and the four 5 ton converters were arranged in pairs over two casting pits, "after the older English fashion," as Holley expressed it; indeed, he went on to comment that the plant was old-fashioned and constricted and, although the works were kept in comparatively good order, the layout called for very hard and hot work, especially on the part of the men about the pit. The converter bottoms, as would be expected on this older plant, were not detachable and this led to further hot work and delays, having to be renewed every 12 heats. The converters were fed with liquid metal from the two blast furnaces, care being taken that half of each charge came from each blast furnace, in order to equalise the composition as much as possible; the charges were of the order of 8,000 lb (= 3.6 metric tons) but each charge was weighed in the ladle before transfer. The standard practice was to blow to remove all the carbon; the blast furnace metal generally had only about 1.5% silicon with under 2% manganese and thus the blow was somewhat on the cold side compared with that at Le Creusot, giving little opportunity for the use of scrap in the converter. Recarburising the blown metal was very much as at Le Creusot, using either the high manganese spiegeleisen or the 50% ferromanganese, depending on the desired carbon in the finished steel. The spectroscope was in use for determining the end of the blow; some use was also made of the colour test for the determination of carbon. Waste in blowing was around 14%. The metal was all top poured, usually into 11 inch ingots for rail production. Annual Bessemer production was stated to be 23,200 tons of rails and 1,800 tons of plate; the 30,000 ingot tons required would call for 13 to 14 blows per pair of furnaces per day. The similar plant at Bessèges, working on the same plan and under the control of Terrenoire, made up to 25,000 tons of rails annually.

The two works of Petin, Gaudet & Cie at Givors and St. Chamond each had a pair of 4 ton converters;[36] those at Givors are known to have been installed in 1869[37] and Holley only visited the works at St. Chamond, because, in his estimation, here was "the newest and best Bessemer plant." It is not clear when this plant was installed but there certainly were original features here, in that the two converters, opposite each other, and their charging platforms, were erected on tall pillars in the centre of the building, quite detached from the walls, the vessel centres 12 feet above ground level with a relatively shallow casting pit. Above the vessels were sheet iron hoods projecting through the roof. The two cupola furnaces for the spiegeleisen were on columns some 18 feet high, at one side of the vessels, with a lift between the two vessels to provide the cupolas with fuel and metal and also to raise the ladle coming from the blast furnace, which was situated some 300 feet away. The converter bottoms were of the detachable type. Ingots were typically 10–11 inches square; one interesting comment is that the ingot moulds "have the bottoms cast in and are largest at the top";[38] the metal was quite obviously, therefore, top poured. The main Bessemer steel product at both Givors and St. Chamond was rails. At the stated production of only 1,500 tons per month from each plant, Holley concluded that the furnaces were not heavily worked. Elsewhere, however, it is reported that the plant at Givors produced 24,000 ingot tons in 1877.[39]

There is a further list of steelmaking furnaces in France in 1879.[40] The earlier list quoted above showed 24 converters and 25 Open Hearth furnaces the later list still shows 24 converters but now there are 44 Open Hearth furnaces. The changes in Bessemer capacity are little confusing. Commentry Fourchambault now have 3 converters instead of 4; Chatillon

36 Clearly some of these converters are not included in the list of 24 given earlier.

37 Jordan, op. cit., p. 358.

38 This provides the single exception to the comment given earlier that, until well after1900, ingots were invariably cast narrow end up. These moulds at St. Chamond in 1875, indeed, would not have looked out of place in many Sheffield works in the 1920s, provided they had been fitted with headboxes!

39 Jeans, op. cit., p. 228.

40 J.I.S.I., 1880, Pt. 1, p. 425.

Commentry likewise have one less furnace at Montluçon but have installed 2 further converters at what appears to be a new site at Saint Montant. Both Terrenoire and the sister plant at Bessèges have reduced their 4 converters to 2; the converter plant at Saint Chamond has been closed; instead there are now 2 at the associated works at Assailly instead as well as the two previously at Givors which are still operating; there is one furnace less at Saint Etienne, Denain et Anzin and Le Creusot being as before.

To conclude this section of the survey, in default of any better information, it seems appropriate to quote the official French statistics for steel production for the period 1874 to 1880 as available from the Iron and Steel Institute surveys of production.[41] They do not, of course, differentiate between Bessemer and Open Hearth production and, as the above survey shows, Bessemer capacity was more or less static over a period when Open Hearth capacity doubled. Open Hearth production would appear therefore to have been a significant proportion of the total even at the beginning of the period and presumably this proportion increased as time went on.

Year	1874	1875	1876	1877	1878	1879	1880
Forged Steel	350	267	175				455
Puddled Steel	17,214	17,858	19,236				18,604
Bessemer & Martin	194,216	231,467	231,998	221,817	304,000	317,000	359,831
Cemented Steel	2,656	2,044	2,780	2,137			
Crucible Steel	7,226	6,143	7,685				7,777
TOTAL	**221,684**	**257,759**	**261,874**				**388,894**

Developments in Austria

Bessemer production in Austria dates from 1863 when Professor Tunner of the Metallurgy School at Leoben collaborated with Prince Schwarzenberg in the setting up of a small converter, believed to be on the Swedish plan, at Turrach.[42] This was followed by a quite remarkable growth in Styria and Carinthia, where the supplies of haematite and, particularly, the spathic ores were well suited for production of blast furnace metal for the Bessemer converter. The situation in Austria, indeed, was not dissimilar to that in Sweden, with plentiful supplies of high quality ore and very poor supplies of mineral fuel, so that all blast furnaces were charcoal fired. It is not surprising, therefore, that the earlier Austrian trials on the Bessemer process were based on Swedish experience.

[41] J.I.S.I., 1875, p. 604; 1877, Pt. 2, p. 239; 1881, Pt. 1. p. 340. The quoted figures are in metric tons.

[42] Jeans, op. cit., p. 241.

Little more is known of the experience at Turrach. There were three 2 ton converters in operation in 1868, but whether they were fixed or tilting type is not known; at the 1867 exhibition they showed anvils and axles in Bessemer steel and they had a good reputation for their harder steels.[43]

In 1864, however, Bessemer steel commenced to be made at Heft in Carinthia.[44] The first converter was a small fixed one, built on the Swedish pattern, just over four feet in internal diameter and just over five feet high – remarkably similar, indeed, to the converter erected the year before at Backa in Sweden. The works soon grew and two further fixed converters were installed, each five feet in diameter and eight feet high internally, each with 23 tuyeres some 0.67 inches diameter set at a slight angle around the periphery of the base and each capable of refining from 3.0 to 3.5 tons of metal run direct from the blast furnace. The lining was said to be of crushed quartzite mixed with clay, the bottom part of the converter generally needing to be replaced after 9 to 11 heats, up to an exceptional 16 heats. The Swedish practice of catching the carbon was generally followed, although some heats had a small amount of blast furnace metal added to the ladle for adjustment of carbon; it is commented that the time of blowing could vary between a minimum of 18 minutes to a very extended 35 minutes, depending on the type of blast furnace metal and the final desired carbon. Ingots were cast uphill in clusters with a central ingot mould as the feeder. In 1867, these two converters, with the smaller one in reserve, were capable of producing some 4,500 tons of ingots per annum. There is a most interesting survey of production for the first three years of operation; in this connection, it should be explained that the steel made was classified according to the grain structure obtained on test samples from each heat which were flattened under a hammer, quenched, fractured and graded into six groups with nominal carbon analyses:

I	II	III	IV	V	VI	VII
1.45% C	1.25% C	1.05% C	0.80% C	0.45% C	0.25% C	0.15% C

The production record is to be found in Table I on p. 112.

43 Jordan, op. cit., p. 377.
44 Jordan, op. cit., pp. 382-386.

TABLE I BESSEMER STEEL PRODUCTION AT HEFT IN CARINTHIA, 1864-1866

	1864	**1865**	**1866**
Number of blows	216	506	1033
Metal to converter, tonnes	339.1	991.5	2875.6
Metal added to ladle, tonnes	2.1	6.9	56.9
Total metal used	341.2	998.4	2932.5
Cast as pig iron, tonnes	nil	2.1	54.7
Dressed ingots, number	576	1597	6981
Steel Grade I, tonnes	1.4	nil	nil
II, tonnes	4.1	4.9	nil
III, tonnes	13.7	29.8	5.7
IV, tonnes	29.0	108.3	130.6
V, tonnes	43.7	193.3	535.7
VI, tonnes	54.8	284.8	1562.8
VII, tonnes	49.7	100.4	144.1
Total Ingot Weight	**196.4**	**721.5**	**2378.9**
Losses:			
Casting runners, tonnes	nil	1.5	9.2
Ladle skulls, tonnes	31.3	102.5	85.8
Runnings, tonnes	26.6	1.5	10.7
Furnaces skulls, etc, tonnes	33.6	33.5	24.9
Production total	**287.9**	**860.5**	**2509.5**
100 kg. blast furnace metal would provide:			
Pig iron, kg.	nil	0.21	1.86
Dressed ingots, kg.	57.57	72.26	81.12
Casting runners, kg.	nil	0.15	0.31
Ladle skulls, kg.	9.17	10.26	2.92
Runnings, kg.	7.80	0.15	0.36
Furnace skulls, kg.	9.85	3.35	0.85
Fire loss, kg.	15.61	13.62	12.57
From this data can be derived the following:			
Average Metal charge weight, kg	1570	1959	2784
Average metal added to ladle, kg	10	14	55
Average number of ingots per blow	2.67	3.16	6.76
Average ingot weight, kg	341	452	341
% of steel in Grades VI and VII	53.2	53.4	71.8

Source: S. Jordan: "Revue de l'Industries du Fer en 1867", Vol. 2, Paris 1869, p. 385.

The improvement in performance over the three years is impressive, the ingot yield going up from 58% to 81%; the latter figure was around good practice average. With the charge weight almost doubling over the period, this indicates more and more use of the larger furnaces; the extended production of the lower carbon materials also implies more furnace time used productively and so less down time for accidents. One interesting feature, however, is the apparent use of larger ingots during 1865; the coincidence between the mean ingot weight in 1864 and 1866 is otherwise hard to explain. The works at Heft did, indeed, exhibit an ingot weighing 3.1 tonnes, which must have represented the whole of one Bessemer blow, at Paris in 1867; this had been forged down in several steps and showed an excellent finish. There was also a highly polished cylindrical driving shaft and numerous examples of steels of varying hardness according to their own scale, to which reference has already been made.

At Neuberg in Styria, operations were commenced early in 1864, also using Swedish practice with a small fixed converter. Shortly afterwards, Baron Rothschild sent his engineers to witness Bessemer's own operations at the Sheffield Works, taking with them a supply of Hungarian pig iron for trials.[45] The outcome was the setting up of a plant with two "modern" converters – of the tilting type – at Neuberg later in 1864.[46] They were lined with a mixture of ground quartz from Olmutz and fireclay from Blanske, near Brno; the vessel would work 60 to 80 blows with such a refractory before requiring relining.[47] The converters were arranged so as to be capable of being fed with liquid metal from the two blast furnaces on the site. The operations were carried out as at Heft, generally using the Swedish method of catching the carbon with no addition, but occasionally adding a little of their own blast furnace metal in the production of the lower carbon grades, possibly in an attempt to prevent blow hole formation in the ingot. It seems that one or two procedures were novel to Neuberg, in that they were in the habit of adding puddling scrap to the metal charge in the Bessemer just before the boil started and that they also had a device whereby powdered carbon could be taken in with the blast to increase the heat on cold blows; the same mechanism was also used on occasion to inject common salt to render slags more fluid.

From documents issued at the Paris Exhibition in 1867 can be quoted details of Heat No. 599 at Neuberg. The charge was 3,516 kg. of what was termed a very grey iron from Blast Furnace No. 2, operating on Altenberg ore (a good quality haematite). The air entered through 49 holes, each 9mm.

[45] Jeans, op.cit., p. 241

[46] There is confusion here. Jordan, op. cit., p. 378 indicates the converters were capable of handling charges from 3750 to 5000 kg.; Jeans, op. cit., p. 241, states they were of 3 ton capacity and that a 1 ton vessel was also erected for trial purposes.

[47] Jeans, op. cit., pp. 241–242.

diameter, under a pressure of 119 cm. of mercury. The first period lasted 28 minutes. The second period lasted only 7 minutes, with the pressure reduced to 109–113 cm. The final period was 3 minutes with 113 cm. pressure. The product was 3,057 kg. of ingots. In the following tables the analyses of the metal and slag are given. A is the original blast furnace metal; B is metal at the end of the first period; C is metal at the end of the boil; D is metal before adding the final blast furnace metal to the ladle and E is the metal before casting. Slag samples b, c, d and e correspond to the metal samples. The analyses quoted are as follows:

	A	B	C	D	E
Carbon	3.93	2.46	0.95	0.09	0.23
Silicon	1.96	0.44	0.11	0.02	0.03
Manganese	3.46	1.65	0.43	0.11	0.14
Sulphur	0.018	tr.	tr.	tr.	tr.
Phosphorus	0.040	0.040	0.045	0.045	0.044
Copper	0.09	0.09	0.10	0.12	0.11
		b	c	d	e
Silica		46.78	51.75	46.75	47.25
Alumina		4.65	2.98	2.80	3.45
Iron oxide		6.78	5.50	16.86	15.43
Manganese oxide		37.00	37.90	32.33	31.89
Lime		2.98	1.76	1.19	1.23
Magnesia		1.53	0.45	0.52	0.61

Also available were details of the production at Neuberg during 1866 and for the first three quarters of 1867:

	1866		**1867**	
		1st.qtr.	2nd.qtr.	3rd.qtr.
Number of charges	607	228	157	172
Metal charged, tonnes	2060.7	844.5	610.8	681.2
Production:				
Ingots	1718.6	717.5	506.0	572.4
Runnings	40.5	10.3	9.4	11.0
Metal ejected	16.0	8.0	5.7	8.0
Chimney skull	16.9	8.4	11.7	9.5
Total	**1792.0**	**744.2**	**532.8**	**601.8**

100 kg. of blast furnace metal would provide:

	1866	1867 1st.qtr.	2nd.qtr.	3rd.qtr
Ingots, kg.	**83.36**	**84.96**	**82.82**	**84.05**
Runnings	1.96	1.21	1.54	1.61
Metal ejected	0.78	0.95	0.93	1.30
Chimney skull	0.81	0.99	1.91	1.38
Fire loss	**13.09**	**11.89**	**12.80**	**11.66**

These figures are evidence of good running of the plant; it is interesting to compare these obtained on tilting converters with the figures from Heft with the fixed converters, quoted above, the major difference being the ladle skulls in the latter case. Comparative costs are quoted as being 308 francs per metric ton for Neuberg as against 312 francs per metric ton for Heft.[48]

Neuberg also published a "Hardness Scale", the details of which are worthy of quotation.

Degree of Hardness	Carbon Content Range %	Breaking Stress kg./sq.mm.	Elongation at Rupture %	Hardening on Quenching
I	1.38–1.58	Not ordinarily used		
II	1.12–1.38	do.		
III	0.88–1.12	89–102	5	Requires care
IV	0.62–0.88	72–89	10–5	Very good
V	0.38–0.62	56–72	20–10	do.
VI	0.15–0.38	48–56	25–20	Less good
VII	0.05–0.15	40–48	30–25	Negligible

Applications of the above materials:

III. Trepanning tools, boring bits, scissor blades, rock drills, fine springs, sheet for saw blades, etc.

IV. Cutting tools, files, scythes, wood drills, saws, etc

V. Wheel tyres, machine parts, springs, plates, angles, gun barrels, etc.

VI. Boiler plates, axles, construction work, thin sheet for pressings, sabre blades, gun barrels, etc.

VII. Axles, machine parts, soft sheet, wire, etc.

[48] All the details on operations at Neuberg, together the table on the "Hardness scale", are from Jordan, op. cit., pp. 378–382.

The Bessemer Works at Graz were established in 1865 by the Southern Austrian Railway Company to provide steel headed rails, crank axles and other axles for the locomotives and tenders. It had two 4.5 ton converters operating on pig iron remelted in reverberatory furnaces together with a spiegeleisen furnace and made up to 4,200 tonnes of steel annually with a workforce of 32.[49]

Bessemer steel production was started in Moravia at Witkowitz (Vitkovice) by the Rothschild family in 1866 with two 4 ton converters and one 2 ton vessel, all of the tilting type. The raw materials mostly came from Hungary: the pig iron from Mariazell and the spiegeleisen from Concordia. To melt these there were three large reverberatory furnaces and a small spiegeleisen furnace, as well as a double hearthed reverberatory to service the small converter. The larger converters had twelve tuyeres, each carrying 8 or 9 holes, each 8 mm. diameter. These tuyeres were made from the clay from Blanske; it had been found useful to grind up old tuyeres and to mix this dust with new fireclay for making fresh tuyeres, which were produced by a moulding machine invented by the manager of the works.[50] The charge was ordinarily 3,600 kg; one charge which was observed, comprising two thirds Hungarian pig and one third Styrian pig, took three hours to melt before being charged; the blow lasted 23 minutes – 12 minutes for the first period, 8 minutes for the boil and 3 for the refine. A 9% addition of spiegeleisen was made. The plant normally produced three to four blows during a twelve hour day shift; it did not work at night. The tuyeres only lasted about 6 blows; it took 36 hours to change them since the vessel was allowed to go cold before the men went in. Trials had been made to use the iron from the blast furnaces on the plant but without success; it was too phosphoric. The steel was mainly used for rails and wheel tyres.

Beyond this date, only sketchy details have been located concerning the Austrian Bessemer Steel production. It is known that Heft changed from the fixed converters to the tilting type in 1870 and that new works were established in Ternitz in 1868; at the time, this was the largest Austrian Bessemer steelworks, with six converters. The only Hungarian works was founded at Resicza the same year. Zeltweg was founded in 1871; a works at Teplitz and a second works in Graz – this one owned by the Austrian State Railways – starting up in 1873, Kladno and Teschen in 1875 and Prevali in 1876. Invariably, these new additions all had two converters; details of the sizes are not known. By 1876, therefore, there were 32 converters in

[49] Jordan, op. cit., p. 386. Operations at Graz are said to be described by Castel in Annales des Mines, 6me. Serie, tome viii, pp. 149ff, together with later contributions in Annales des Mines (Tomes xii and xiii) by Gruner and Clerault.

[50] This information on tuyeres is from Jeans, op. cit., p.242. The remaining information on Witkowitz is from Jordan, op. cit., pp. 387–388.

operation in the Austro-Hungarian empire with an annual production of around 90,000 metric tons. Table II gives annual outputs from the individual Bessemer works in the Austro-Hungarian empire for the period 1864 to 1877.[51]

Developments in Germany

Bessemer failed to obtain a patent in Germany. The Prussian authorities stated that a patent essentially similar to his had been registered many years earlier. Pressed to state what this patent was, they eventually reported that the similar process to which they referred was that of Nasmyth.[52] Bessemer protested that the two processes were completely different and even persuaded Nasmyth himself so to inform the Prussians; nevertheless the decision stood and the Bessemer process, therefore, could be used by any German firm without let or royalty. This led to Germany becoming the Continental country which made the most use of the Bessemer process. Such is the story as generally accepted in Britain.

Research into the records held by the Krupp Archives, however, gives a rather different version of this story.[53] Alfred Krupp's agent in London in the 1850s was Alfred Longsdon, brother of Robert Longsdon, who was Bessemer's friend and partner. Thus it was that Krupp had knowledge of Bessemer's activities and appreciated that Bessemer's efforts to produce liquid steel could be competition for the crucible process. Krupp, therefore, tried out the procedures detailed in Bessemer's 1855 patents in his own works and became convinced of the practicability of the idea. That Bessemer was aware of these activities is confirmed by the survival of a letter to Alfred Longsdon from Bessemer dated 25th January 1856, as follows:

> I shall be much pleased to hear that Mr. Krupp has made a trial of the principles of my invention and I have no doubt that his practical knowledge of the whole subject will be of great service in doing so. . . . I shall be happy to hear anything Mr. Krupp may have to say on the subject and, in the meantime, I would suggest that if Mr. Krupp feels desirous of entering into a negotiation that he takes the necessary steps to secure Patents in Prussia for this invention in order to prevent any accident from their publication.

51 Table II is a direct copy of that provided in Jeans, op. cit., p. 243.

52 J. Nasmyth, British Patent 1001, 4 May 1854. This claim was clearly in error; Nasmyth's patent covered the introduction of steam into the molten metal in the puddling furnace.

53 H. Dickmann:"Alfred Krupp – Henry Bessemer", Stahl und Eisen, Vol. 82 (1962), pp.911–913. The author is indebted to the Historisches Archiv of Fried Krupp GmbH for bringing this information to his attention.

TABLE II BESSEMER STEEL PRODUCTION IN THE AUSTRO-HUNGARIAN EMPIRE, 1864–1877

Area	----Bohemia----		Moravia	Silesia	Lower Austria	--------------------Styria--------------------					---Carinthia---		Hungary	
Name	Teplitz	Kladno	Witko-witz	Teschen	Ternitz	Turrach	Neuberg	S.R. Graz	Zeitweg	Graz State	Heft	Prevali	Resicza	TOTAL
Opening Date	May 1873	Sep 1875	May 1866	Jan 1875	Jun 1868	Nov 1863	Feb 1865	Jan 1865	Sep 1871	May 1873	Jul 1864	Mar 1876	Dec 1868	
No. of Fces.	2	2	3	2	6	3	2	2	2	2	2	2	2	32
Year						**Production in Metric Tons.**								
1864						109					197			306
1865						211	794	1,895			695			3,545
1866			524			308	1,718	2,906			1,379			6,835
1867			1,585			724	2,824	2,396			1,236			8,765
1868			2,727		3,346	547	3,090	3,294			1,304		287	14,495
1869			3,230		7,144	464	3,112	3,166			1,217		2,389	20,722
1870			3,200		6,055	607	3,980	3,878			764		3,628	22,112
1871			3,278		10,565	1,145	5,261	4,577	1,327		3,102		6,257	35,512
1872			3,800		20,514	901	5,697	5,063	7,142		5,189		7,098	55,404
1873	2,934		5,570		29,849	1,432	4,435	4,503	9,480	1,990	7,588		9,040	76,821
1874	9,963		6,630		34,127	897	3,607	3,659	11,358	7,255	10,160		9,302	96,958
1875	9,270	921	7,197	3,163	19,395	1,081	3,238	3,710	7,480	11,335	10,613		13,203	87,443
1876	11,835	7,635	2,734	5,113	15,606	1,419	3,834	5,196	6,653	3,350	9,192	340	22,132	89,926
1877	12,235	8,912	12,280	5,252	14,403	1,267	3,602		10,530		10,801	5,793	17,646	97,470

Source: J. S. Jeans: "Steel: Its History, Manufacture, Properties and Uses".

On 23 April 1856 Krupp applied for such a patent; the Minister refused to assign it. On 18 June 1856 the application was renewed with an amended, clearer description of the process. The letter of refusal dated 25 August 1856 considered that the patents of Nasmyth and Martien[54] showed prior usage of the Bessemer principle. A final application made 29 September 1856 met with the reply that the method had been openly published by Bessemer's own presentation at the Cheltenham meeting, followed by printing of the text; so it was that the element of novelty was lost.

It was a further five years before Krupp began to construct the first Bessemer steelworks in Germany; it is presumed that the delay was due to the troubles which Bessemer met in the late 1850s. The operations were put in hand with the full cooperation of Bessemer and Longsdon, two 2.5 Ton converters being ordered from Sheffield with an agreement dated 5 August 1861 for Krupp to pay £1,000 annually in exchange for technical detail and, more importantly, an exclusivity agreement on the process for Krupp throughout Germany. The first £1,000 was paid in two instalments in 1861–2; there is a strong probability that payments were then discontinued. The first steel was made from the converter on 16 May 1862. It is said that Krupp kept the Bessemer operations secret, the Bessemer shop being referred to as a tyre mill (Raederwalzwerk C) and the steel as "C-steel". These operations proved so successful that Krupp ordered two 5 Ton converters before the second 2.5 Ton vessel had been delivered from Manchester. In 1864 he added two small converters and four 5 Ton vessels made in his own works; the total equipment at this date, therefore, was six large and four small converters, together with eighteen reverberatory furnaces for remelting the pig iron. Initially, the low phosphorus pig iron was obtained either from England or Sweden; later it was a domestic product coming from the Georgesmarienhutte at Osnabruck; eventually he had his own blast furnace plant operating on imported Spanish ores. The liquid steel, available in quantity for the first time, was used mainly for rails and tyres. Demand was such that a further five 5 ton converters were added in a second shop, used for the production of rails; this plant operated continuously from 1865 to 1874 and then intermittently until 1883, when it was closed down and the plant used as spares for the main shop. By 1876, the pig iron was all melted in cupolas. Annual production figures for the Krupp Bessemer Steelworks are available for the period 1862 to 1925, when they finally went out of operation. Figures up to 1880 are quoted as follows:

[54] J. G. Martien, British Patent 2082, 15 Sept 1855. This patent covered the running of molten blast furnace metal along a channel through which air or steam could be fed. It was not really an application of the Bessemer principle. It was this patent to which the work of Mushet was linked by the Ebbw Vale Company, much to Mushet's subsequent regret.

Year	Ingot Tons	Year	Ingot Tons
1862	423	1870/1871	40,956
1863	4,032	1871/1872	60,573
1864	10,021	1872/1873	65,394
1865	17,905	1873/1874	71,102
1866	19,283	1874/1875	51,046
1867	19,043	1875/1876	69,355
1868	18,346	1876/1877	77,338
1869	22,462	1877/1878	89,600
		1878/1879	108,380
Jan–Jun 1870	20,640	1879/1880	124,051

The second Bessemer plant to be established in Westphalia was that at Hoerde, installed in 1864 by Hermann Wedding; he had studied the Atlas Works of John Brown in Sheffield and he built his new plant on very much the same plan. The first Prussian furnace was that at Konigshutte, which commenced operations in January 1865.[55] This was the only German establishment to exhibit Bessemer steel at the Paris Exhibition in 1867; the display was said to be disorderly with little of note, the Bessemer ingot and the Bessemer rail exhibited being of very mediocre quality.[56]

A list of the converters in being in Germany in 1868[57] reads as follows:

Essen, Westphalia (F. Krupp)	18 @ 2T to 5T
Bochum, Westphalia	6 @ 3T to 5T
Hoerde, Westphalia	5 @ 3T to 5T
Oberhausen, Westphalia	2 @ 4T
Dusseldorf, Westphalia	2 @ 3T
Konigshutte, Silesia	2 @ 4T
Zabose, Silesia	4 @ 4T

The works at Hoerde at this date had two Bessemer shops; one had two converters, together with 4 reverberatory furnaces for melting pig iron and two smaller ones for the spiegeleisen; the other had three converters together with cupolas. Hard steel was made using pig iron with 3% manganese and 1.0–1.5% silicon; the blow lasted only ten minutes or so. It was reported

[55] O. Johannsen: "Geschichte des Eisens", Dusseldorf, 1953, pp. 464–466. According to Jordan, op. cit., however, the works at Konigshutte originally worked on local pig and proved unsuccessful. It was not, therefore, until 1867 they were in production commercially; by this time they were using imported English Bessemer pig iron.

[56] Jordan, op. cit., p. 390.

[57] Jordan, op. cit., p. 391.

that the end of the refining was judged from the appearance both of the flame and of slag samples taken out on the end of an iron bar; it was said that the dense brown fumes precluded the use of the spectroscope.[58] Softer steels were produced using haematite pig iron imported from Cumberland. Final additions of spiegeleisen were made in either case. There is an interesting comment that the ingots were stripped whilst still very hot and were hammered direct without reheating.

Figures are available for production in the years 1875,[59] 1876[60] and 1877:[61]

Year	No. Converters in being	No. Converters working	Output, tons.	Productivity tons/converter
1875	. .	35	235,391	6,728
1876	78	32	270,804	8,496
1877	81	39	390,434	10,011

A complete list of the 78 converters in 1876 is available; unfortunately no idea of capacity is provided:[62]

Prussia

Konigshutte, Upper Silesia	4	
Osnabruck Steel Works	4	
Hermanshutte, Hoerde	5	
Hoesch, Dortmund	2	
Union, Dortmund and Hattingen	4	
Bochum Steel Works	19	
New Bochum Works	7	
Krupp, Essen	18	
Gute Hoffnungshutte, Oberhausen	4	
Pheonix, Ruhrort	2	
Rheinische Steel Works, Meidereich	8	
Ponsgren and Giesbers, Dusseldorf	2	
Rothe Erde, Aix la Chapelle	2	
Steinhaufer Hutte, Witten	2	66 Total

[58] Jordan, op. cit., p. 392, was not convinced his was true. A typical analysis of the fume was quoted as MnO 45%, SiO_2 20%, FeO 10–15%, CaO + MgO 20–25%.

[59] Jeans, op. cit., p. 178, quoting Dr. H. Wedding in "Reichsanzeiger", 4 Oct 1877. Wedding compared this figure with the 22 converters in America, producing 375,517 tons, or 17,069 mean annual tons per converter, putting this down to the superior mechanical appliances in the American works.

[60] Jeans, op. cit., p. 174

[61] J.I.S.I., 1880, Pt. 1, p. 243.

[62] Jeans, op. cit., pp. 174–175

Saxony		
Konigin Marienhutte, Zwickau	4	4 Total
Bavaria		
Maxhutte, Regensburg	2	
Gienauth Bros, Kaiserslauten	2	2 Total
Alsace-Lorraine		
Dietrich and Co, Niederbron	2	
De Wendel and Co, Hayange	2	4 Total

Germany at this date was the largest Continental producer of Bessemer steel. The number of converters shown above to be out of service is significant, however. Germany, and France and Belgium too, had large markets to supply and the simple truth was that they were limited by the supplies of suitable blast furnace metal they could provide. The resources of Continental Europe in low phosphorus ores were very largely limited to Austria and Sweden; there were vast resources of iron ore in Alsace-Lorraine which just could not be utilised on account of their high phosphorus content and many of the German works either imported iron ore from Spain, North Africa and elsewhere or relied on imported Bessemer pig iron from Cumberland. All these resources were in demand elsewhere for the same reason – it may be remembered that John Brown had secured ore supplies from Spain and built blast furnaces in his steelworks so as to achieve self sufficiency in his raw material supplies. The situation was soon to be changed dramatically, and we shall discuss this in a later chapter.

Developments in Belgium

The story of the Bessemer process in Belgium is in reality the story of the Cockerill works at Seraing. They took out a licence from Bessemer in 1863 and started operations in 1864. It is not clear what the original establishment was; the table which is reproduced below[63] indicates that there was only one furnace until 1870, whilst another source[64] clearly indicates the operation of two 4–5 ton furnaces in 1868. Seraing exhibited rails, locomotive tyres and axles, all in Bessemer steel, at Paris in 1867. At that date the works was operating entirely on bought pig iron, mainly on Bessemer pig from Cumberland.

Whatever the early history, it is clear that, by 1875, Seraing had become an important centre of Bessemer steelmaking: it was described[65] as:

[63] Jeans, op. cit., p. 267.

[64] Jordan, op. cit., p. 394.

[65] Holley, op. cit., pp. 65–80 gives his own report on a visit to Seraing from which these details are taken.

> a vast establishment, than which [of the Continental works] Krupp's and the Creusot works alone are larger.

Six miles down the Meuse from Liege, it had four collieries, thirty iron ore mines, seven blast furnaces (including two working on Bessemer iron) and two Bessemer plants – the old one, with four 5 ton furnaces which had produced up to 34,000 tons of steel in the year,[66] and the new one, with four 7 ton furnaces, fed directly from the blast furnaces. There were also a forge, with 7 hammers, a steel rolling mill with a 30-inch blooming train and a 24-inch rail train (as well as the old iron plant with puddling furnaces, hammers and its own roll trains), a foundry, machine shops and a bridge and boiler works; in addition the firm had its own shipyard at Antwerp.

The Bessemer blast furnaces were fed with a mixture of Algerian and Spanish ore and produced a metal with the following composition:

Carbon	4.00 – 4.50%
Silicon	2.25 – 2.50%
Manganese	3.00 – 3.75%
Sulphur	0.025 – 0.040%
Phosphorus	0.060 – 0.075%

The new plant had two 7 ton vessels which commenced operations in February 1874; the other two were commissioned six months later. They were both originally supplied with metal from cupolas; one pair was fed directly from the blast furnace when Holley visited the plant. With the facility with which the operations were performed and the ample handling power, he classed this as the best Bessemer plant in Europe. The vessels were set at one side of the pit which was reasonably shallow since they were raised above it and each pair of converters was serviced by three cranes. The vessels themselves had shells 8 feet in diameter; the internal diameter was 6ft. 9in. and the internal height 11ft. 6in. The bottoms were detachable but the custom was to replace worn tuyeres at intermediate stages rather than remove the bottom completely every time. The time for changing the vessel bottom or replacing tuyeres was in any case around 12 hours. The refractories used were of the following compositions:

	Silica	Alumina	Lime	Moisture & Volatiles
Vessel Lining	87.0	8.5	1.0	3.5
Vessel Bottom	67.5	26.5	1.0	5.5
Tuyeres	69.0	23.5	2.0	5.0
Nozzles	70.0	24.0	1.2	4.5
Setting clay	78.0	17.0	1.6	3.4

66 There is a footnote to the effect that the old plant at Seraing has been described in Engineering, September 1873.

The charges to the furnaces averaged 14,350 lb (6.4 Tons). The blast pressure was around 21–22 lb. per sq.in. One unusual feature was the introduction of cold rail end scrap, to the extent of 15–20% of the original charge weight, in the middle of the decarburising period, turning down the converter for the purpose. Using the spectroscope, from 30 seconds to up to a minute after the carbon lines appeared, when there was a thick brown manganese smoke apparent, the vessel was turned down, a paddle dipped into the metal, drawn out and plunged into water. The thin sheet of slag adhering to the paddle was then compared to a standard slag test colour; the indications were as follows:[67]

Lemon yellow:	0.75% C or over
Orange yellow:	0.60% C or over
Light brown:	0.45% C or over
Dark brown:	0.30% C or over
Bluish black:	0.15% C or over

The small globules of metal trapped in the slag were also tested by flattening them cold on an anvil: the edges would crack if the metal was too hard and would flatten completely if too soft. It was said that the operators soon became very proficient in determining the temper of the metal. If too hard, the blowing would be resumed; if too soft, a little crude iron would be added. No spiegeleisen was added after blowing and generally there would be something of the order of 0.5% manganese retained in the steel, together with a silicon content of 0.10–0.15%. This latter had no detrimental effect in the presence of the higher manganese content. The experience had been that the loss in the process was 10–11% with metal direct from the blast furnace but 14–15% when cold pig was remelted in the cupola.

The classification of the main types of steel produced, with the mechanical tests obtained, was reported as follows:[67]

Carbon, %	Breaking load, lb/sq.in	Elongation, %	Uses
0.55–0.65	98,000–156,000	5–10	Springs, etc
0.35–0.55	80,000–98,000	10–20	Rails and tyres
0.25–0.35	68,000–80,000	20–25	Arms and cannon

For tyre ingots, Swedish, Cumberland and Seraing pigs were mixed; it was reported that, out of 25,000 tyres made in one year, only 28 had been

[67] Jeans, op. cit., p. 271 quotes an extended steel classification at Seraing:
0.25–0.35% C Extra Soft
0.35–0.45% C Soft
0.45–0.55% C Half Hard
0.55–0.65% C Hard
0.65–0.75% C Extra Hard

returned as unsatisfactory. For gun barrel steel[68] the same mixture was used, but some spiegeleisen was added in the cupola and the quality kept up by the addition of Fagersta charcoal iron as required.[69]

Top pouring was general practice; some larger forging ingots were bottom poured. The general rail ingot was 12 in. square. The ingots were stripped hot and charged direct to the reheating furnaces in the rolling mill. The cast number of the heat was stamped on all ingots and a 3in. test ingot was taken from each heat, forged down to 0.75 in. bar and bend tests carried out, both hot and cold, and carbon contents determined on the bar using the Eggertz colour test. All these details were then recorded against cast number.

In addition, it is reported that Seraing did produce plate for its shipbuilding activities and four analyses of typical heats are given:[70]

Carbon	0.08	0.10	0.12	0.12
Silicon	0.02	0.01	0.01	0.45
Manganese	0.43	0.30	0.30	0.15
Sulphur	0.050	0.050	0.065	0.070
Phosphorus	0.065	0.045	0.045	0.085

The plates from these casts gave test results not inferior to comparative samples from British and Swedish plates.

This would seem to be a most appropriate way in which to close a technical discussion of Bessemer steel in Europe. It is necessary, however, to add that a second Bessemer plant in Belgium was added in 1873: the Angleur works with two converters. This is included in the production table below:[71]

Year	No. of Works	No. of Converters	Production in Ingot tons
1864	1	1	296
1865	1	1	969
1866	1	1	1,460
1867	1	1	1,767
1868	1	1	2,509
1869	1	1	3,699

68 Holley reports that some 500 tons of rifle steel were being made for Remington of New York at the time of his visit to Seraing.

69 A typical analysis of the Fagersta pig iron is quoted as C 4.95%, Si 0.77%, Mn 4.49%, S trace, P 0.027%.

70 Jeans, op. cit., p. 274.

71 Jeans, op. cit., p. 266.

Year	No. of Works	No. of Converters	Production in Ingot tons
1870	1	2	5,977
1871	1	4	10,854
1872	1	4	14,985
1873	2	7	21,268
1874	2	10	36,584
1875	2	12	53,500
1876	2	12	71,758

There will be found certain discrepancies between this list and the earlier evidence. Holley did report, however, that the plant he saw in 1875 had space allotted for a third pair of converters which were to be added in the near future. This would account eventually for ten converters at Seraing plus the two elsewhere, giving the Belgium complement of 12, as here stated.

Postscript

In Russia it is known that there was a 5 ton converter at the Aboukowsky Works in 1876;[72] it is also reported that there was some early work on the Bessemer process in Italy but it seems to have come to nothing.[73] Our survey has, therefore, covered most of what can be reported on the growth of the Bessemer process in Europe in the first twenty years after it bcame a viable operation. A useful summary of the output of Bessemer ingots in the various European countries at around the end of the period under survey is available:[74]

Year	Country	Converters available	Converters working	Ingot tons	Productivity tons/converter
1879	Gt. Britain	104	68	834,511	12,272
1876	France	26	17	218,000	12,821
1877	Germany	81	39	390,434	10,011
1877	Austro Hungary	32	17	97,470	5,733
1877	Belgium	12	8	76,000	9,500
1879	Sweden			20,820	

72 Jeans, op. cit., p. 255

73 Jordan, op. cit., p. 394.

74 J.I.S.I., 1880, Pt. 2, p. 243..

6 The Pneumatic Steel Process in the United States of America

Bessemer's presentation at the Cheltenham meeting of the British Association, as we have seen, was printed verbatim in *The Times* on 14th August 1856, three days after the meeting. This report, being considered of importance in the United States, was quoted in *Scientific American* on 13th September 1856, resulting in rather a surprising letter to the editor, published in the issue of 18th October, which reads as follows:[1]

> In November 1851, I commenced a series of experiments with a view to converting fluid pig metal into malleable iron, with the aid of a strong blast of air, and without the use of fuel, which process I term "air boiling". My object was to drive off the carbon in the iron and to make powerful blasts of air do the work of the fire and the manipulation of the puddler's bar in the puddling process. My first efforts were quite satisfactory, as with a blast taken from my furnace I produced a fair article of malleable iron. I found when using grey iron cold blast answered my purpose, but when the metal was white I found hot air had a better effect. I therefore had a small furnace erected to heat the air in blast pipes.
>
> My experiments were conducted publicly at this establishment; hundreds of persons called to see the trials I had made, and the subject was discussed amongst the iron masters, etc. of this section, all of whom are perfectly familiar with the whole principle and object I had in view, as discovered by me nearly five years ago.
>
> I was surprised to notice in *Scientific American* of the 13th September an account of a similar process of converting pig iron into malleable iron, claimed as the discovery of Mr. Bessemer of London and made within the past two years, the process not differing in the slightest from that I had in practical operation nearly five years since.
>
> I have reason to believe my discovery was known in England three or four years ago, as a number of English puddlers visited this place to see my new process. Several of them have since returned to England and may have spoken of my invention there. A charcoal furnace such as I have – using cold blast – produces various grades of metal, that I found had to be treated in the air boiling process with some variation; this caused difficulties which I have succeeded in removing and expect shortly to have the invention perfected and bring it before the public.
>
> *William Kelly,*
> *Suwannee Iron Works, Eddyville, Kentucky*
> *30th September 1856.*

[1] This is copied from J. McHugh: "Alexander Holley and the Makers of Steel", Baltimore, 1980, pp. 112–113.

William Kelly did, indeed, apply for a Patent early in 1857. Meanwhile, as was his practice, Bessemer had applied for patent protection in most countries where his invention might be used and accordingly he was granted U.S. Patent No. 16,082 on 11th November 1856, based on British Patent 2321, dated 17th October 1855. Kelly's application was accompanied by numerous affidavits confirming usage of the method by Kelly prior to the date of Bessemer's application for the British patent;[2] these implied that Kelly, indeed, started his work as early as 1847. The outcome was that Kelly's priority was acknowledged by the commissioner; he authorised the granting of a patent for the process to Kelly and withdrawal of such patent rights from Bessemer, who was, however, still to retain protection for his machinery. All this was subject to appeal by Bessemer; Bessemer, who was then in deep trouble with his home licencees and must have had grave doubts as to the real validity of the process, though fit not to contest these arrangements and Kelly was duly granted U.S. Patent No. 17, 628 on 23rd June 1857. The comment of the leading American steel historian on Kelly's success puts the situation rightly in context.[3]

> The validity of this claim cannot be impeached. But it must also be said that Mr. Bessemer, with the aid of Mr. Mushet, successfully employed this principle in the production of steel and that Mr. Kelly did not. The Kelly process produced refined iron of good quality. Furthermore the machinery with which Mr. Kelly operated his process was not calculated to produce rapidly or at all large masses of even refined iron; whereas Mr. Bessemer's machinery was successful after it was perfected in producing steel in large quantities and with great rapidity.

Kelly continued rather desultory efforts, none of them even reasonably satisfactory.[4] In 1861, with the Civil War rendering home production of metal more vital, the Cambria Ironworks in Pennsylvania undertook to provide Kelly with facilities for a trial of his method and for this purpose obtained a small tilting converter — clearly in contravention of the Bessemer patent!

> The first Bessemer converter ever erected in America was built at Cambria, by Mr. Kelly, and still remains there, a cherished relic. It was calculated to convert about half a ton of metal and received its blast from a foundry blowing engine. But I never heard even a tradition of a perfect conversion made in this vessel.[5]

[2] Details of the affidavits and the Examiner's report are to be found in J. D. Weeks: "The Invention of the Bessemer Process", Trans. Amer. Inst. Mining Engineers, Vol. xxvi, (1897), pp. 980-991.

[3] J. D. Swank: "The Manufacture of Iron in All Ages", 2nd. ed., Philadelphia, 1892, p. 399.

[4] McHugh, op., cit., pp. 124–129, provides details of these trials.

[5] R. W. Hunt: "A History of the Bessemer Manufacture in America", Trans. Amer. Inst. Mining Engineers, Vol. v, (1876–7), p. 210.

The importance which the Bessemer process was by then assuming in Britain and in Europe became clear to the Americans and, despite the failures with the Kelly trials, a group including the Cambria Iron Company, the steelmakers Park Brothers and influential industrialists such as E. B. Ward and Z. S. Durfee entered into an arrangement with Kelly, forming the Kelly Pneumatic Process Company early in 1862. The value of the contribution made by Mushet to the development of the original Bessemer invention in Britain was clearly recognised. The situation in America, however, was different from that in Britain in this respect. Whilst British Patent No. 2219, dated 22nd September 1856, taken out by Mushet on treatment of the blown metal with his "triple compound" of iron, carbon and manganese, had been allowed to lapse, thus making the principle freely available to Bessemer steel makers in Britain, the parallel U.S. Patent, No. 17,389, granted 26th May 1857, was still in force. Accordingly, Z. S. Durfee was dispatched to England to secure the rights on Mushet's American patent and in this he was successful. He was also given the task of endeavouring to persuade Bessemer to sell his U.S. patent; in this, however, he was not successful.

Meanwhile, Holley had been in Europe and had visited Sheffield, where he saw the Bessemer operation being worked both at Bessemer's own works and, more impressively as far as Holley was concerned, at John Brown's Atlas works. Being convinced of the future of the process, he persuaded the two business men, John Griswold and John Winslow, both of Troy, New York, to join him in trying to secure the Bessemer machinery rights in America. Holley returned to England in the summer of 1863; at first it seemed that he was to be as unsuccessful as Durfee had been, but at the end of the year he had agreed a licence on a royalty basis and a three year option to purchase the Bessemer patent rights in America. It could well be that Holley's success with Bessemer was related to the fact that the competition came from an organisation which had connections with Kelly and that Bessemer was still unhappy with the treatment he had received at the hands of the American Patent commissioners seven years previously.

There were now, therefore, two interested groups in America at this time, each possessing part of the necessary licensing to clear production in the United States and each determined to control the future of the, as yet, undeveloped American steel industry which up to that date had been largely dependent on imports across the Atlantic.[6]

Meanwhile, whilst Durfee was in England in the spring and summer of 1862, the Kelly group made the decision to erect an experimental converter adjacent to the Eureka blast furnace at Wyandotte, Michigan on premises

[6] Pennsylvania, which produced almost all the steel in the U.S.A. at that time, had 14 steelworks in 1864, producing a total of only 9,771 tons of blister and crucible steel. See J. B. Pearse: "A Concise History of Iron Manufacture in Pennsylvania", New York, 1876, p. 257.

owned by Captain Ward, one of the subscribers. So it was that William F. Durfee, cousin of Z. S. Durfee, was:[7]

> invited to design and superintend the construction and working of the machinery and apparatus necessary to test, by an experiment on a large scale, the merits of a process for the production of steel invented by William Kelly, of Eddyville, Kentucky

Durfee relates how it was decided to install blowing equipment which would be suitably powerful for subsequent commercial production, should the experiments prove successful and that likewise that the same conclusions should apply to the vessel itself; otherwise the equipment should be designed with economy in mind. To avoid complications with a cupola, it was also decided to take the metal direct from the blast furnace. Durfee also makes it clear that he had never seen any apparatus for the proposed method of steelmaking, and had been furnished with no plans. He was confident that his cousin would obtain the patent rights and he, therefore, obtained copies of all the relevant Bessemer patents, together with copies of all available textbooks. On this basis he designed and built a converter; he was very pleased when his cousin, on returning from England in the autumn, said it looked very like converters he had seen abroad! Incidentally, when it became clear that the Bessemer rights were not to be achieved by the Kelly syndicate, a politically motivated plan was put into operation; a stationary converter of about the same capacity was installed. As things turned out, this converter was not put into operation until some months after Durfee left Wyandotte. This diversion, as well as delays in the construction of the blowing engine, occasioned both by strikes and the Civil War, meant that it was not until the spring of 1864 that operations could be put in hand at Wyandotte. The first public demonstration of the converter took place on 6th September 1864:[8]

> Although Durfee and the workmen had conducted many trial conversions successfully, they all know that the converter might fail at a critical moment. Thus, all were nervous and tense. The visitors, although they did not share that knowledge, were aware of the importance of the preparations being made for the blow. Ward had little to say, but as he watched he slowly shifted silver dollars from one hand to the other – always a certain sign with him of either nervousness or rising temper.
>
> As the group watched, the workmen slowly tipped the converter on its axis and the vessel received its charge of molten metal.[9] Then it was swung

[7] W. F. Durfee: "An Account of the Experimental Steel Works at Wyandotte, Michigan", Trans. Amer. Soc. Mech. Engs., vol. 6, (1884–5), pp. 40–59.

[8] B. C. Korn: "Eben Brock Ward, Pathfinder of American Industry", Ph. D. dissertation, quoted in McHugh, op. cit., pp. 179–182.

[9] Two and a half tons, according to Hunt, op. cit., p.202.

> back to its upright position, the blast turned on and, almost immediately, the familiar roar could be heard. At the proper moment, as the men stood motionless and silent, the brilliant flame shot from the top and the process began its familiar cycle. When the flame reached the proper colour and the turmoil within the converter subsided, the vessel turned slowly on its trunnions, the blast was shut off and the carburiser run in. A violent reaction quickly took place within the vessel and, when all was quiet, the molten metal was poured into the waiting moulds. Whether or not the watching men realised it, during that half hour an historic event had taken place. Under the direction of William Durfee in this experimental converter at Wyandotte, Michigan, they had seen the first steel produced in America by the pneumatic process.

Durfee was anxious to see ingots of his steel rolled into rails. The roller at the mill feared that the steel would prove too strong for his mill, which had been designed for rolling wrought iron. Overriding his reasonable objections, Ward insisted that it should be tried; this resulted in a broken roll. It was not until the following May that another trial was made, this time in the Chicago Rolling Mills. Durfee was unable to be present, but he received a letter from the Manager reporting on the excellent working of the steel, saying that six good rails had been produced and complimenting him:

> You have done what you set out to do and done it well and I am glad to congratulate you and rejoice with you, for I can appreciate some of your difficulties and wanted you to hear some of the praises bestowed upon your labors, as you richly deserve.[10]

Durfee resigned on 1st June 1865. His place at Wyandotte was taken by a Mr. Hart, who had acquired some experience at the works of Petin, Gaudet et Cie. He decided to build a reverberatory furnace for melting the iron, so as to be able to use a blend of English and American pig irons. He was succeeded by R. W. Hunt who removed the reverberatory furnace, relying once more on direct metal from the blast furnace; subsequently, however, he erected a small anthracite fired cupola and it was in this cupola that Missouri Iron Mountain pig was melted and blown in the converter in October 1865, providing exceptionally good steel. Hunt left in May 1866; the works had by then been bought by Captain Ward who ran them for his own requirements until they were abandoned in 1869.[11]

Having observed closely the operations in Sheffield and armed with working drawings from Bessemer, Holley undertook to erect a two ton converter plant at Troy, New York, at an old flour mill site on the banks of the Hudson River. This site had a water wheel and in the initial stages this was used to actuate the blowing cylinders. The first blow was on 16th

[10] Durfee, op. cit., p. 59.
[11] Hunt, op. cit., pp. 202–203.

February 1865. Before this, however, samples of American pig iron had been despatched to Sheffield and trials had been done on them in the 1.5 ton vessel in Bessemer's own works. As a result Crown Point iron was used in the first heat and remelted in a reverberatory furnace. The record reads:

> 2,497 lb pig; 175 lb. New Jersey Zinc Co. Franklinite[12] for recarbonising. Blast 5 to 9 lb. Blew well. Three taper ingots cast, weighing 482, 491 and 561 lb respectively: 1,534 lb product and 706 lb. scrap. Casting 54.4%, scrap 26.4%, loss 16.2%. Scrap mostly a large skull due to slow handling. Test piece (1¼" sq.) bent double cold. Hammered as well as possible. Finer fracture than the charges made from the same brand of iron at Bessemer's Works in England on 30th. November. Welds pretty well and hardens pretty well.

This record of a hopeful beginning is in Alexander Holley's handwriting. The next trial was made 27th February using the same grade of iron; only 50% ingot yield was obtained. Holley's notes are as follows:

> Blast 10 lb. Blew 22 minutes. Vessel not hot enough. Ladle nozzle too small – 1⅛ "diameter.

The next day came the third trial, again with the same raw material:

> 77.0% ingots, 3.8% scrap, 19.2% loss. Vessel hot and blew well with 8 to 10 lb. blast. Metal came through bottom by side of tuyere; stopped it with water. Steel all poured out of ladle. Nozzle 1⅞" diameter.

On 17th May, after several more heats, the water wheel shaft was found to be rotten. Whilst this was being replaced, the height of the vessel was increased, the pit was enlarged, new tuyeres and nozzles were substituted and the Franklinite furnace was raised 6" to give a better flow into the vessel. Meanwhile a cupola was being installed. On 20th July the cupola was tried out:

> Iron melted in cupola in one hour after stopping tap hole; run out very hot from ladle. Coal consumed 1,226 lb. Iron melted and blown 2,997 lb. Result: ingots 85.4% scrap 2.0%, loss 14.4%

The water wheel continued to give trouble and was last used on 8th. December 1865. On 10th March 1866 work recommenced with blast from a steam engine. After some teething troubles, this worked well and during

[12] Franklinite is a mineral found in New Jersey, generally considered as a source of zinc. According to J. Percy: "Metallurgy: Iron and Steel", London, 1864, p. 199, it is a magnetic iron ore with the iron partially replaced with zinc, containing approximately 45% iron, 20% zinc and 10% manganese. The residue after the extraction of the zinc could be smelted in the blast furnace to produce "Franklinite Pig", which was a form of spiegeleisen with up to 12% manganese. It seems that the mineral itself had been used as an addition during the refining of pig iron to wrought iron, with beneficial results.

March some 65 tons of ingots were made with 2.4 tons scrap.

It is recorded that between November 1865 and April 1866 three attempts were made to produce chromium steel, using either chromium ore or a "triple compound of iron, carbon and chromium"; in all cases, the ingots crumbled on hammering and the trials were then discontinued. This trial is of interest in that John Baur, with works in Brooklyn, took out a patent in August 1865 for a steel "greatly improved, toughened and hardened by the addition of chromium".[13]

An octagon ingot was cast on 25th April 1866. This was 15" across flats; the mould was filled to within 12" of the top and then a 6" diameter bar, 20" long was lowered into the metal to a depth of 4" to act as a handle for forging.

In May 1866 a total of 80 charges was made, providing 118.8 tons of ingots, a yield of 81.5%, scrap being 1.8% and loss 16.7%. Whilst the old plant continued to run with increasing success, the decision was taken to install a new plant with two 5 ton converters and, with this almost ready for operation early in 1867, Holley left to complete the Pennsylvania Steel Company's plant at Harrisburg, Z. S. Durfee taking over at Troy.[14] He built the Forge at Troy, to cog the ingots into blooms which were then sent out for rolling into rails. With both Bessemer plants in operation and improvements both in equipment and technique, output was steadily rising. On 19th October 1868, however, the roof of the new plant caught fire and much of the machinery was destroyed. Holley came back to supervise the rebuilding, incorporating many of the improvements which he had instituted at Harrisburg. The small plant was kept running and the new shop commenced operations on 12th January 1870. At that date, a 30 inch mill was commissioned in the Forge shop and the ingots were rolled into blooms. The mill was further improved and in 1875, when the title of the works was changed to "The Albany and Rensselar Iron and Steel Company" with R. W. Hunt as general manager, the output of the two Bessemer plants was 1,300 ingot tons per week, sufficient to keep the rail mill in operation on double shifts. The largest monthly output in that year was 5,498 tons.[15]

Following the survey of history of operations at Troy, it is now appropriate to return to the situation which arose between the two competing groups: the Kelly Process Company, holding the Kelly and Mushet patents, and Winslow, Griswold and Holley, with the Bessemer rights which they

[13] J. Baur, U.S. Patent No. 49,495, 22 August 1865. His chromium steel was used in the stiffening ribs of the Saint Louis Bridge in 1868. A later analysis showed this to contain around 0.75% C and 0.60% Cr; this material was produced by the crucible process. See also British Patent 370, 1876.

[14] The above information on the operations at Troy is derived from R. W. Hunt: "The Original Bessemer Plant at Troy", Trans. Amer. Inst. Mining Engineers, Vol. v (1876–7), pp. 60–70.

[15] R. W. Hunt: "A History . . .", pp. 203–207.

had purchased outright in July 1865, together with several valuable Holley patents on Bessemer engineering. Both parties came to the conclusion that, if matters between them came to court, the outcome would be expensive and most likely unsatisfactory; they therefore decided early in 1866 to combine their interests. Accordingly, a new organisation, with the somewhat cumbersome title of "The Trustees of the Pneumatic or Bessemer Process of Making Iron and Steel", was formed, Winslow, Griswold and Holley taking 70% of all royalties and the Kelly Process Company 30%; it was also a unanimous decision of the Trustees that Holley should be the engineering adviser. The new plant at Troy was, in fact, the second undertaking of the new organisation.

Their first, however, and the third Bessemer plant to be built in America, was the Pennsylvania Steel Works at Harrisburg, to which some reference has already been made. The company had been organised by the collaboration of a number of prominent railway men, essentially as a rail making plant to free them from the necessity of importing British rails. The first engineer was William S. Butcher, a prominent Sheffield steel man, and he saw the foundations laid in 1866. The plans were then changed and it was arranged to lay out the works in accordance with the advice of Holley. With construction well in hand, Holley left Troy at the beginning of 1867 to go to Harrisburg, where he assumed sole charge for the construction. The two 5 ton converters started operation in June 1867. Until the rail mill was completed in May 1868, the ingots were shipped to the Cambria Iron Works at Johnstown, Philadelphia. In June 1868 an official publication stated that the annual capacity of the Bessemer plant was about 10,000 tons, whilst that of the rail mill was 30,000 tons. It was commented that additional converters would be erected as required. No such additions had been made up to 1875, but the earlier figures were soon proved to be a gross underestimate and by 1875, the maximum daily output had been 281 ingot tons, the maximum weekly output 1,291 tons and the maximum monthly output 5,455 tons.

The fourth American Bessemer works was the Freedom Iron and Steel Works, near Lewistown, Pennyslvania. This works was set up by John A. Wright who purchased the complete machinery in England, with the exception of the blowing engine. The original intention had been to produce boiler plates, but the plate mill, with a Ramsbottom reversing engine, was soon changed to a rail mill. The first blow was made on 1st May 1868. The company only ran for a year, due largely to the unsuitability of the iron produced by the associated Logan Iron Company for Bessemer steel manufacture; the works was closed and much of the machinery sold. This is the isolated instance of failure in the period of Bessemer steel growth in America.

The Cleveland Rolling Mill Company installed a pair of 5 ton converters, on the same lines as the Pennsylvania Works in 1868; the first blow took place on 15th October in that year. Shortly afterwards a second pair of 5 ton vessels was added and all four remained in operation until 1875, when the second pair was removed and replaced by Siemens Martin furnaces. This company made a speciality of wire, screws and so on.

The Cambria Iron Company, of Johnstown, Pennsylvania, had considerable experience of handling ingots in Bessemer steel made by the Pennsylvania works and the chief engineer, George Fritz, had assisted Holley in the designing of the rolling mills at Pennsylvania and Troy. In 1871 he called on Holley to reciprocate in assisting him in the erection of a Bessemer plant at Cambria. To this, Fritz added a blooming mill with roller tables and hydraulic devices for turning over the pieces on the tables. This plant was looked upon as one of the finest Bessemer plants in the United States; in the month of March 1876 this one melting shop produced 6,051 ingot tons from two 5 ton converters.

This type of plant with two nominal 5 ton converters set the pattern for all succeeding installations. It should be noted that they all tended to operate with around 6 ton charges. These were the Union Iron Company at South Chicago, which commenced operations on 26 July 1871; the North Chicago Rolling Mill Company, commencing 10 April 1872 and reputed to be making the highest ouput of any in the world by 1875 (a figure of 6,457 tons in the month having been recorded); the Joliet Iron and Steel Company, Illinois, commencing 13th March 1873; the Bethlehem Iron Company, where the plant jointly designed by Holley and Fritz started up 4th October 1873; the Edgar Thomson Steel Company of Pittsburgh which started up in August 1875; the Lackawanna Iron and Coal Company of Scranton, Pennsylvania which started up 23rd October 1875; finally the Vulcan Iron Company of St. Louis, which was being erected in 1875.[16]

It was in February 1865 that the first production Bessemer plant in the United States was commissioned at Troy. A little over ten years later, there were eleven plants, each with two 5 ton vessels, capable of turning out, in aggregate, over half a million tons of Bessemer steel ingots per annum. The official statistics for Bessemer steel ingot production in U.S. tons per annum over this period are as follows:[17]

1867	**1868**	**1869**	**1870**	**1871**	**1872**	**1873**	**1874**	**1875**
3,000	8,500	12,000	42,000	45,000	120,108	170,652	191,933	375,517

16 The bulk of the information on the later plants is derived from Hunt: "A History ...", pp. 207-215.

17 American Iron and Steel Institute: Official Statistics.

Less detail is known of the further installation beyond 1875. At the beginning of March 1880, however, there were still eleven works operating 22 converters. Under construction, there were said to be ten further converters as under:

Pennsylvania Steel	3
Bethlehem	2
Lackawanna	1
Pittsburgh Bessemer	2
North Chicago	2

The same source also quotes the following figures for Bessemer steel production, again in U.S. tons:[18]

	1874	1875	1876	1877	1878	1879
Pig Converted	204,352	395,956	539,474	562,227	739,765	923,717
Ingots made	191,933	375,517	525,996	560,587	732,226	928,972
Rails produced	144,944	290,863	412,461	432,169	550,398	683,964

[18] J.I.S.I., 1880, Pt. 1, pp. 283–284.

7 Siemens and the Development of the Open Hearth Processes

William Siemens (1823–1883) was born near Hanover, the seventh of fourteen children of a German farmer; he was christened Karl Wilhelm. His elder brother, Werner, was destined to found the engineering firm of Siemens and Halske in Berlin and it was he who arranged for Karl Wilhelm to be trained in science at the University of Gottingen, prior to becoming an apprentice in his brother's works. At the age of twenty he came to England to sell an electroplating process which his brother had invented; the process was purchased by Elkingtons of Birmingham and this was to be largely responsible for the demise of "Sheffield Plate".

In the following year he returned to England; as it turned out, he was to stay for the rest of his life. By 1845 he was learning the new science of thermodynamics from Joule in Manchester. Two years later, his younger brother Frederick joined him and together they applied their energies to making improvements in furnace efficiency. By 1856, Charles William Siemens, as he was now known, had become a British citizen and was an engineering consultant in London, working on the application of the regenerative principle in furnaces. The idea of preheating the air for combustion followed naturally from Nielson's application of hot blast in the coke fired blast furnace. In 1857 Frederick patented the idea of passing the spent combusted gases from a furnace through a chamber containing firebrick checkerwork to absorb their sensible heat; a subsequent reversal of the gas flow caused the air to be preheated prior to the combustion of the fuel, the spent gases then being passed through a similar regenerator chamber at the other end, so heating this and rendering it capable of preheating more air on once more reversing the flow.[1] The early trials soon demonstrated that, using solid fuels, the ash was carried over with the flue gases and this led to the gradual choking of the checkerwork, reducing the efficiency of the combustion and also proving very difficult to clean out. This was obviated by the provision of a gas producer to convert coal into a relatively dust-free

[1] F. Siemens, British Patent 2861, 2 December 1856. Reference to Fig. 21 will clarify the means employed to arrive at the somewhat complicated gas flow patterns.

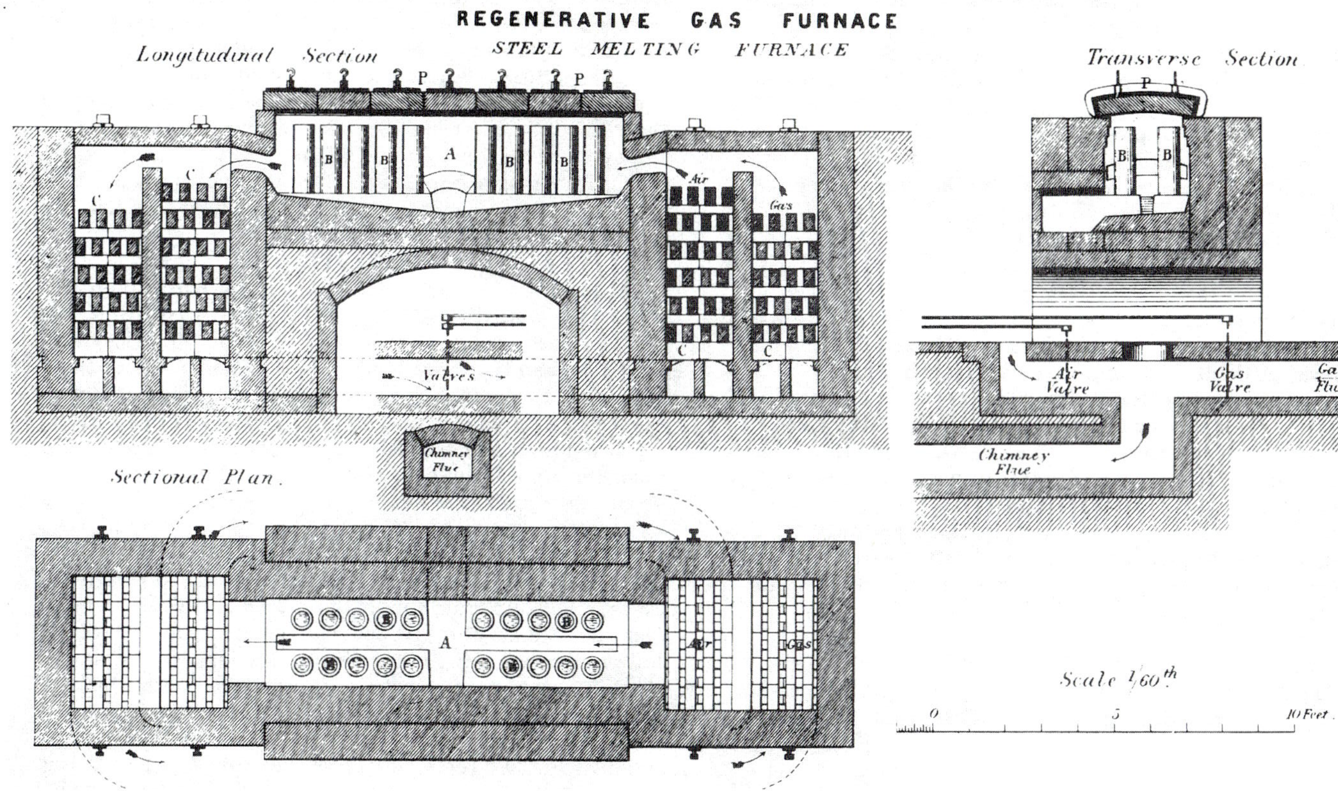

Fig. 20. Siemens Regenerative Gas Fired Crucible Steel Melting Furnace. This illustration is taken from the Proceedings of the Institution of Mechanical Engineers, 1862. It indicates a regeneratively heated gas fired hearth on which was placed a layer of coke breeze to support a number of crucibles. There was a series of covers in the roof of the chamber through which the crucibles could be loaded or removed. The melting process was precisely as in the conventional coke fired crucible holes; by 1862 the crucibles would generally hold from 50 to 60 lb. and a normal charge would be a mixture of bar iron and cast iron, generally both of Swedish origin.

gaseous fuel.[2] By providing two sets of checkerwork at each end, one for the preheating of the air and a smaller one to preheat the gas, each fitted with appropriate reversing valves, very effective combustion could be realised and temperatures as high as 1650°C provided in the furnace chamber; such an arrangement was patented at the beginning of 1861.[3] The first use to which solid fuel fired regenerative furnaces had been put was the reheating of billets for rolling, at the iron works of Lloyd and Foster in Wednesbury and at the steelworks of Marriott and Atkinson in Sheffield.[4] The first full scale application of the producer gas fired regenerative furnace, however, was in 1861 at the Flint Glass Works of Lloyd and Summerfield in Birmingham and, saving 50% of the previous amount of fuel used for the melting of glass, was stated still to have been at work in 1867. Other glass makers followed, including Chance Brothers in Birmingham; here it was that Siemens met Michael Faraday who was sufficiently impressed to invite Siemens to give a lecture on his furnace technology at the Royal Institution.[5] In this lecture Siemens drew attention to the fact that trials in steelmaking were in hand at the Brades Works near Birmingham and appeared promising. Later the same year he reiterated this:[6]

> Another important application of the regenerative gas furnace is as a steel melting furnace, in which the highest degree of heat known in the arts is required, presenting consequently the greatest margin for the saving of fuel. Plate 18 [here reproduced as Fig. 20] represents a regenerative steel furnace which has been in satisfactory operation in Germany for a considerable length of time, being worked with lignite, a fuel little superior to peat in heating power. The application of the regenerative gas furnace is indeed rapidly extending in Germany, but has not yet practically succeeded in Sheffield where it was also tried; it is, however, in course of application at the Brades Steel Works near Birmingham.

This furnace was, in effect, a modernised version of the Huntsman crucible furnace. The flat hearth of the furnace chamber carried some twenty conventional steel making crucibles; the roof of the chamber was provided with a number of holes, fitted with covers – these were removable to allow the taking out of the crucibles for teeming in the conventional manner. This

[2] Other fuels than coal were used elsewhere; Siemens himself referred to the use of lignite in one of his early furnaces in Germany and wood sawings and other waste were used by Lundin in connection with the Open Hearth furnace which he erected at Munkfors.

[3] C. W. Siemens and F. Siemens, British Patent 167, 22 January 1861.

[4] W. Pole: "The Life of Sir William Siemens", London, 1888, p. 102.

[5] Siemens is said to have burned his lecture notes inadvertently and only a short abstract appears in the Proceedings of the Royal Institution, Vol. iii, p. 537.

[6] C. W.Siemens: "On a Regenerative Gas Furnace as Applied to Glasshouses, Puddling and Heating", Proc. Inst. Mech. Eng., 1862, pp. 35–36.

form of furnace gradually replaced the conventional coke fired melting holes in a number of the larger Sheffield firms; there was a prejudice in favour of doing things the old way, nevertheless, as well as considerable capital tied up in existing plant. On the Continent and in America, however, the Siemens crucible melting furnace had a major part to play in special steel production.[7]

The metal melted in the crucibles originally was blister steel; the process was a refining of the only steel previously available. At the time of the introduction of the Siemens gas-fired crucible furnace, however, following the release of Swedish cast iron to the Sheffield market,[8] considerable quantities of crucible steel were produced by melting together the cast iron and the bar iron. The logical step, as far as Siemens was concerned, was to provide a furnace with a dished hearth on which the mixture could be melted without the use of crucibles and a tap hole through which the finished metal could be run out into ingot moulds or a ladle.

The first attempt at this appears to have been at Tow Law, in County Durham, in 1862. Charles Attwood had patented a method for producing steel by the melting together of bar iron with cast iron or speigeleisen.[9] The local Weardale iron ores are essentially carbonates which have undergone some oxidation; they are rich in iron and virtually free from phosphate minerals. Moreover they are relatively rich in manganese and the blast furnace metal produced from them would resemble the continental spiegeleisen. At Attwood's request, Siemens designed a small Open Hearth furnace for him which included a roof which was depressed in the centre and a removable iron pan on which the furnace bed could be constructed. Some steel was certainly made using the hearth. According to Attwood's patent, when melting on the Open Hearth, he had found it advisable to cover the metal with "a thin or superficial coating of molten glass", indicating this could be obtained by adding cullet or broken common black bottles. In the event, however, it seems that he

[7] For further details on the Gas Fired Crucible Furnace refer to K. C. Barraclough: "Steelmaking Before Bessemer", Vol. 2, London, 1984, pp. 76–81, 201–202, 209–214, 228–231, 236–239.

[8] With no coal available and only charcoal as fuel, official Government policy in Sweden had been to convert all available cast iron into bar iron, thus benefitting the economy from the large value addition obtained in this way (of the order of £30 per ton as against £6 per ton in Sheffield). The growth in demand for iron, however, had gradually brought about an imbalance and in 1854 the law was changed to permit the export of such cast iron as could not be converted into bar iron; this was followed in 1858 by permission for the export of Swedish iron ore, hitherto prohibited.

[9] C. Attwood, British Patent 1473, 5 May 1862. This covered the melting of steel in crucibles, in larger pots "which may be tapped at their bottoms instead of being removed from the furnace" or "in an open hearth or chamber of the furnace itself".

reverted to the use of crucibles; the quality of the material from the hearth itself had not proved to be to the desired standard.[10]

In 1863, Le Chatelier, the French Inspecteur General de Mines, had worked out a process for the manufacture of steel by melting puddled blooms in a molten bath of cast iron on an Open Hearth lined with bauxite; this material was extremely infusible and contained no hurtful impurities. In view of his official position, Le Chatelier could not participate in a patent; an association was therefore formed with the firm Boigues, Rambourg et Cie, with works at Montluçon, who patented the use of bauxite[11] and entered into an arrangement with Siemens. He reported that a furnace "of great heating power" was constructed under his supervision and:

> would certainly have accomplished the desired object if the company had displayed the least determination to succeed.

The furnace bottom of bauxite proved unsatisfactory, since it was not solidified by the heat and rose to the surface of the liquid bath. This defect was rectified by the substitution of a fritted white sand bottom and some good steel was subsequently obtained. One night, however, through lack of attention, the covering arch of the furnace was partly burned down. Siemens commented that:

> . . . this slight accident, which proved nothing except an ample sufficiency of heating power, sufficed to deter the company from pressing on with the attainment of that success which was so nearly within their reach.[12]

Siemens also commented on the difficulties of finding a suitable refractory for the roof of his furnace:

> Welsh Dinas brick, consisting of nearly pure silica, is the only material, of those practically available on a large scale, that I have found to resist the intense heat at which steel melting furnaces are worked; but though it withstands perfectly the temperature required for the fusion of the mildest steel, even this is melted easily if the furnace is pushed to a still higher heat.[13]

[10] C. W. Siemens: "On the Regenerative Gas Furnace as Applied to the Manufacture of Steel", J. Chem. Soc., 1868, p. 287. See also J. S. Jeans: "Steel: Its History, Manufacture, Properties and Uses", London, 1880, p. 90.

[11] W. E. Newton (on a communication from Guillaume Boigues and Paul Rambourg) British Patent 708, 16 March 1863. This specifically refers to the use of "the Siemens patent regenerative gas furnace".

[12] C. W. Siemens, ". . . Manufacture of Steel", p. 288.

[13] ibid., p. 297. A footnote indicates the composition of the Dinas rock from Pont Neath Vaughan as 98.3% silica, 0.72% alumina, 0.18% protoxide of iron, 0.22% lime, 0.14% alkalies and 0.35% combined moisture. The bricks were reported to be made from the crushed rock with an admixture of 1% lime, fired to a high temperature.

Fig. 21. Early Siemens Acid Open Hearth Furnace. This drawing is based on that given in *The Making, Shaping and Treating of Steel*, published by the United States Steel Corporation. It shows a structure not dissimilar to that shown in Fig. 20, but with the metal now being melted direct on the furnace hearth instead of in crucibles.

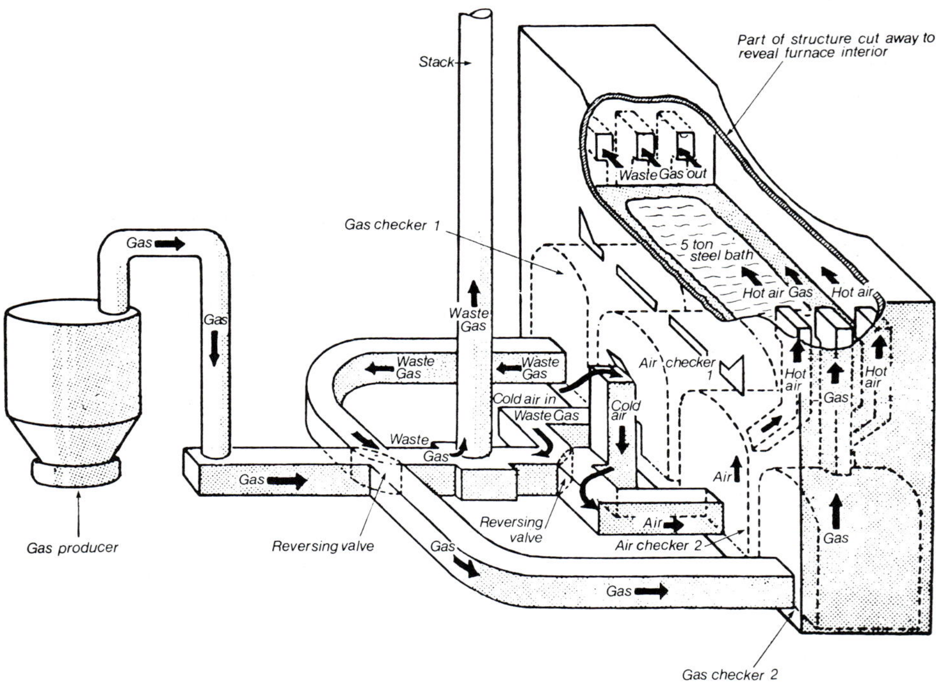

A schematic drawing indicating the general features of the early Open Hearth furnaces is provided in Fig. 21.[14] From this may be appreciated the means employed to arrive at the somewhat complicated gas flow patterns. With minor modifications, this remained the pattern of all succeeding furnaces,

[14] This drawing, an adaptation of one which originally appeared in "The Making, Shaping and Treating of Steel", published by United States Steel Corporation, was prepared for a contribution by the author to the 1983 Conference of the Historical Metallurgy Society. See K. C. Barraclough: "Sir William Siemens: 1823–1883", Historical Metallurgy, Vol. 18, No. 1, 1984, p.3.

whether the capacity was 2 tons, as at Sireuil, or up to 20 tons by 1880 (or, indeed, up to 200 or more tons eventually). There were alterations to the arrangements of the furnace ports to provide better combustion; the roof in some early furnace construction was brought downwards in the centre and the angle of the ports adjusted to deflect the flame towards the bath (for instance at Le Cruesot and Bochum in the mid 1870s). One very important improvement [15] was the incorporation of an enlarged space immediately before the checker work at each end of the furnace, so that the exhaust gas flow lost some of its velocity and suffered a change in direction, so dropping a large proportion of the solid particles entrained in it. These were known as "slag pockets" and they collected considerable quantities of matter which would otherwise have tended to block the channels in the checkerwork; they had access doors through which the debris could be removed at regular intervals. The features of the dished roof and slag pockets may be seen in Fig. 22, which may be taken as representative of furnaces in service around 1880. The placement of the regenerator chambers became rather a matter of individual preference; there was a tendency to move them out somewhat from under the hearth of the furnace and so make their cleaning easier, as well as restricting the damage which could be caused by a metal breakout.[16]

The next stage, and a most important one, was the setting up of the Siemens Martin process at Sireuil. Here in 1864 Siemens installed a two ton furnace, with the roof and walls constructed in Dinas brick and with a fritted white sand hearth and it was here that the two brothers, Pierre and Emile Martin, successfully made steel by melting a bath of pig iron and feeding into this puddled steel or iron, with or without admixtures of steel scrap. Their original operations at Sireuil are embodied in a patent taken out in this country on their behalf by Brooman:[17]

> The direct manufacture of cast steel many be effected in a reverberatory furnace, preferably a Siemens gas furnace, in which cast iron is melted to form a bath, wherein cold or heated pieces of iron, natural steel, iron or steel turnings, scrap or preferably puddled balls cut into pieces are dissolved until a bath of steel is obtained. A portion only of this bath is then run off, and a further quantity of iron or steel and, if needful, cast iron, is added to replace the metal run off and so the operation is continuously carried on. The bath must be raised to a temperature of from about 1500 to 1800°C and it is needful to remove from the bath the black scum charged with oxide of iron and substitute for it a clear scum free from oxide. The use for this purpose of a flux, such as the dross from blast furnaces heated with wood, facilitates raising the temperature and, by

[15] C. W. Siemens, British Patent 43, 5th Jan 1875.

[16] C. W. Siemens, British Patent 3370, 5th Sep 1877.

[17] R. A. Brooman (a communication from Francois Marie Emile Martin and Pierre Martin), British Patent 2031, 15 August 1864. The quotation is from the official abridgement.

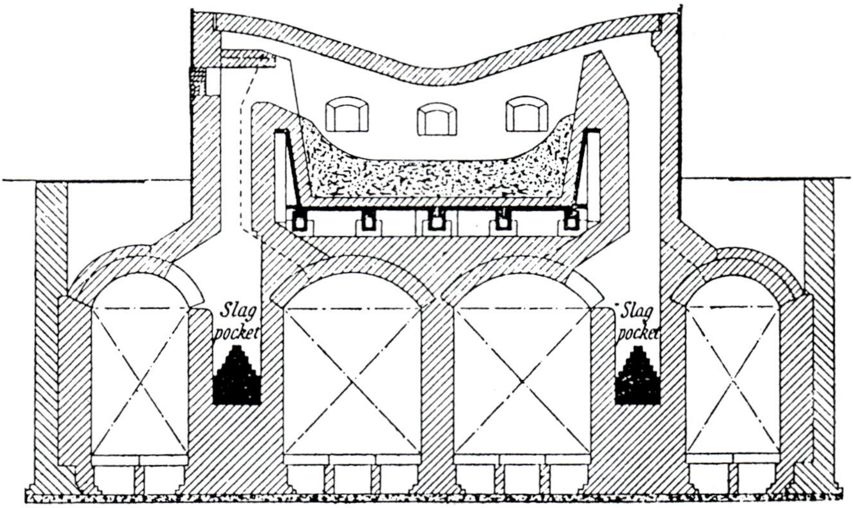

Fig. 22. Typical development of the Siemens Open Hearth Furnace by 1880. This drawing is reproduced from Fig. 127 of D. Carnegie and S. C. Gladwin: *Liquid Steel: Its Manufacture and Cost*, London, 1920. In particular it illustrates the dipped roof, allowing the flame to be directed towards the hearth, and the provision of so called "slag pockets" which collected much of the dust from the combusted gases, so preventing the clogging up of the checker work.

dissolving the oxide reduced in this clear scum, the matter is purified and its quality assured. Manganese, chlorides or fluorides or nitrates may be added to the bath of cast iron. Carbonaceous or other reducing matters may sometimes be used to remove oxide from scum containing it.

Operating in this way, a useful quality of steel was achieved. It is indicated that the furnace could be kept running continuously for a week or more. It soon became evident, however, that the production of steel with a regularity of temper (that is to say, to the desired carbon content) was difficult and almost precisely a year later Brooman took out a second patent on their behalf which described a batch process rather than a continuous one:[18]

> In a reverberatory furnace . . . heated by gas according to Siemens' patent and raised . . . to a continuous white heat, I place about 700 lb. of cast iron in pieces, each piece weighing about 4 lb., and a bath of liquid cast iron at high temperature is obtained; into this bath I introduce, by degrees, pieces of puddled steel in blocks of 2 to 4 lb. until about 200 lb. has been added. Fifteen to twenty minutes suffice for the fusion of this 200 lb; . . . I then add a second charge of puddled steel heated to white heat and always by degrees; after a third charge of steel has become fused in, I take off the black oxidising scum on the bath and replace it by a clear vitreous scum, such as that from a wood heated blast furnace, and add an equal weight of siliceous sand. The scum preserves the bath from oxidation while the sand prevents the metal from becoming short or brittle. I continue to add to the bath the doses of 200 lb. of blocks of puddled steel as before stated as fast as their fusion will admit, for instance 3,400 lb. arising from 700 lb. of cast iron 2,300 lb. of puddled steel and 400 lb. of debris or pieces of steel from previous fusions. At this stage the whole is stirred, the metal tested by portions drawn from the bath and, according to its quality, I add from 40 to 100 lb. of similar cast iron, previously heated to white heat, which fixes the exact grain of the cast steel. To raise the temperature of and purify the bath at the end of the operation, if necessary, I introduce a flux of manganese and sea salt and fluorspar melted together, in the proportion of two parts by weight of sea salt to one of fluorspar. The metal, stirred and brought to the right degree of quality is run into moulds or ingots . . .

There follows a description of the casting operations using moulds on a roundabout so that they might be filled in succession from the furnace spout; no ladle was used at this stage. The patent specification then continues:

> For a . . . furnace of about 4,000 lb. capacity the duration of the operation is from five to six hours, which allows two successive fusions in twelve hours and therefore twenty four fusions per week; it is necessary, however, to have two furnaces since each requires repair after ten to twelve fusions. The waste is about ten to twelve per cent, according to the nature of the materials employed. The consumption of coal is about thirty three bushels in twelve hours, but anthracite, lignite, peat or similar materials may be substituted. It is also necessary to add about thirty three bushels in twelve hours for the furnace for reheating the pieces of cast iron or puddled steel before introducing them into the bath. . . . The proportions of cast iron and puddled steel may be varied according to the nature of the cast steel to be obtained; all pure steely cast irons, that is to say, those which give puddled steel of good quality,

[18] R. A. Brooman (a communication from Emile and Pierre Martin), British Patent 2137, 18 August 1865.

will produce, by this process, good qualities of cast steel. Likewise good steel is produced from natural irons and steels refined by wood"

The patent indicates that the example quoted would produce a "semi-hard" steel; by using puddled iron rather than puddled steel in the fusion, a soft steel would be obtained which "would not temper". This insistence on the use of puddled materials is interesting; an explanation might be that the value of the puddling process in removing some of the phosphorus had been recognised – for certain, the availability of good low phosphorus material in France was not great. One point to be noted quite clearly, however, is that the addition of iron ore to the "scum", as the slag is called in the patent, is not in any way a feature of the process – the only reference to the use of this material, indeed, is as an aid to the removal of any small pieces of cast iron which might become trapped in the "scum". In this way, the process, which became generally known throughout Europe, where it was widely used for the next fifteen years, as the **Siemens-Martin Process**, the product often being known simply as "Martin Steel", was referred to by Siemens as the **pig and scrap process**. This was in distinction from the **pig and ore process** which he later developed and which was to play an important role in British steelmaking, simply being known here as the **Siemens Open Hearth Process**.[19]

Hitherto Siemens had supplied the furnaces, leaving the manufacturer to organise the steelmaking procedure. In 1865, having been disappointed by the results, he took the decision to carry out steelmaking experiments himself. It so happened that a furnace he had erected previously in Birmingham became disused and the premises were offered to him; he accordingly rented the premises for a year at the beginning of 1866. His operations there commenced in April and by August he could write:[20]

The steel melting furnace which I erected in Birmingham has now been in full operation for some time and is a most complete success. It has been working day and night for five weeks in melting extra soft steel for wire and the furnace lining is still intact. The body of the furnace will last for several years and the lining when it fails can be replaced in a day. The consumption of fuel does not exceed one and a half tons of slack per ton of cast steel of the softest kind produced . . . I can produce a ton of cast steel for less money than Bessemer and of superior quality.

His experimental work continued and he eventually extended his lease, adding a further furnace:[21]

19 As time went by, the distinctions became more and more blurred and eventually pig, scrap and ore were used in both processes as occasion demanded. The confusion is confounded by vagaries in the available production statistics – in particular, the relevant French figures which combine Bessemer and Open Hearth production.

20 W. Pole, op. cit., pp. 146–147.

21 C. W. Siemens: ". . . Manufacture of Steel", p. 289.

> The second, erected in 1867, is an open bath furnace, capable of melting a charge of 24 cwt. of steel every six hours. Although these works have been carried out under every disadvantage, inasmuch as I had to educate a set of men capable of managing steel furnaces, the result has been most beneficial, in affording me an opportunity of working out the details of producing cast steel from scrap iron of ordinary quality . . . and in proving these results to others.

In this way he familiarised himself with the capabilities of his furnaces, as well as working for prospective clients on their own materials. Among these were the Bolton Steel Company, the London and North Western Railway and the Great Western Railway Company. The Bolton Steel Company proved to be a disappointment to him. He had designed a puddling furnace that could be converted to an experimental steel melting unit; after only two months trials, however, even though these were conducted under the superintendence of Siemens' assistant, William Hackney, they reverted to puddling. A furnace at the works of Rowan and Company in Glasgow was worked experimentally for only a few days in May 1867; this however, eventually led to the establishment of the Steel Company of Scotland. Trials at Barrow Haematite Steel Company, about the same time, were also abortive. His experience at the Crewe Works of the L.N.W.R., on the other hand, was gratifying and they became the first British establishment, in 1868, to adopt the production of steel in the Open Hearth furnace, using a variant of the Siemens Martin Process; this was a consequence of his demonstrating in Birmingham that he could take old rails and convert them into good steel.[22]

The work done with the Great Western Railway was to have far reaching consequences as far as Siemens was concerned. They sent a truckload of old wrought iron rails, originally produced by Dowlais Iron Works, to Birmingham. Siemens had some difficulties, but eventually produced steel ingots from some of them. The ingots were sent to John Brown's Atlas Works in Sheffield for rolling into rails which were subsequently laid down at Paddington, where they would be subjected to extensive wear. They survived their trial (indeed, it is reported that they were not taken up until 1878 – eleven years later) and Siemens was asked if he could continue the supply.[23] This led him to set up a production steelworks and the site chosen was at Landore, near Swansea. On 6th August 1868 he wrote:

> It really is the utmost time to move and I am quite satisfied to adhere to Landore and get to work. I have commenced setting out the works upon your plan and hope to see you very soon or we shall lose the whole year.

22 J. S. Jeans, op. cit., p. 91.
23 W. T. Jeans: "Creators of the Age of Steel", New York, 1884, p. 157.

By the middle of 1869 the works at Landore were producing 75 tons of steel per week. According to Siemens himself:[24]

> The process chiefly employed at the Landore works consists of introducing on the bed of an intensely heated regenerative gas furnace . . . about 6 tons of pig metal, which may be No. 3 or No. 4 haematite pig. When a fluid bath has been formed, oxide of iron (which should by preference have been melted before hand with such proportions of lime or other fluxing materials as to form with the silica in the ore and in the pig metal a fusible slag) is added, or natural ores may be used in their raw condition if they contain lime and manganese, as, for example, the African Mokta ore. When about 30 cwt. of this ore have been dissolved (with ebullition) in the metallic bath, it is found that a sample taken from it contains only about 0.1 per cent of carbon; a point which can easily be detected by the eye of the workman owing to a peculiar bright appearance of the sample when chilled in water and broken by the hammer. In order not to pass beyond this point, samples are taken out from time to time during the latter part of the operation. . . . The requisite point of decarburisation being reached, the supply of ore must be stopped and from 8 to 10 per cent of ferro-manganese or spiegeleisen added to the bath. When this has all been well incorporated by stirring, the metal is ready to be tapped into a ladle mounted upon wheels, which is afterwards propelled into the foundry and discharged either into ingot moulds, to be hammered and rolled, or into dried clay moulds for the production of steel castings.

This is quite clearly the **pig and ore process** or **Siemens Open Hearth Process**.

There is a mention above of the addition of lime with the ore. It soon became the practice to add lime towards the end of the boil. Although there is no mention of this in any of the process descriptions which are available, Siemens himself referred to it in the discussion on dephosphorisation at the Spring Meeting of the Iron and Steel Institute in 1880; he reported that he had always used lime in considerable quantities in connection with his charges, without doing much injury even to the silica bottom usually employed. He said that the effect of the lime was to make the slag yield up a portion of the iron to the bath.[25] This procedure continued right to the end of the working life of the Siemens Open Hearth furnaces in Sheffield in the 1960s; an important feature was that, in the replacement of some of the iron in the slag with calcium, the oxidation potential of the slag as regards carbon removal was reduced and this treatment with lime introduced a means of better control of the process; it did not, however, materially assist the removal of phosphorus or sulphur, the slag still remaining siliceous and so not corrosive to the furnace lining.

2 C. W. Siemens: "On Smelting Iron and Steel", J. Chem. Soc., 1873, pp. 662-663.

25 J.I.S.I., 1880, Pt. 1, p. 123.

8 Open Hearth Steelmaking in Britain to 1880

The spread of the Open Hearth manufacture in Britain is not easy to trace. It seems that the Bolton Steel and Iron Company eventually utilised their one small furnace, since it was reported working in 1869. In the same year, the North Yorkshire Iron works were operating a furnace at Stockton on Tees endeavouring to work with the Cleveland ores; this proved an impossibility and the works closed down within two years at a considerable financial loss.[1] There is a mystery concerning what was evidently the first establishment of the Open Hearth process in Sheffield. It is known that Vickers and Sons, of the River Don Works, had installed a large range of Siemens gas-fired crucible melting furnaces by 1871.[2] It is clear, however, that they also operated Open Hearth furnaces in the early 1870s; Siemens, in his lecture to the Chemical Society in March 1873, from which the previous description of the melting process was taken, exhibited a series of slag and metal samples:[3] "kindly supplied me by Messrs. Vickers and Company", whilst Holley, in a report published at the end of 1874,[4] states:

> The tyres of Vickers, Sons and Company, popularly supposed to be in Crucible steel, are all Open Hearth steel and in uniformity are proved superior, and in other qualities at least equal, to any crucible tyres ever made by that firm.

Beyond these two scraps of evidence nothing else is known of such activities at the River Don Works until the list of British Open Hearth furnaces in 1877[5] which indicates that there were then 10 furnaces at Vickers works; a lengthy report on the River Don Works dating from 1867 certainly gives no indication of any melting facilities other than coke and gas fired crucible furnaces.[6]

[1] J. C. Carr and W. Taplin: "History of the British Steel Industry", Oxford, 1962, pp. 34–35.

[2] J. D. Scott: "Vickers; A History", London, 1962, p.19.

[3] C. W. Siemens: "On Smelting Iron and Steel", J. Chem. Soc., 1873, p. 663.

[4] A. L. Holley, Private Publication No. 3: "The Siemens Steel Manufacture from Pig and Ore", December 1874, New York, p. 32.

[5] J.I.S.I., 1878, p. 596.

[6] Engineering, 25 October 1867, pp. 383-385.

The first establishment founded specifically to operate the Open Hearth process in Britain was the Hallside Works of the Steel Company of Scotland, on the outskirts of Glasgow. The origin of these works is interesting.[7] Charles Tennant operated the St. Rollox Works near Glasgow, the largest chemical works in the world in the 1860s. To produce sulphuric acid, iron pyrites from his mine in Spain was roasted to extract the sulphur; the solid remaining was treated to remove the copper, which was a valuable by-product, leaving a final residue of "Blue Billy", mainly iron oxide, representing almost three quarters of the original pyrites. There was no use for this and eventually thousands of tons piled up outside the works. Tennant discussed the problem with Siemens, who considered it might be used for steelmaking and after some preliminary trials announced that, whilst he had not so far been able to make steel using Blue Billy, he felt confident in the outcome. As a result, he installed a rotating gas fired furnace[8] for Tennant and this eventually produced a rather inferior wrought iron, for which little use was found and eventually the process was abandoned. Tennant meanwhile had found considerable interest in a steelmaking enterprise among his business friends and commissioned Siemens to build it, using conventional furnaces and materials until the trials with Blue Billy were brought to a successful conclusion. In the event, the steelworks was found to be an admirable investment using the Siemens pig and ore process and

[7] A. Borthwick: "Hallside: One Hundred Years, 1873–1973", Private Publication, B.S.C. Special Steels Division, 1973.

[8] A. L. Holley: Private Publication No. 1: "Siemens Direct Ore Process", October 1874, London, 12 pp. The details of the furnace are as in C. W. Siemens: British Patent 4075, 11th December 1873. Holley quotes the composition of "Blue Billy" as 94.11% peroxide of iron, 0.40% lead, 0.12% sulphate of lime, 4.40% silica, 0.10% copper, 0.62% sulphuric acid and 0.13% chlorine. This was mixed with a Spanish brown haematite ore from Camorata, with 54.9% metallic iron and 9% lime, reduced in the vessel with powdered coke, the vessel being lined with bauxite bricks and heated with the usual regenerative system to a "bright yellow heat" and being rotated on its axis 40 times within an hour. Figures for the production between 13th August to 12th September 1874 were recorded by Holley as follows:

		t.	c.	q.	lb.
Charge:	Hammer Scale	4.	0.	2.	10
	Blue Billy	18.	10.	1.	27
	Camorata ore	9.	5.	0.	24
	Red Antrim ore		9.	0.	0
	TOTAL OXIDE	32.	5.	1.	5
	Foundry Coke	7.	17.	1.	23
Product:	Balls	19.	6.	0.	7
	Slag	4.	11.	2.	1

Siemens had suggested that by incorporating manganese ore or spiegeleisen with the charge a form of puddled steel might be obtained but there is no evidence to indicate this was tried. According to Borthwick, op. cit., p. 24, there were two of these rotating furnaces at Hallside, both being taken out of commission in 1875.

no more was heard of Blue Billy as a steelmaking addition. So it was that the first real commercial plant in Britain was launched. The site at Hallside was carefully chosen; it had an area of nearly 100 acres, was virtually level, had access by rail to coal and iron ore supplies and to the port, had ample water supplies, was free from objection by local landowners to the loss of amenities and had ground available for dwellings for the workforce. The works were built during the early part of 1873 and four 6 ton Open Hearth furnaces were in operation and rails were being rolled in the 26 inch mill before the end of the year. By the middle of 1874 a further four 6-ton furnaces had been installed.

Without doubt, however, the major early commercial developments were made by Siemens himself at his works at Landore. Their history is not clearly recorded but it is known that by the middle of 1874 there were twenty four furnaces there and they were turning out over 1,000 tons of steel per week. Most of the furnaces were of 6 tons capacity, as at Hallside.

Holley visited both Landore and Hallside in the latter part of 1874 and prepared a report for the information of his American clients.[9] This is a most valuable historical document since it gives an independent view of early commercial development of the Siemens process (pig and ore) as against the continental Siemens-Martin process (pig and scrap). His preamble reads:

> The manufacture of cheap steel from pig-iron and ore, with or without scrap, as devised by Dr. C. W. Siemens and worked out and successfully practised by "The Landore Siemens Steel Company" . . . and later by "The Steel Company of Scotland" . . . is a matter of very great importance to American iron-masters and of comparative novelty to metallurgists, as the details of the process have been kept more or less secret.

He pointed out that the sizes of furnace and the plants themselves were very similar and accordingly considered them both together. Details from the report are as follows:

> **The Furnace**. The Siemens regenerative gas-furnace . . . does not differ greatly from the Siemens Open Hearth furnaces used in this country [U.S.A.] although the new form at Landore is much improved in detail. The roof pitches down at a sharper angle from the ports to the centre, so as to throw the flame more upon the bed. The plates upon which the sand-bed rests are so arranged as to better prevent the metal from getting into the regenerators . . . the side plates are largely cut away below the bed, so that air can circulate freely between the bottom of the bed and the ports and roof of the

[9] A. L. Holley, Private Publication No. 3: "The Siemens Steel Manufacture from Pig and Ore", New York, December 1874, pp. 3–35. It should be noted that Holley refers throughout this report to the works of the Steel Company of Scotland as being at Newton, near Glasgow; these quite clearly are those normally referred to as Hallside Works.

regenerators. The regenerator arches are strengthened . . . The Newton [Hallside] furnace is 10 feet by 6 feet 11 inches on the bed. The Landore improved furnace is 11 feet 8 inches by 7 feet 4 inches on the bed . . . There are three doors on each side of the furnace.

The Furnace Bottom and Bricks. The sand for bottoms is heated to redness before it is used, in order to burn out such impurities as would tend to flux too much or impair this solidity. This is a very important matter and has much to do with the excellence of the bottoms. One Open Hearth furnace is constantly employed at Landore in calcining sand for the others. Three parts of silica sand – in this case, a crushed rock containing at least ninety-eight per cent of silica – and one part of common sea-sand, which is chiefly silica but contains some fluxing ingredients, are mixed and shovelled some eighteen inches deep upon the brick bed of the calcining furnace and heated some two days by the gas flame, not hot enough, however, to melt the sand together. The steel furnace has a three inch brick bottom laid on the plates and a fourteen inch brick lining all round the sides. A melting heat is gradually put on and then twelve inches of calcined sand are thrown in to form the bottom. On the slopes of the furnace, sand is laid one inch at a time, each layer being partially fused on by a melting heat. The whole operation of making a bottom and its side slopes occupies a week. The bottom lasts two years and the slopes, being renewed with sand layers after every heat, endure indefinitely.

The slopes run up the side walls so high that the bath does not come into contact with the brick sides of the furnace. Otherwise the bricks would be rapidly destroyed, especially in the ore process. Experiments on the pig and ore process, in this country [that is, U.S.A.] and elsewhere, have failed because the ore so rapidly destroyed the bottom, especially at the upper level of the metal bath. Extraordinary cutting of the bottom and slopes at this point does not appear in the furnaces under consideration and this is doubtless due to the very refractory character of the sand and to its purification by previous burning.

The silica bricks used at Landore, besides many sold to other companies, are made by the Landore Siemens Company. The silica rock, which is rather hard, but more like gannister [sic] than quartz in structure, is broken by hammers until it will enter between crushing rollers. These are 22 inches in diameter and 20 inches long and make eleven revolutions per minute. The two top rollers are geared together; the two bottom ones, between which the partially crushed stone falls, are not geared, so that there is a rubbing action between them. There is also a Carr's disintegrator – a pair of rapidly revolving toothed disks – which grinds a part of the material finer than the rollers leave it. The sand, as ready for use, contains some few pieces one eighth to three sixteenths of an inch square and from this down to fine flour. It averages nearly ninety nine per cent silica and one per cent oxide of iron, lime, alumina and other impurities. The sand is thoroughly mixed with one and a half per cent of lime, previously wetted so as to form a lime cream. The mixing is done by repeatedly turning over the mass on the floor with shovels and seems to be pretty thorough, although a mortar mill (revolving roller in a pan) should do it much better. The bricks are hand moulded. The material is rammed into iron moulds, not hard, but by three or four blows of a hand rammer; then smoothed off and laid on a warm drying floor. After twelve hours drying, when the moisture seems to be driven off, the bricks are put in "beehive kilns", about 35,000 at a time, and burned in the ordinary manner. They are charged to the steel works at £4 per 1,000

The silica bricks are very tender when unburned and require careful handling; and they are not very hard after burning, but they are very pure and therefore resist heat remarkably well. The roofs and ports of the furnaces made of them should stand 200 heats and often endure 250 . . .

Materials Charged. The charges, when pig and ore alone are used, average about 5 tons 10 cwt. of pig and 30 cwt. of ore. The theory is to use enough or a little more than enough ore to make good the waste in the Open Hearth process, just as fettling ore makes good the waste in the best puddling practice. A higher percentage of such ores as have been used has tended to impair the toughness and malleability of the steel. The ore averages 20 per cent of the whole charge and, the waste being 10 per cent, a 50 per cent ore should just make good the waste.

Steel scrap or any form of good malleable iron may be used in any proportion up to 75 per cent of the charge. Usually scrap is employed in much smaller proportions – only such forms as sculls, bad and short bloom ends, etc., which are unfit for drawing into merchant bars. Long rail ends and bloom ends are worth more for rerolling than for remelting.

When pig and scrap are used together, their proportion is generally a little greater, or about 5 and three quarter tons to 30 cwt of ore. I have seen pig and ore used without any scrap, but the ordinary practice is as follows:

		Percentages of Materials Used			
		Pig	Scrap	Iron in Ore	Speigel
Average Practice:	Landore	61.2	21.4	10.4	7.0
	Newton	66.0	15.0	12.0	7.0
New Schedule:	Landore	68.0	12.5	12.5	7.0
	Newton	55.0	26.0	12.0	7.0
Actual Practice Observed at Newton, 2 weeks, Summer 1874		67.4	15.6	9.2	7.8

Any red or brown haematite ore sufficiently free from phosphorus and sulphur is used. The phosphorus in the whole bath should not exceed 0.10 per cent. The same rule holds for these as for Bessemer mixtures. At Newton and Landore chiefly Spanish and Algerian haematites are employed. The Cumberland ores of England are said to be equally good, but more costly. The following analyses exhibit the general characteristics of the ores I saw tried:

	Camorata	Mokta
Manganese protoxide	1.65	2.15
Lime	9.00	1.43
Magnesia	trace	trace
Silica	7.86	6.80
Sulphur	0.07	trace
Phosphorus	0.025	0.012
Metallic Iron	54.90	59.30

The Marbella ore much used at Newton averages 56 per cent of iron. Artificial oxides of iron are, to a small extent, substituted for ore. All the heating furnace cinder and a little of the rotator cinder[8] are used at Newton. At Landore, Mr. Riley prefers to use most of the heating furnace cinder and hammer and roll scale in the blast furnace, as it is found to cut the Open Hearth bottom more than ore.

The pig employed is almost all No. 2 and No. 3, made from Bessemer ores. One advantage of the Open Hearth process over the Bessemer is that its clean working is not dependent on a certain percentage of silicon in the pig. The heat is got by other means . . . The spiegel used at both these works, for all products, is the ordinary material such as is used in Bessemer works. But few experiments have been made with ferromanganese because speigel has proved good enough for the comparatively hard products such as rails and tyres. No plates are made in these works at present.

The Process. The pig is charged into the furnace and melted, as in the usual Open Hearth process. When scrap is used it is generally charged with the pig. A number of experiments have been made as to the best period for charging the scrap and Mr. Siemens tells me that, in the case of heavy scrap certainly, time is saved by melting it by the direct flame rather than by the hot bath. The pig seems to carburise it and so enable it to melt, just as well in the one case as the other. It is, of course, more oxidised when charged with the pig, but the oxide, although more costly than ore, facilitates the process. When scrap is charged after the pig bath is partly melted (and this is still considerably practised at Landore) the whole of the scrap is charged together instead of a little at a time. The reason is obvious. In order to save both furnace bricks and fuel, the higher temperatures required as the metal becomes decarburised should be of the shortest possible duration and should therefore not be checked and prolonged by throwing in cold materials. When ore is used, this must be charged throughout the whole process, so that getting the scrap melted with the pig is the more important. There are no auxiliary heating furnaces in these works. If much scrap were used, they would undoubtedly pay. Only a little cheap scrap is generally to be had in a Siemens steel works not running in conjunction with a Bessemer plant.

The pig and scrap having been melted in from two and a quarter to two and a half hours, the ore is charged, a few shovelfuls at a time, until the bath gets white hot, when as much as three barrowfuls are charged at once. As I repeatedly observed it, the remainder of the process is divided into two distinctly defined periods of about three hours each. The first is the period of the rising of the slag. The ore is shovelled in, coarse and fine together (the coarsest not larger than the fist) and some of it quite damp. It instantly sinks under the slag previously formed from the impurities and oxidation of the melting bath. The ore crackles as it sinks, from the explosive formation of steam, and when the bath gets very hot, it makes a strong ebullition for two or three minutes after charging. As the charges of ore increase, the slag increases in amount, but more largely in frothiness and volume. It expands and rises, and its surface is covered with large, heavy moving bubbles or rather blotches. This rising is not constant and the surface action is not continuously active until the last hour of the period. The bulk of the ore is put in during the latter part of this period and the beginning of the next. The exact times and proportions are a matter of judgment. The supply of oxygen must be as rapid as possible to save time but not so rapid as to prevent the bath from constantly getting hotter. The combustion of the carbon in the pig certainly adds heat to the bath but the cold ore absorbs heat in larger proportion.

After two and a half to three hours from the charging of the first ore, the slag rises to its highest point and begins to flow towards the gas. A decided surface current seems to be developed towards the inlet ports. This action indicates the coming on of the next period – the slag, although hotter and thinner than before, loses volume and within ten minutes settles several inches. Another sort of ebullition then commences and increases during the early part of this period, that is, the boiling of the metal. The whole surface is covered, not with large and sluggish slag bubbles, but with the lively spouting of the metal and slag, sometimes of the metal through the slag. The bubbles are smaller and more active than in the former period, the difference being similar to that observed in the boiling of water as compared with the boiling of heavy liquids.

The sudden settling of the slag and its flowing towards the gas could be more fully explained by means of many analyses and tests of temperatures during this part of the operation. The silicon in the pig is of course burned out first, partly by the flame, which is always oxidising when the temperature is high, and partly by the oxygen of the ore. This operation quickly forms a slag which is still further mechanically thickened and increased by the release of silica and other earthy impurities in the ore. The carbon of the pig has not yet begun to burn freely so there is no rapid disengagement of carbonic oxide gas to produce ebullition and the temperature does not increase rapidly enough to keep the slag thin as fast as it forms. These conditions appear to account for the sluggish movement and frothy rising of the slag. When the silicon is nearly consumed, at a period corresponding with the somewhat sudden appearance of the carbon flame in the Bessemer vessel, the carbon of the pig all at once begins to burn and to throw off carbonic oxide, which accounts for the bubbling of the second period. The rapid combustion of the carbon – a large quantity of ore now being mixed with it and highly heated or melted – must considerably increase the temperature of the slag and thus promote its fluidity and its consequent settling into a more compact state.

The flowing of the slag towards the flame may perhaps be explained in this way. The flame at the entrance end of the furnace is at the point of fiercest combustion; it has lost little or no heat and it is beginning to undergo compression under the dipping roof of the furnace. Hence the slag at this end may be so much more highly heated than at the exit end that it settles, causing the higher, because more porous, slag at the exit end to flow towards the entrance end by force of gravity.

Ore is added until within an hour, usually, of the charging of the spiegel; but the quantities and times are regulated by the character of the test pieces, which now begin to be taken, and by the appearance of the furnace, that is to say, the character of the boiling, to show that decarburisation is going well, and the various indications of temperature, to ensure the fluidity of the metal.

One sampler for eight furnaces examines all the test pieces and decides when to put in spiegel and when to tap. One reverser at regular intervals reverses as many furnaces as he can attend to. Thus the actual making of the steel is put out of the hand of individual melters.

From this description an expert in the regular pig and wrought iron process will readily observe that there will be no great difficulty in working into the pig and ore process, although a vast amount of money was sunk at Landore before the general features, as I have described them, were ascertained. In the old process, where 75 to 80 percent of decarburised iron are added to 25 or 30 per cent of pig, dilution plays a larger part. In the pig and ore process the oxidation of the carbon in 60 to 80 per cent of pig is the principal feature and is performed, in somewhat less time, by the greater and more intimate mixture of the oxidising agent. There should seem to be little difficulty in working up from the less active to the more active addition, especially as the proportion of scrap is very largely varied at these works, without at all modifying the amount or treatment of the ore.

The speigel – averaging 7 per cent – is charged cold, about twenty minutes before tapping, and when the carbon in the bath has been reduced to about one tenth per cent. It is placed, through the side doors, on the slopes under the ports, by means of a peel. Some of it melts and runs into the bath, but the greater part of it, when it gets red hot, is pushed into the bath, where it is rapidly taken up; and in about five minutes from this time the furnace is tapped. The bath is still bubbling when the spiegel is charged; but the decarburisation is now nearly complete and the agitation it causes gives place to the spiegel reaction.

The time of operation for five or six tons heats . . . was from eight and a quarter to nine hours. Mr. Siemens says it averages ten hours for six ton heats at Landore and that he does not want to make it faster; the product is all the better for going slow. One charge that I particularly timed contained 14,375 pounds [6.42 tons] of pig and scrap and 3,696 pounds [1.65 tons] of ore and was tapped in eight hours twenty minutes from charging of the pig.

The coal used in the producers for the Open Hearth process alone averages about 1,750 pounds [0.78 tons] and coal for all purposes in the ingot manufacture averages about 1,850 pounds [0.825 tons] per ton of ingots.

The Casting. Rail ingots are usually 12-inch in both these works. From ten to twelve two-rail ingots are run at a heat. The moulds at Landore are 3 inches thick at the top and 4.5 inches at the bottom. The taper averages 1.5 inches for two-rail ingots. The moulds last ninety to one hundred heats with top casting.[10] The taphole is stopped in a manner which I judge to be not generally known or employed and which is certainly very successful and highly valued at these works. The breast is made in the usual manner, with a taphole some 5 inches in diameter formed in it. This hole is stopped with small anthracite (Welsh) coal, averaging about pea size. The coal is rammed, not very compactly, clear up flush with the bottom of the bed. It does not, apparently, burn away. The outside of the hole is well rammed, for a depth of six inches, with sand and a little clay. To tap the furnace, it is only necessary to remove the sand and haul out the coal with a hooked bar. It is not at all fused together, except at the very bottom of the bath and here it is so soft that one stroke of a bar pierces to the metal. At the same time, the coal is compact enough to prevent the metal from bursting out. There is rarely any trouble with the tap hole and opening it is a quick and easy operation.

The tapping spout is double – two wrought iron spouts, one laid in the other with a 2 inch space between. This space is rammed with sand and clay

[10] There are subsequent references to bottom casting in the report; it appears that, at the data of writing, this procedure was in the development stage. Holley indicated that he would discuss the subject later in another of his confidential reports. In the cost figures (q.v.), however, there is an item for "bottom casting repairs".

> to form a refractory channel, in case the upper spout should burn through.
> At both Landore and Newton, each furnace . . . is provided with a casting ladle capable of holding 6 or 7 tons, and mounted on a car . . . As soon as the steel has all run out of the furnace, the ladle is moved from under the tapping spout and the slag (more than half as much in volume as the steel) is allowed to run into a concave bottomed pit some five feet in diameter. A piece of bent iron bar, to form an eye for the crane hook to grapple, is cast in this mass of slag, which when cooled is lifted upon a car and dumped.

The inclusion of a quotation of such length seems completely justified since it gives a first hand account of operations in the two melting shops where Siemens process had been brought to a viable commercial operation, at the time when the development had been carried to its logical conclusion and when the expansion henceforth was, initially, to be the increase in numbers of furnace installations, each very much like those here described at Landore and Hallside and, eventually, by increase in size of the individual installations. The modus operandi was to remain virtually unchanged for almost a century. Moreover, this account was prepared by a man skilled in the art of steelmaking who was casting a practised eye over what he found to be important developments, thereby providing an extremely valuable contribution to technological history.[11]

Holley went on to report that the major product at Landore and so far the only product at Newton was rails; the ingots were forged into 7.5 inch square blooms under 8 ton hammers; these were then reheated for rolling in 26 inch mills into rails. There was a 30 inch cogging mill being erected at Landore to obviate the necessity of hammering the ingots. Landore was also making tyres, the production of castings was increasing and some material was also taken to bars and wire. Rail steel usually contained about 0.40% carbon; some of the soft steel at Landore contained well under 0.20% carbon. Landore was producing, on average, 56 tons per week per furnace; this included all down time, a furnace in operation making between 68 and 80 tons per week. Results at Newton were essentially similar.

The cost schedules for the production of ingots both at Newton and Landore are included in the report; these are the official cost account figures, using the standard procedures for usage of raw materials and also incorporating lower labour costs and raw material prices which had been agreed in November 1874. The quoted figures are in dollars, at the rate of five dollars to one pound sterling; those for Newton, recalculated back to British currency are quoted in Table III. As a comparison, the previous ingot cost at Newton was quoted as £7.10.0 per ton, with a pig iron cost of £5.16.0 per ton. The figures for Landore are essentially similar.

[11] The unique value of these Private Publications of A. L. Holley will have become abundantly clear to the reader at this stage.

TABLE III COST OF INGOTS AT STEEL COMPANY OF SCOTLAND - NOVEMBER 1874

		£.	s. d.
Cost per ton of Ingots:			
Pig	1,863lb @ £4.5.0 per ton	3.	10. 6
Ore	585 lb @ £1.12.6 per ton		8 .6
Spiegel	196 lb @ £8.2.0 per ton		14. 4
Scrap	341 lb @ £4.16.4 per ton		14. 8
Furnace Cinder	2 lb @ 3.0d. per ton		0. 1
		5.	7. 1
Credit: Scrap	112 lb @ £4.7.6 per ton		4. 5
		5.	2. 8
Coal for all purposes	1,848 lb @ 8.0d. per ton		6. 7
Refractory materials, not for furnaces			2. 0
Stoppers and nozzles			2
Stopper plates			6
Stores, oil, etc			8
Moulds			2. 10
Bottom casting repairs [10]			3
Repairs to furnaces			2. 6
Water			4
Cartage, locomotives, horses			2
Wages, Open Hearth shop			12. 7
Wages, gas producer			11
Testing			1
Loading and unloading			4
Yard Labour			4
Disposal of ashes and debris			1. 0
Salaries			1. 3
Interest, rent and taxes			1. 8
Royalty			4. 6
	Cost of one ton of ingots	£ 7.	1. 4
Summary Cost per Ton of Ingots:			
Materials		£ 5.	2. 8
Fuel			6. 7
Repairs of all kinds			9. 3
Labour			15. 5
Sundries			7. 5

Source: A. L. Holley: Private Publication No 3: "The Siemens Steel Manufacture from Pig and Ore", New York, December 1874.

It is reported that three 10 ton Open Hearth furnaces were added at Hallside in 1875.[12]

Detail on the growth of Open Hearth furnace capacity in Britain in the period 1870–1880 is difficult to find. There are statistics of production (in thousands of ingot tons per annum) available from 1873 onwards;[13] the comparative figures for Bessemer steel production are also included:

	1873	1874	1875	1876	1877	1878	1879	1880
Open Hearth	78	91	88	128	137	175	175	251
Bessemer	495	539	620	700	750	807	834	1,034

The production for 1879 is also summarised according to district:[14]

	Open Hearth Furnaces available	Ingot Tons
South Wales	48	85,000
Scotland	16	50,000
Sheffield	15	21,000
Lancs & Cheshire	12	15,000
North East Coast	5	1,000
Others	6	3,000
TOTAL	**104**	**175,000**

There is also a list of the steelmaking companies which had installed Open Hearth furnaces by 1878;[15] there is, unfortunately, no indication of the furnace size:

Landore Siemens Steel Company	24 furnaces
Vickers and Sons	10
Bolton Steel & Iron	2
London & North Western Railway	4
Charles Cammell	6–8
Dowlais Iron Company	6–8
Panteg Steel and Engineering	12
Joseph Whitworth and Company	2
John Spencer, Newcastle	2
W. T. Henley and Company	2
Steel Company of Scotland	12
Railway Steel and Plant	2

[12] Borthwick, op. cit., p. 24.
[13] British Iron and Steel Federation Statistics.
[14] J.I.S.I., 1878, p. 596.
[15] J.I.S.I., 1880, Pt 1, p. 244.

Albion Steel and Wire	2
Steel Casting Company	1
White and Challingsworth	1
Farnley Iron Company	1

This catalogue gives a total of 89 to 93 furnaces, as against the 104 furnaces quoted a year later. The order in which these establishments is listed is interesting. There is elsewhere a comment that the formation of the works at Landore gave an impetus to the process which was "soon afterwards" adopted by Dowlais and Panteg[16] and the early establishment at Vickers and Sons in Sheffield has already been noted, as have Bolton Steel and Iron and the L.N.W.R. It appears not unlikely, therefore, that the order of this list is chronological, at least in recording the first installations at the various works. It is also worthy of comment that three of the firms on this list as having Open Hearth furnaces are also known to have installed the Siemens regenerative gas fired crucible melting furnaces during this period: Vickers and Sons (216 crucibles), John Spencer and Company (96 crucibles) and Steel Casting Company (36 crucibles). It could be that their involvement in Open Hearth steelmaking arose from their contact with Siemens in connection with the installation of crucible furnaces.

If, indeed, this is a chronological record, the early establishment of Open Hearth furnace capacity at the Sheffield works of Charles Cammell, as well as at Vickers, is a matter which has not previously been appreciated.

Some information, on the state of the art at the end of 1878 is available from a later Holley report.[17]

It has already been indicated that the date of the foundation of the Open Hearth plant at Panteg, in South Wales, is not definitely known, but it was obviously extended subsequently, since in 1878 it was described as the latest Open Hearth installation in Britain; nevertheless the layout is described as similar to that at Landore and having many disadvantages. The main building is described as:

> a great pit, some two hundred feet long, from which side ventilation is cut off by lines of furnaces. Handling hot ingots and moulds and the operations of casting are thus done between two fires. Such an arrangement would be quite impracticable in a hot country; I found the heat in the middle of the Panteg pit, when a set of ingots was stripping each side, insufferable even in winter.

[16] Jeans, op. cit., p. 103.

[17] A. L. Holley, Private Publication, Second series No. IX: "The Open Hearth Steel Manufacture in Europe", New York, January 1879, pp. 36–50.

There were ten 10 ton furnaces operating in September 1878, with two 14 ton furnaces nearly completed. The 14 ton furnaces had beds 18 feet from port to port and 7 feet 6 inches wide. The flame was deliberately directed down away from the roof by the ports in the Panteg furnaces and this had a beneficial effect on roof life, even though the pig and ore process in general use here was said to be much harder on roof and ports than the scrap process. A gradual change was being made to the pig and scrap process, however, for rail production; the pig and ore process was retained only for their more special production of blooms and forgings, and for tinplate sheet. This was largely because the available scrap was not so uniform as the pig. A 10 ton charge on pig and ore normally took up to 14 hours; on one third pig and two thirds scrap it would be tapped in 10 to 11 hours. With nine furnaces in operation, up to 1,000 tons of ingots were made in the week. The normal routine operated on No. 3 or No. 4 Bessemer pig, with 12–20% scrap; the iron in the ore, which constituted up to 12% of the charge, did not quite make up the oxidation loss. Spiegel was used with 15–16% manganese. Pig and scrap were charged together cold; only the spiegeleisen was preheated.

Ladles were on cars, running under the furnace spout on rail tracks parallel to the lines of furnaces. The spout was some 10 feet long, with a removable section which was pulled up after all the metal had entered the ladle, so as to run the slag into a pit. It was becoming general practice, however, to let as much slag as possible run on top of the metal in the ladle, since it had been found that the slag entrained a considerable amount of shot steel, which drained out of the slag into the metal in the ladle, thus improving the yield. A casting method was being developed to run eighteen ingots – the total yield from the furnace – on one bottom plate, the moulds having solid tops with a ventilation hole; this was giving some problems with short run ingots; it was also noted that there was a large "mould" available to receive the metal in case of emergency.

Dowlais was described as having one of the most conveniently arranged and well managed Open Hearth plants in Britain; the furnaces were, however, small. There were six 6 ton furnaces,[18] arranged in pairs with their own casting facilities. The production was all by the pig and scrap process; the pig used was largely Bessemer pig which turned out too high or too low in silicon for the Bessemer plant; the scrap came from their own Bessemer plant, from that at Ebbw Vale and from the many railways in the region.[19] The proportion was of the order of 15 to 25 cwt. of pig to 6 tons of scrap.

[18] The previous list indicates 6 to 8 furnaces; this presumably indicates some expansion was in hand at this time.

[19] Holley comments that he saw scrap railway wheels piled up in the yard. The were charged whole to the furnace: cast iron hub, wrought iron spokes and steel tyre. He remarks:

> This is a commentary on the capacity of the Open Hearth furnace for digesting heterogeneous and cheap materials.

Only the spiegel was preheated. Each furnace averaged 70 tons of ingots per week for eleven months in the year. Interestingly, the cost of the Open Hearth steel was said to be some months above that of the Bessemer steel and some months slightly below; it should be noted that Dowlais was one of the few plants where a valid comparison could be made.

The bulk of the steel was made into rails. Here, as at Panteg, there is a comment on the production of "phosphoric steel" rails, some thousands of tons having been produced at Dowlais, whilst at Panteg they had "within the last two years, made some twenty thousand tons of phosphoric steel rails, with phosphorus over 0.25%, for the Great Western Railway."

This is a topic which will be discussed more fully in the next chapter – the technique was developed at Terrenoire in France. It was essentially making use of the fact that phosphorus, like carbon, has a hardening effect; on the other hand, the two together have an embrittling effect on the metal. This is the reason that the hardest steels produced in Sheffield by the crucible process were always made using the low phosphorus Swedish iron. In both the Bessemer and Open Hearth processes it had been found necessary to restrict the phosphorus content by using haematite ores to obtain optimum results. These ores, however, were not plentiful in Europe and attempts to use cheaper and more plentiful sources eventually led to the compromise of restricting the carbon content to about 0.15% with phosphorus contents of 0.25–0.30% and at the same time incorporating manganese to the value of almost 1%. Ingots of these compositions could be rolled into rails which were not unduly brittle and would wear as satisfactorily as the lower phosphorus steel with 0.3–0.4% carbon. It is interesting to note that this European practice was used in South Wales, where scrap wrought iron rails, with relatively high phosphorus contents, were plentifully available.

There is also an interesting list of the labour complement in the Open Hearth shop at Dowlais:

Three chargers	per furnace
One pit man	per furnace
Two labourers	per furnace
One ladle man	per six furnaces
One foreman	per six furnaces
Twelve gas man above	for 12 blocks of producers
Six gas men below	for O.H. and heating furnaces.

At Landore, by 1878, four of the old 6 ton furnaces had been replaced by 12 ton furnaces since Holley's last visit whilst "a number" in the "new plant", which had sixteen 6-ton furnaces originally, had been replaced by 10 ton furnaces; the remainder were now considered to be 7 ton units. Here, as at Dowlais, it was noted that the pig and ore process was gradually being superseded by the pig and scrap process – not that the quality was other than excellent, but since it was wasteful and long. The current practice varied between pig with 12–20% scrap and 20% of ore (containing 60% iron) to 20% pig with 80% scrap. Hammer scale and roll scale was mixed with a little clay, pressed into blocks and these then charged as an adjunct to the ore addition. The regular pig and ore product gave 80 tons per week on the 7 ton furnaces; one run of 94 tons was produced in 13 heats – 7.23 tons per heat. In one particular week, fifteen 7 ton furnaces made 1,180 tons of ingots, or 78.5 tons per furnace, compared with the norm of 68 tons per week in 1874. Landore had recently been approved for the supply of ship plate by the Admiralty, the material having 0.16–0.20% carbon, 0.6% manganese and phosphorus below 0.030%; the ingots were hammered into slabs, six inches thick, which were then rolled into plates. The plates were annealed after shearing. This was one application where the Open Hearth furnace was found to give a more acceptable product than the Bessemer converter.

At Newton [Hallside] Holley reported that four 10 ton furnaces had been added to the line of the original eight furnaces – these, as at Landore, were now classed as 7 ton furnaces. The furnace technique had remained very much the same as in 1874. To some extent, however, the spiegeleisen was being replaced by higher manganese ferroalloy; again, this is a matter which had Continental origins and will be discussed fully in the next chapter. A form of bottom pouring similar to that at Panteg was being investigated; in all these reports, it would seem that the method of casting was going through a transitional stage and that bottom pouring was recognised as being advantageous but that the technique was still to be fully developed. In 1878 the output at Newton was of the order of 1,000 ingot tons per week, 10% of these being made into plates and forgings, and the remainder into rails.

9 The Open Hearth Process in Europe to 1880

The European situation

The pioneering work carried out by the two brothers, Pierre and Emile Martin, at Sireuil in France between 1864 and 1866 has already been discussed in Chapter 7. The Siemens Martin process was taken up by a number of works on the Continent but it is extremely difficult to evaluate a chronological pattern. There is a list of Open Hearth furnaces and production for 1878 and also a further list showing growth up to 1882[1] and these may be summarised as follows:

	1878 No. of Fces.	Ingot tons	1882 No. of Furnaces
Great Britain	102	175,000	163
France	46	72,000	48
Germany	42	65,000	47
Russia	17	22,000	30
Austria	15	25,000	15
Sweden (incl, Finland)	6	6,000	8
TOTAL	**228**	**365,000**	**311**

There is also a comment that the annual potential of the Open Hearth furnaces in existence in 1882 was of the order of 1.25 million tons.

It will be seen therefore that somewhat less than half of the Open Hearth production was on the continent and this must be largely due to the lack of suitable indigenous ore supplies. Indeed, of the rising industrial nations, only the United States had an adequate supply of suitable ore. Even in Britain, the Cumberland and Furness ores had to be supplemented by imports from Spain during the 1870s; the ores from the Bilbao region were the major European source. France and Germany were singularly short of

[1] J.I.S.I., 1880, Pt., 1, p. 425; J.I.S.I., 1883, Pt. 1, p. 326. The information on Swedish furnaces does not agree with the detail given by Odelstjerna; see Section on Swedish Open Hearth Steelmaking and Footnote 45.

low phosphorus ores and this restricted their growth in steelmaking, both by Bessemer and Open Hearth methods. Germany imported ore from Spain and Austria, as well as from Britain, whilst France called on supplies from Corsica and Algeria. A further complication was that much of the available scrap was wrought iron, mainly in the form of rails, which itself was generally high in phosphorus. Of the Continental steelmaking countries only Austria, Sweden and Russia had significant supplies of suitable iron ore.

This shortage of low phosphorus raw material led quite early to serious investigations into the properties of steel with a moderate phosphorus content. Indeed, as early as 1868, Bessemer steel at Zwickau in Saxony, and at the Maxhutte works in Bavaria, was made from pig containing between 0.10 and 0.20% phosphorus but with up to 4% manganese; by catching the carbon at 0.10 to 0.15%, a residual manganese of up to 0.5% was obtained and many examples of such steel rolled well and gave reasonable ductility when cold.[2]

This appeared to indicate two features. Firstly, the deleterious effect of phosphorus in steel could be minimised by keeping the carbon as low as possible. Indeed, it was eventually worked out that, whilst the best steel for any purpose, combining the maximum toughness for any particular strength, would have no more than 0.02% phosphorus, a steel containing a higher phosphorus content could be tough enough and strong enough for ordinary purposes, provided that the percentage of carbon was diminished as the proportion of phosphorus increased; to quantify this, with 0.7 to 1.0% carbon the phosphorus content must not exceed 0.03 to 0.04%; with 0.05 to 0.06% phosphorus steel for rails should not exceed 0.45% carbon; with 0.10% phosphorus the safe carbon limit was 0.3% carbon and, with 0.15% carbon, steel with 0.25 to 0.30% phosphorus was said to make serviceable rails.[3]

Secondly, manganese appeared to be beneficial to the working of such steels. It will be remembered that Mushet remelted the over oxidised or "rotten" Bessemer steel in crucibles with spiegeleisen and thereby rendered it forgeable and that this became standard practice in Bessemer steel production.[4] With the introduction of Open Hearth steelmaking, the same

[2] A.L. Holley, Private Publication, Supplement to No. 2: "The Value of Manganese" Brooklyn, April 1875. It is to be remembered that phosphorus has the tendency to render steel "cold short" — that is, brittle when cold.

[3] W. Hackney: "The Manufacture of Steel", Proc. Inst. Civil Eng., Vol xlii, 1874-75, Pt. iv, p. 37.

[4] The argument was that the addition of the spiegeleisen removed the oxide of iron and thus the brittleness. Blown Bessemer metal with 0.08% C was found to contain 0.35% oxygen by reduction with hydrogen; similarily, a calculation based on the carbon and manganese losses by oxidation from a known addition of spiegeleisen to blown metal with 0.10% carbon gave an oxygen content of 0.278%. See Hackney, op. cit., p. 31.

principle was applied, spiegeleisen being used to stop the boil prior to tapping the furnace. The low carbon material, either fully blown Bessemer metal or fully decarburised Open Hearth metal was undoubtedly "red short" – that is, difficult to forge or roll without breaking up – but according to one of the leading steelmakers of the time[5] this could either be due to the presence of oxygen or want of manganese. He pointed out that the addition of spiegeleisen to such materials gave rise to a violent evolution of gas which signified the removal of the oxygen; if the bath contained 0.4% carbon, however, such additions were tranquil and he took this to indicate that such metals did not contain appreciable amounts of oxygen. If, however, the residual metal with higher carbon contained only traces of manganese, it was still generally red short to some degree and the conclusion was that a residual manganese content was essential, whatever the carbon content.

The effect of sulphur on forgeability was also studied and it was reported that at Landore steel with less than 0.04% sulphur and a minimum of 0.2% manganese would forge well; 0.08% sulphur required a minimum of 0.4% manganese.[6]

All these aspects pointed to the value of a reasonably high residual manganese in the metal. In the pig and ore process in Open Hearth furnaces the retention of manganese in the bath was soon found to be a virtual impossibility; the time scale was different as compared with the short but violent oxidation in the Bessemer converter; with the more gradual oxidation conditions under a slag cover it was found that the manganese and the silicon appeared to oxidise virtually simultaneously; in one particular heat at Landore, by the time the carbon had been reduced to 1.5% there were only traces of silicon and manganese in the bath. Any manganese in the final steel, therefore, had to be added at the end of the process. It was soon found that the use of spiegeleisen was unsatisfactory, in that its carbon to manganese ratio was too high for satisfactory production of low carbon materials. The production of alloy additions with a more favourable composition was, in fact, pioneered by the French works at Terrenoire[7] and alloys of manganese and iron, which became known as ferromanganese, containing up to 50% or even 60% manganese, with around 5 to 7% carbon, became available. With these, steels with 0.5 to 1.0% manganese could be produced with carbon contents of 0.15% or less.

[5] Hackney, op. cit., pp. 29-31.

[6] Hackney, op. cit., pp. 32-33. It was from such evidence that a requirement for a manganese to sulphur ratio of at least 5:1 became part of steelmaking tradition. The author remembers applying this even in the making of free cutting stainless steel in the 1950s.

[7] A Browne, British Patent 1574, 1st May 1873: "a communication from the Foundries and Forges Company of Terrenoire, la Voulte and Bességes in the French Republic". This refers to furnace linings suitable for the making of ferromanganese with from 25-75% manganese.

Terrenoire and Phosphoric Steel

This preamble is to put in context the work which was done at Terrenoire from about 1873 onwards. Six 4 ton Open Hearth furnaces had been installed at Terrenoire, in an adjacent shop to the four Bessemer converters. By experience, it was found that the Open Hearth furnaces were more adaptable for controlling the analysis of the product and the production of the rail steel with relatively high phosphorus content and low carbon content was soon the role of these furnaces, the Bessemer plant being operated on the more pure pig irons smelted from Algerian ores. One of the early production heats of the type of material which became known as the "Phosphorus-Manganese Steel" was witnessed on 20 July 1874.[8] Details were as follows:

8.40 am.	2,000 lb Solsinora (Corsica) pig charged
9.40 am.	400 lb Tredegar rails charged
10.20 am.	400 lb Tredegar rails charged
10.45 am.	400 lb Tredegar rails charged
11.20 am.	400 lb Dowlais rails charged
11.35 am.	400 lb Dowlais rails charged
12.05 pm.	400 lb Dowlais rails charged
12.30 pm.	400 lb Dowlais rails charged
1.00 pm.	400 lb Dowlais rails charged
1.25 pm.	400 lb Dowlais rails charged
1.50 pm.	400 lb Dowlais rails charged
2.15 pm.	400 lb Dowlais rails charged
3.00 pm.	130 lb red hot Ferromanganese charged
3.05 pm.	Furnace tapped

Half the pig was preheated to a good red heat; the remainder was charged cold. All the rail scrap came from the French Northern Railway ("Nord"); the rails were cut to 4-5 foot lengths and were all preheated to a good red heat. The furnace worked very hot towards the end of the process. The time of operation, 6h.25m, was short on account of the low charge; typical heats were 8,000 lb in seven to seven and a half hours. The bath was thoroughly stirred after the addition of the ferromanganese and the comment was made that the reaction appeared to be a little stronger than that which would have been anticipated using spiegeleisen. The metal was cast into 9 inch ingots, the total ingot yield being 5,880 lb, the yield being 90.1%.

[8] A. L. Holley, Private Publication No. 2: "The Value of Ferro-Manganese", London, October 1874, pp. 6-12. British Patent 370, applied for 29th January 1876, by the patent agent A. Browne on behalf of Terrenoire, covers the production of "Phosphoric Steel".

During the latter part of the melt, metal samples were taken from the furnace with a spoon and cast into a 3 inch diameter mould, 1 inch deep. When solid but still red hot, the samples were taken to a 2 ton hammer, struck lightly at first and then with full force to flatten them to a disc 5-6 inches in diameter and just over a quarter of an inch thick. Any hot shortness showed up a cracking round the edge of the disc. The disc was then fractured and the fracture observed. All samples taken during the melt gave cracked edges, the fractures becoming less granular as the melt proceeded until the final sample showed a partly fibrous structure, indicating the carbon was largely removed. Two samples taken from the furnace after the addition of the ferromanganese, one just as the metal was being tapped and the other at the end of the tapping both showed sound edges, the manganese having removed any tendency to hot shortness. Analyses of the various raw materials and the test samples were as follows:

	C	Si	Mn	S	P
Solsinora Pig	....	0.32	0.80	0.06	0.11
Ferromanganese	6.50	0.20	47.60	0.01	0.15
Dowlais Iron Rail	0.06	0.15	trace	0.14	0.348
Tredegar Iron Rail	0.12	0.14	nil	0.25	0.455
Sample at 12.55pm.	0.32	0.03	nil	0.13	0.283
Sample at 1.45 pm.	0.12	0.03	nil	0.13	0.280
Sample at 2.30 pm.	0.05	0.03	nil	0.13	0.280
Sample at start of tap	0.16	0.03	0.75	0.13	0.286
Sample at end of tap	0.12	0.03	0.75	0.13	0.281

The ingots were reheated, rolled to blooms of 7" x 5.5" section, which were reheated and rolled into rails. Some ingots were overheated; the others rolled satisfactorily into rails which subsequently withstood loading tests and drop tests in a very similar manner to the standard Bessemer steel rails with normal low phosphorus content.

This work was the prelude to a widespread acceptance of many thousand tons of such material for rails in Western Europe; the production of phosphorus-manganese steel rails at Panteg, Dowlais and Landore has already been noted. Indeed, trials were also carried out in America.[9] As experience was gained, the composition of the steel for rails was modified to provide a somewhat stiffer rail but one which would still roll satisfactorily (although

[9] A.L. Holley: "Chemical and Physical Analysis of Phosphoric Steel," Proc. Inst. Civ. Eng., vol liii, 1877-78, Pt. iii, pp. 3-28

it was noted that the practice was to roll these from somewhat lower temperatures) and produce sound rails:[10]

	1874	1876	1878
Carbon	0.12-0.15	0.15-0.20	0.30-0.35
Silicon	0.03	0.02-0.03	trace
Manganese	0.70-0.75	0.90-1.00	0.90-1.00
Phosphorus	0.27-0.29	0.21-0.25	0.20-0.25

The six 4 Ton furnaces which were reported at Terrenoire in 1874 had, by 1878, been upgraded to 7 Ton capacity and two 20 Ton furnaces added. These latter were, at that time, the largest Open Hearth furnaces in operation. It was reported that 15,000 tons of phosphoric steel rails were produced annually for the Northern Railway. The only change in practice appeared to be the inclusion of around 10% of phosphoric steel scrap along with the old iron rails. By this time, better quality raw materials were being used in the Open Hearth furnace for the production of boiler plate steel, using a charge of around 35% good Bessemer pig, together with plate scrap and blooms puddled from Bessemer pig. The average analysis of such steel was 0.15% carbon, a trace of silicon, 0.25% manganese and 0.06% phosphorus. A specially pure grade was also produced to meet the requirements of the French Admiralty, with considerably lower phosphorus content, giving a 30% elongation and a strength of 70,000 p.s.i. (31.2 tons per square inch). A further speciality of the Terrenoire Works was their production of sound castings; this will be dealt with in some detail later.

Saint Chamond and the Pernot Revolving Hearth Furnace

Mechanical agitation of metal during ironmaking and steelmaking processes has always been found to speed up the processes; the rabbling by hand done by the puddler had lately been mechanised in the Danks furnace with largely increased outputs and one only has to think of the violent short time eruption in the Bessemer converter in contrast with the leisurely, comparatively calm occurrence of virtually the same chemical reaction in the pig and ore process in the Open Hearth.[11] It was with a view to speeding up the

[10] A.L.Holley, Private Publications, 2nd Series, No IX: "Open Hearth Steel Manufacture in Europe", New York, Jan 1879, pp. 19-20.

[11] Hackney, op. cit., p. 15, quotes an American writer, "In pudding the iron is agitated by the power of one man: in Bessemerising it is torn into spray by a 500-horse engine!".

reactions in the Open Hearth that in 1874 Charles Pernot of the Saint Chamond Works devised a furnace with Siemens regenerators but with a rotating hearth [Fig 23]; this became known as the Pernot Furnace and provoked much interest. As one commentator put it:[12]

> The Pernot furnace differs from the ordinary form in having the bed circular and arranged so that it may be rotated on an axis inclined at an angle of 5° to 6° to the vertical. The rate of rotation is about two revolutions per minute.
>
> The effect is that, as the charge of liquid metal always lies in the deeper part of the bed, every part of the bottom and all unmelted pieces of steel or iron are alternately in the course of each revolution exposed to the full heat of the flame and dipped into the liquid bath. The increased rapidity of working that must result from such an arrangement and the saving of the brickwork that follows from the lower heat that will suffice to do the work are evident when it is borne in mind that in an ordinary steel melting furnace, with a fixed bed, the bath of metal is often covered with slag one inch or two inches thick and that all the heat that is got into it must pass through this slag blanket. With a rotating bed, not only is the liquid metal kept constantly stirred and each part of the bottom exposed in turn to the direct heating action of the flame but the slag, being lighter and less perfectly liquid than the metal, is dragged back by the rotation of the bed and carried part of the way up the ascending side. The entire bed is mounted on a carriage and may be drawn out from under the roof to facilitate the repair of the one or the other. The weak points of the arrangement would seem to be the risk that the cast iron plate carrying the furnace may be damaged by heat and that slag or steel may get over the edge of the rotating bed and jam it and the ventilation space at the ends between the bed and the fixed brickwork also appears narrow. Furnaces of this design are, however, reported to have been in regular work at Saint Chamond for more than a year and one is also working at Allevard, so they would seem to have had a fair trial and proved sufficiently durable.

The original furnace installed at Saint Chamond in 1874 had a hearth 7 feet in diameter, 20 inches deep at the centre and could produce 5 tons of steel at one heat. A second furnace with a hearth 8 feet 9 inches in diameter and a depth of 21 inches, with a capacity of 10 tons, was installed early in 1875. Holley was at Saint Chamond just after the installation of the second furnace. His description, however, relates to the 5 ton furnace:[13]

> **The Furnace...** The roof is supported at its extremities by the walls of the gas and air conduits and on the sides by a cast iron plate in the centre of which is a circular opening corresponding to the hearth. This plate is enclosed in the walls

[12] Hackney, op. cit., p. 23, The Pernot furnace was covered by four British Patents, taken out on behalf of Charles Pernot by the London agent J.J. Johnson: 2819, 27th August 1873:4291, 31st December 1873;1317, 16th April 1874; 1997, 9th June 1874. It should be noted that there is lack of agreement on the speed of rotation of the furnace; Holley (see footnote 13) quotes it as being three or four revolutions per minute.

[13] A.L. Holley, Private Publications No. 5:"The Pernot Revolving Hearth Furnace and Practice for Siemens Martin Steel", Brooklyn, 1875, pp.5-14.

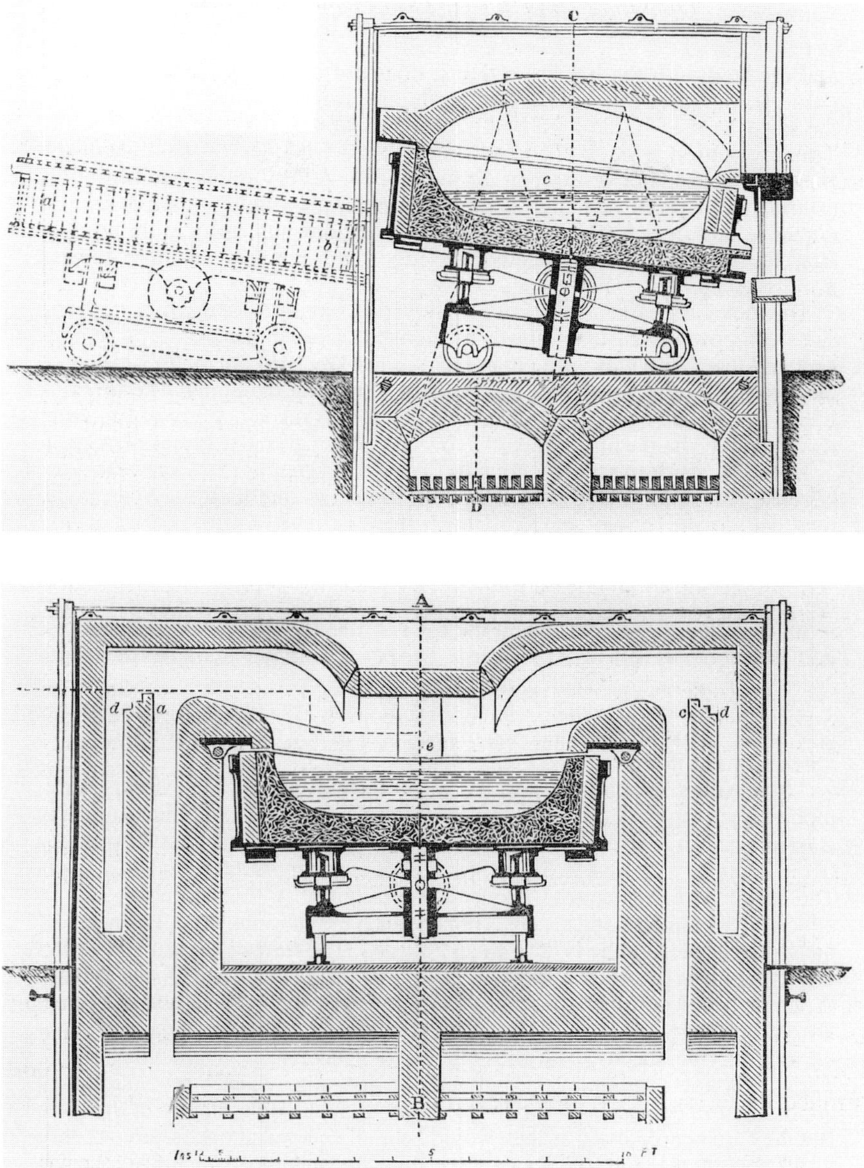

Fig. 23. The Pernot Revolving Hearth Furnace at St. Chamond. This modification of the Open Hearth furnace was designed and installed at the St. Chamond Works in France by Charles Pernot in 1874. The revolution of the central hearth at a tilt angle of around 5° to the horizontal ensured that the liquid metal always lay in the deepest part of the hearth. Any as yet unmelted portions of the charge were alternately immersed in the molten bath and subjected to flame action some two or three times a minute, thus ensuring rapid melting and thorough mixing of the bath. This illustration derives from the same *Encyclopaedia on Chemistry* as Fig. 15.

and covered with the bricks so that it does not come into contact with the flame. It is inclined the same as the hearth. The movable hearth barely touches the lower face of it, the joint being closed as well as may be with dry sand. This precaution is quite sufficient; there is no tendency on the part of the flames to escape nor of exterior air to come in. A very stable equilibrium is established, a proof of which is that, if sometimes a light smoke is seen, it does not continue but disappears rapidly without any observable draught from the hearth. The sand placed on the edge of the hearth is not drawn inward nor disturbed.

The movable hearth consists of a shallow pan, having a plate iron bottom and sides composed of cast iron staves or segments fixed to the plate by key bolts and held together with a wrought iron band. The key bolts are on the outside, so that broken staves can be renewed without disturbing the lining. The firebrick lining wall is laid against the staves. The bottom is made entirely of fire sand; this is rammed hard while the hearth is out and is then glazed in the usual manner.

Rotating Mechanism. This consists first of bevel gearing, the large gear being keyed to the bottom of the hearth in segments, so that they may be readily removed. Inside of this geared segment there is keyed to the bottom plate a four-armed spider, the centre of which is the base for the centre pin or king bolt, and the arms of which form journal boxes for the four conical rollers that carry the weight of the hearth. The whole apparatus is borne by a car which may run it out from under the roof. Upon the car is a flanged track which carries and guides the conical rollers of the hearth. The bevel gearing that rotates the hearth is driven by an inclined shaft and steam engine.

The arrangement by which the hearth may be quickly uncoupled from the driving gear is very good ... the connection may be broken and the car drawn out in five minutes or less if necessary.

The Process. When the furnace is ready for the charge, the hearth is put in motion; then the pig iron, previously heated to redness in an auxiliary furnace, is introduced. It is spread evenly over the bottom, without difficulty, by means of the rotation of the hearth. Then the rail ends are charged, cold, and all at once. The entire charging requires about fifteen minutes. A small movable car on which the workman stands enables him to reach the charging door. By means of the same car the taphole is dressed and opened. The furnace is then allowed to work; the materials become heated and the pig iron soon commences to melt. When melted, it constantly remains, by force of gravity, on the lower part of the hearth, in front of the charging door; the solid materials, on the contrary, follow the motion, enter the iron bath and come out of it, keeping up a relative agitation. These materials acquire a very high temperature, since at each revolution they are brought in contact with the flame; hence they dissolve rapidly in the bath. But they are not oxidised, as they would be if all charged at once in an ordinary furnace where the bath could not cover them; they are dipped under the cast iron and protected by a cover of it each revolution. None of the material sticks to the bottom; it is always completely fused when the furnace is working normally.

While the melting is going on, the motion brings new surfaces in contact with the atmosphere of the furnace and the refining of the bath is accomplished much more rapidly than if this motion did not occur, and without necessitating the stirring of the mass. The agitation, as a result of this motion, tends also to make the metal homogeneous, which is never so surely reached in a furnace with a fixed hearth.

The operation is continued until the refining is complete and the metal has become soft iron. In order to ascertain when this is accomplished, tests taken from time to time are hammered, tempered and broken. The quantity of spiegel necessary to obtain the proper hardness is added at the time indicated by the tests in the usual manner. The spiegel is charged in pieces heated to redness; in a few minutes it is completely melted, which fact is ascertained by running a rod through the mass; tests are then taken and the furnace remains in motion until these tests indicate the desired temper.

When this result is reached, the bottom is stopped so that the taphole comes at the lowest point. The system of tapping a small hole with a pointed rod, and of making up the taphole, is very cleverly worked out in these French establishments. ... A ladle is used ... Bottom casting is not employed...

In order to clean and repair the bottom, the opening in the stave where the taphole is situated is completely cleared and the remaining slag and pieces of broken bricks are scraped out; then a few shovelfuls of sand are thrown on the bottom, the taphole is closed and the engine is set in motion for a short time so as to uniformly heat the furnace, when all is ready for a new operation.

Five charges are regularly melted in twenty four hours in the five-ton furnace. The ten-ton furnace was making four heats in twenty four hours when I saw it.

Repairs. It is the custom at St. Chamond to repair the ports and other internal parts once a week. Whenever it is necessary to make such repairs, the gas is stopped and the car is drawn out. The furnace then cools very rapidly and in a few hours the workmen can get at the roof and work without any inconvenience from the heat. Thus repairs are made with great facility and require no protracted stoppage. Sixteen hours is about the usual time necessary to cool the furnace and do any ordinary amount of brick-work. It is stated, and it seems probable, that an entire roof can be put in between Saturday night and and Sunday night. The solidity of the furnace may suffer a little from this rapid cooling; however it has not been noticed at Saint Chamond that its stability has been endangered. With a strong shell and simple and careful construction, there is little danger in this respect.

The facility with which the bottom can be repaired is even more striking. In case the steel gradually works through the bottom, it can be detected at once by the heating of the bottom plate; if it cannot be stopped by cooling the plate by a stream from a hose, the flame can at once be stopped and the charge tapped; or, if the charge is not fully melted, a part can be tapped and the hearth can then be withdrawn, so that the remainder can be cooled and lifted out with a crane. In case of a "bear" it is not necessary to tear the furnace down, or at least remove the roof, in order to get it out. The bottom can be run out under a crane and the entire contents and lining can be lifted out of the conical pan which contains them. The sand bottom is ... rammed in layers, as in a Bessemer vessel, which greatly contributes to its endurance. The feature about which I have made very particular enquiries is the roof – its stability, when enough depressed in the centre to throw the flame well down upon the bath. There seems to have been a compromise here; the roof does not dip as much as in ordinary Open Hearth furnaces and it is not as strong as if it had been dome shaped. However, I cannot ascertain that these roofs have failed by settling and so long as the furnaces turn out five heats a day, the flame must be pretty well directed.

A table presenting data for twenty five consecutive heats in a week's campaign may be summarised as follows:

	Total	Mean per Heat
Givors Pig, lb	56,800	2,272
Crops Ends, Steel Rails,lb	158,000	6,320
Corsican Spiegeleisen, lb	20,640	826
Total Charge, lb	235,440	9,418
Coal, Producers, bushels	1,226.88	49.08
Coal, Auxiliary Furnace, bushels	252.76	10.11
Ingot Product, lb	221,576	8,863
Ingot Yield		94.11%

It was noted that, using the Pernot furnace instead of the conventional Open Hearth furnace, the daily production could be virtually doubled whilst the coal consumption was reduced by something more than half. This resulted in costs calculated as being the equivalent of $44.87 per ton in the Pernot furnace and $52.85 per ton in the ordinary furnace, using the same raw materials. In addition it was also claimed that the Pernot steel was more uniform in properties.

Developments at the establishment at Saint Chamond involved the modification of the furnace roof and the arrangements of the ports. Fig 24 shows the original arrangement on the left with the improved design on the right.[14] This provided a much stronger roof with the flame more effectively directed upon the revolving hearth. By 1878, the two 5 ton furnaces had been modified and were nominally 8 ton units. A summary of the operations on one of these furnaces in 1876-77 was recorded:

Total number of turns (12 hour shifts in 4 weeks)	44
Average number of heats in 24 hours	3.27
Average ingot tons in 24 hours	21.729
Average ingot tons per heat	6.639
Coal for producers per ingot ton	645 lb.
Coal for preheating furnace per ingot ton	118 lb.
Coal for steam engine (estimated) per ingot ton	100 lb.
Total coal per ingot ton	863 lb.
Average total metal charge per ingot ton	2,373 lb.
Average metal loss in process	5.94%

14 A.L. Holley, Private Publications, 2nd Series, No IX: "The Open Hearth Steel Manufacture in Europe". New York, Jan 1879.

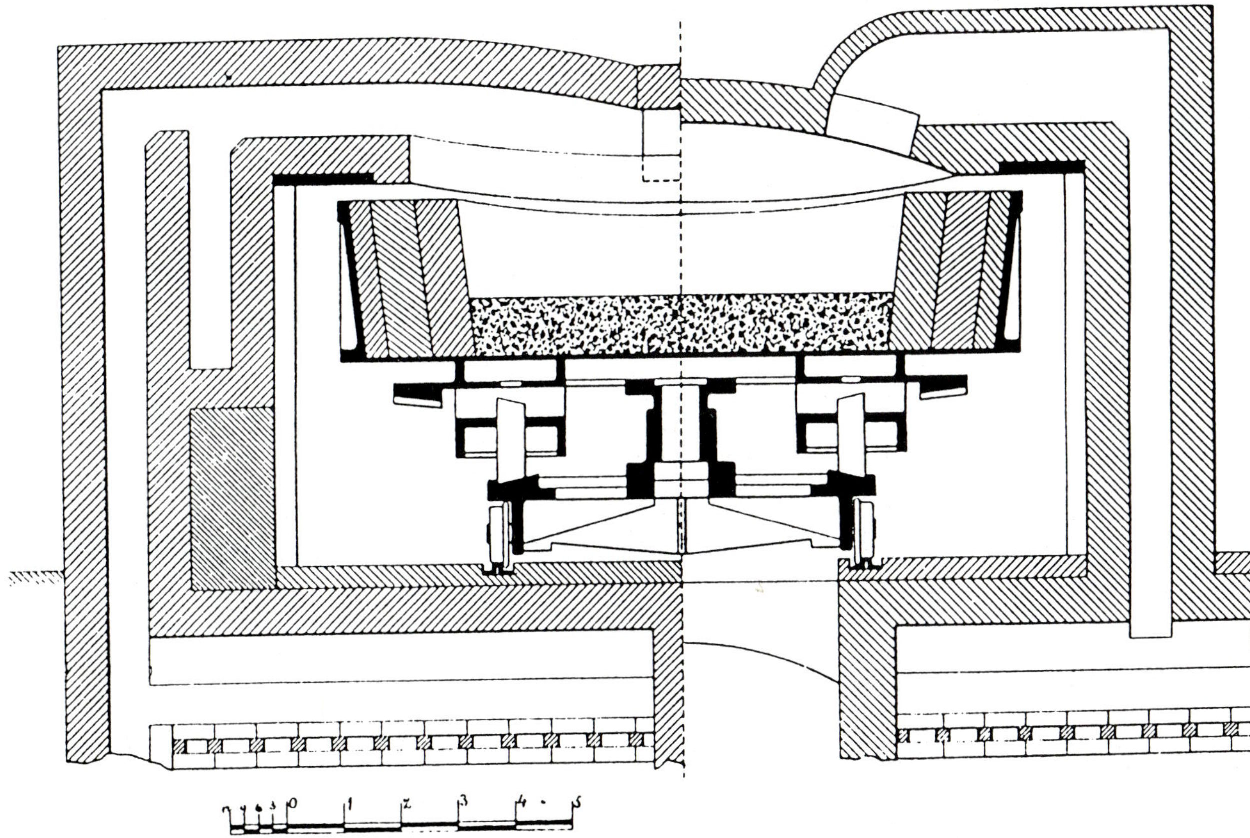

Fig. 24. Development of the Pernot Furnace. This illustration, from the Holley Private Report, 2nd Series, No. IX, illustrates the original design on the left and the improved design of roof line and port design, current in the furnaces in 1878, on the right.

In addition to the two smaller units, the third furnace, originally a 10 ton unit, had been rebuilt as a 20 ton unit. This latter, at the end of 1878, had only been run at the rate of two heats in in twenty four hours; it was known it could be driven harder, but it currently was producing all the steel needed; the major requirements were large ingots for cannon, armour plates and shafting. An 80 ton hammer was being installed and the foundations for three further 20 ton Pernot furnaces were visible at this time, indicating that the performance of the furnace had proved satisfactory.

Despite the apparent advantages, the spread of the use of the Pernot furnace was slow. By the end of 1878, in addition to the three furnaces at St. Chamond there were two others in France (one at Allevard and one at Denain), two in Vienna and three in Russia.[15]

There is no evidence of any interest being showing in Britain in the Pernot furnace; this may well have been due to the fact that a great proportion of the British Open Hearth melting was by the pig and ore process; Siemens himself[16] had asserted that, whilst the Pernot furnace had proved satisfactory for the making of steel in France, it would not work satisfactorily with ore additions:

> since in that process, considerable ebullition ensued and he should be afraid that the boiling mass would go between the rotating lip and the standing part of the furnace and cause it to stop.

There was considerable interest shown in the Pernot furnace by steel-makers in the United States, however — largely, it is felt, because of the publicity given it by Holley. The patent rights in America were purchased by Cooper and Hewitt in 1876;[17] and they licensed the Cambria Iron Company to erect two furnaces in 1878;[18] it was reported in 1878 that the Open Hearth plant there was under construction and would have a pair of 15 ton Pernot furnaces, expected to produce about 25,000 tons of steel per annum.[19] There is, however, no subsequent information available on the use of the Pernot furnace in America.

[15] Coal fired puddling furnaces with the Pernot hearth, however, were much more widespread, with 7 in France, 2 in Belgium, 3 in Germany, 2 in Austria, 1 in Italy, 1 in Spain, 2 in Sweden, 4 in Russia and even one in Hong Kong in 1876.

[16] In the discussion of the paper presented by Hackney, op. cit., pp. 94-95.

[17] J. McHugh: "Alexander Holley and the Makers of Steel", Baltimore, 1980, p. 298.

[18] McHugh, op. cit., p. 308.

[19] A.L. Holley and Lenox Smith, "The Works of the Cambria Iron Company", p. 11; this is a collected reprint of a series of articles which appeared in "Engineering" from May to September 1878.

The Terrenoire "Solid Steel" Process

Work which was carried out at Terrenoire between 1867 and 1877 in the production of ingots and castings in steel free from blowholes provides the first fundamental investigation into what was eventually, in steelmaking terms, to become generally known as deoxidation.

Huntsman had found that, if he cast an ingot from liquid steel which had just been melted, it would emit numerous sparks during the teeming operation, the metal on solidification would tend to rise slightly in the mould and the solid ingot would be honeycombed with small holes in the interior. Indeed, a number of broken ingot pieces showing such effects were found when the Attercliffe site was being cleared for building purposes in 1899, subsequent to the move of Benjamin Huntsman and Company to their new premises further from the city centre. We also know that some of the early crucible steelmakers threw a handful of moistened sand into the top of the ingot mould immediately after casting and then rammed a cast iron plug into the sand to prevent the steel from rising.[20] In fact, this procedure was perpetuated by the early Bessemer steelmakers; the use of "stoppers" to be placed into the tops of moulds in this context had been patented in 1867.[21]

The later crucible steelmakers arrived at the procedure known as "killing by fire" to obviate this phenomenon; this involved heating the steel, when once completely molten, for a further period of up to an hour at the maximum possible temperature the furnace would allow. During this period, as later work made quite clear, there was a reaction in the crucible between the small amount of slag present and carbon within the steel, resulting in a small amount of silicon metal being introduced into the liquid metal; at the same time any oxygen present in the metal would essentially be tied up as silica rather than the much more reactive iron oxide. On casting, this was sufficient to prevent the oxygen present in the molten metal from reacting quickly enough with the carbon in the steel to produce carbon monoxide. By these means, the tendency for this gas to produce blowholes in the solidifying metal was virtually blocked.[22]

Two aspects in connection with crucible steel and the manner in which

[20] G. Broling: "Anteckningar under en Resa i England, Aren 1797, 1798 och 1799", Vol. III, Stockholm, 1816, p. 66. (See translation in the author's "Steelmaking Before Bessemer", Vol. 2, p. 258).

[21] A.L. Holley, British Patent 546, 27 February 1867.

[22] At the high temperature of the "killing by fire" the reaction $SiO_2 + 2C = Si + 2CO$ proceeds relatively slowly, followed by the rapid reaction $2FeO + Si = 2Fe + SiO_2$. On casting, the slow reaction between silica and carbon becomes even slower due to the fall in temperature and the time interval between casting and solidification is relatively small. The opportunity of gas formation is thus minimal and the ingot solidifies before any bubbles collect. The presence of excess silicon also tends to suppress the reaction further. Thus the higher carbon crucible steel will be solid with 0.10% silicon, lower carbon materials need 0.2-0.4% silicon

it differed from Bessemer and Open Hearth steel have to be made clear, however. In the first place, crucible steel was invariably high in carbon – a crucible steel melt lower than 0.5% carbon was a rarity and many crucible melts had in excess of 1% carbon. This is an important feature in this respect since the oxygen content of the metal, other things being equal, is related to the carbon content of the steel by a formula of the type:

$$[C] \times [O] = k$$

where [C] and [O] are the concentrations of the two elements and k is a constant. This leads to the type of curve shown in Fig. 25. Examination of almost any analyses which have so far been quoted, either for Bessemer steel or Open Hearth steel, will show that the normal silicon content in the period under discussion was of the order of 0.03% or less and in many cases was merely a trace: on the other hand, contemporary crucible steel melts would generally show silicon contents of 0.05-0.10%. The steel ingots from Bessemer and Open Hearth processes, therefore, made by normal procedures, would have a tendency to show blowholes and this would be particularly the case with the lower carbon casts.

In addition, it has to be remembered that the crucible steel ingots were generally of the order of 3" square section and that they would be solid in two or three minutes after casting; the bulk steels were not normally cast in ingots of less than 8" to 10" square and these would take from 15 to 25 minutes to solidify,[23] allowing much more time for gas evolution to occur and for blowholes to form in the centre of the ingot. Furthermore, the almost universal practice of casting into moulds with the wide end down,[24] so as to facilitate the removal of the moulds after casting, tended to give rise to areas of shrinkage along the central axis of the ingot and these would form a natural location for such honeycomb formation due to gas evolution during the solidification process.

The initial work at Terrenoire was directed to producing strong and tough castings for projectiles.[25] Work on cast iron had shown that lower carbon materials would withstand a drop test[26] better than the higher carbon materials. Logically, therefore, further lowering of the carbon could be of

[23] As a rough guide, the solidification time for a square ingot is given by the formula $t=0.25d^2$ where t=time and d=distance across the flats in inches.

[24] The only recorded exception to this prior to 1900 was at Saint Chamond. Queried as to why they had adopted this, for its time, unusual practice the answer was "because it is better" (Holley, Private Publication, No. 5 p. 55).

[25] The history here given is taken from A.L. Holley, Private Publication, 2nd Series, No VII: "Solid Steel Castings by the Terrenoire Process", New York, Nov 1877.

[26] 225mm. diameter cast bars were placed on supports 50cm. apart and a weight of 400 kgs allowed free fall upon the centre. The height of fall could be varied and the height of fall required to cause fracture was the assessment of the relative "strength" of the material.

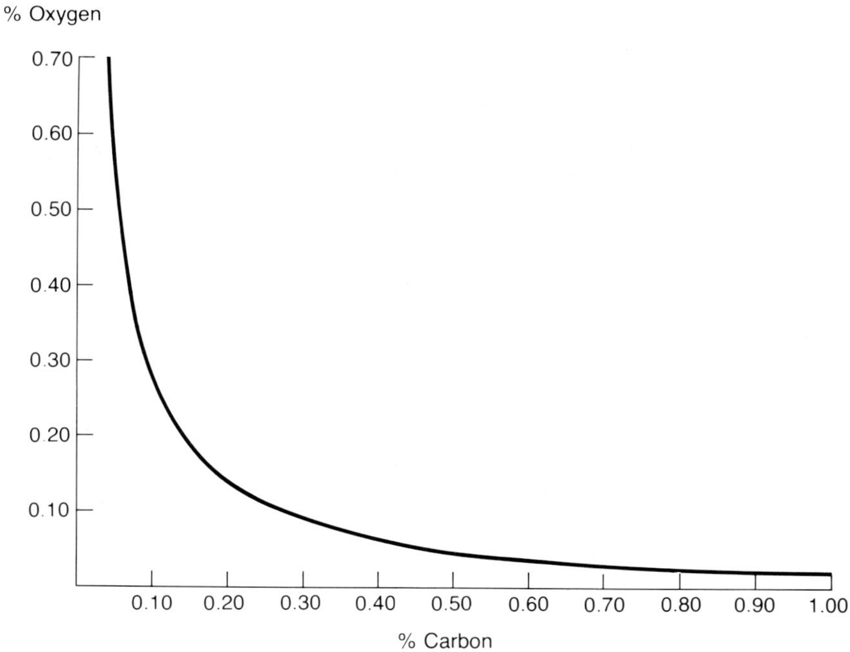

Fig. 25. Notional plot of the oxygen content of blown Bessemer metal prior to deoxidation. This is based on the figures determined by contemporary workers of 0.35% oxygen in metal containing 0.08% carbon and 0.278% oxygen in metal with 0.10% carbon and the expression [C] x [O] = k; in this case k = 0.028.

advantage. Accordingly casts were produced in the Open Hearth furnace with differing cast irons with scrap additions. With the same relative proportion of scrap, the "strength" as measured by the drop test appeared to be related to the amount of silicon in the cast iron; when the scrap proportion was increased beyond a certain limit, bubbling during the casting operation began to appear but, again, the effect of silicon was noted in that the time of bubbling appeared to vary with the amount of silicon in the cast iron used. This was in 1870.

The next stage was to utilise a cast iron with 8% of silicon and some manganese and with they found they could add considerable amounts of scrap and provide a solid casting, but the results were not consistent. Further work was, therefore, focussed on operating the standard Siemens Martin procedure and finally adding a sufficiency of silicon by the use of 11 to 12% of a pig iron containing 3.5 to 4.0% silicon as the final blocking addition to the bath. This regularly gave a product without blowholes; the metal, however, lacked fluidity on casting, it ran badly and the product did not stand up well to forging. This was felt to be due to the presence of silica particles within the cast structure. Work was, therefore, instituted with the aim of combining the silica in such a way that it would become fluid and rise from the bath into the slag before tapping the furnace. Eventually it was found that this could be accomplished by the incorporation of sufficient manganese along with the silicon in the final addition.[27] Combined additions of ferromanganese and high silicon cast iron were, therefore, made at the end of the heat prior to tapping.[28] There remained a further problem, in that it was not always possible to assess the degree of oxidation of the bath at the end of the process; this was important, since it was necessary to be able to add both silicon and manganese in quantities which would provide the required residual amount of each element in the final steel. This was eventually solved by adding a certain amount of manganese to the initial charge so that a slag rich in manganese, and therefore one which allowed better control of the oxidation, was formed.

It was not until 1875 that Terrenoire felt they had mastered the technique; by that time they were producing castings and ingots in both "hard" and "soft" grades of their "acier sans soufflures" (which may be translated literally as "steel without blowholes" or more tacitly as "solid steel"). The provision of these two grades required somewhat different techniques.

In the manufacture of Hard Steel, it was pointed out that the furnace had to be operated so as to minimise the oxidation; it should, therefore, be kept as hot as possible with as little excess air as possible. The initial bath had to contain 6-8% manganese, using spiegeleisen and additions of ordinary cast iron or ferromanganese as required, and was normally 11-13% of the total charge; this was charged cold, spread on the bottom of the furnace and generally melted in about an hour. If, however, large pieces of scrap, such as casting runners or ingot ends had to be used, these would be charged on the furnace hearth with the spiegeleisen on top of them; in this case, a charge of 3,000-4,000 lb would take up to four hours to melt. As soon as the bath was molten, the scrap would be fed, in batches of around 450 lb. each; this scrap would be runners from previous similar charges, ingot scrap and rail ends. Feeding would continue until it was considered time to test the bath.

The degree of oxidation of the bath was checked by taking slag samples. A flat iron bar (about 35 mm. x 10 mm.) was thrust into the furnace and,

[27] Holley reports information from A. Pourcel, Chief Engineer at Terrenoire, to the effect that steel samples, one with silicon only and the other with silicon and manganese, had been heated in a stream of chlorine gas to volatilise the iron. The sample with silicon only had left behind a network of silica and silicate preserving the shape of the original metal, whilst the other only left a small powdery residue. (p. 36).

[28] The co-addition of silicon and manganese was the subject of British Patent 120, 12th Jan 1875, taken out by A. Browne on behalf of Terrenoire. As a final refinement, an alloy with 12-15% silicon and 17-20% manganese, said to be lower in carbon than the normal spiegeleisen, was developed at Terrenoire as a blocking agent to be used at the end of Siemens Martin heats. (A. Pourcel, J.I.S.I., 1882, Pt 2, p. 515). Addition of ferromanganese to a bath of soft metal was followed by a violent evolution of gas; with the ferro-silico-manganese alloy the reaction was quite calm, the gas evolution being completely prevented by the addition of 0.3-0.4% silicon as the mixed alloy. It was also commented that metal samples taken from the bath prior to the addition were full of blowholes; those after the addition were perfectly sound.

clearing away the top slag, was moved around in the under layer near the metal. The rod was then quickly withdrawn; the slag was knocked off as soon as it had set and was left to cool. As it was only a few millimetres thick this took about two minutes. The fracture of the slag revealed its colour and texture. The technique was based on the effect of manganese on the slag texture. With little manganese present in the slag, it would show up the black colour due to iron oxide; manganese lightened the colour to a light olive green. After about a third of the scrap had been added the effect of manganese was beginning to show; from being originally black and porous, the slag turned green and vitreous. At first the whole thickness was the same colour; as time went on, the outside of the specimen, oxidised in the furnace, showed a black film, whilst the interior remained olive green, the inner part against the metal rod retaining a still lighter colour, the three strands being quite distinct. As further scrap was melted in, the streaks grew darker; the outside became black, the middle a brownish green and the inner a dark olive green. When all the scrap has been added, the dark bottle green colour tended to pervade; should this happen before all the scrap has been added, a little extra ferromanganese would be added to restore the manganese level. Only in this way could the level of oxidation be maintained so that the desired levels of residual manganese and silicon could be assured in the final metal.

It was equally important to test the physical state of the metal towards the end of the melt. A spoonful of metal was therefore taken at this stage from the bath and poured into a 75 mm. diameter mould to cast a block 30-35 mm. thick. This was knocked out when set and hammered whilst still hot into a disc about 150 mm. diameter and 10 mm. thick. The disc would invariably show cracks around the edges. It was then cooled until black and quenched in water, prior to being bent to show its softness; if at the correct stage, it would bend almost double without cracking; if then placed under the hammer it would flatten and crack across without coming apart. If too hard, it would break at a smaller angle; if too soft it would hammer double without cracking. The remedies were to add a further scrap charge in the first place or a controlled addition of spiegeleisen in the second.

When the slag and metal were both in the right condition, the bath was left to heat up for a little while before making the final additions. These consisted firstly of an addition of from 11-12% of the bath weight of "special pig", charged hot. This had the following composition:

C	Si	Mn	P
3.0	4.2-4.6	3.5	0.10

As the pig dissolved, the bath, previously bubbling gently, gradually quietened until the surface was smooth, broken only by the occasional large bubble. At this stage, an addition of from 1.0-1.8% of the bath weight of a ferromanganese with 50-60% manganese was made. The bath was then rabbled vigorously for a minute and the furnace tapped. Metal samples taken before tapping would flatten under the hammer and give sound edges.

In the production of Soft Steel the furnace had to be even hotter. The initial charge was, in general, larger, since the added materials were lower in carbon. They were also much lower in manganese – often virtually manganese free. For these reasons, the initial charge was of the order of 14% of the total and was made up to contain 12% manganese. When this was melted, the furnace was generally held for a short period to pick up sufficient heat. Then the "refining materials" would begin to be charged, in small batches as previously. In this case, however, they would be good quality soft steel scrap together with puddled blooms from Bessemer quality pig and all preheated to a bright red heat prior to charging. Both slag and metal tests were taken as before. In this case, the slag would be allowed to come to a somewhat darker level before concern was shown. The metal samples would be softer; they would, however, still show cracks around the edges but not opening up so far into the body of the disc. The correct stage had been reached when the bend could be closed under the hammer without cracking across. The final addition was normally a special Terrenoire pig iron with the following nominal composition:

C	Si	Mn	P
1.60	7.50	14.00	0.125

The usual addition was 3.5-4.0% of this alloy, depending on the appearance of the slag and the desired residual manganese content. In general, the addition of 0.80% manganese would leave about 0.60% in the final steel. Any extra manganese required would be made up by using a 75% ferromanganese. The pig was preheated to bright redness and charged to the hot bath; if ferromanganese was also needed, it was added when all the pig was remelted. After rabbling the bath thoroughly, the furnace would be tapped.

This, then, was the technique perfected by 1875 at Terrenoire. By the end of 1878[29] the process was in use for the manufacture of castings to the extent of 200 tons per month at Terrenoire. It had also been licensed to Krupp's works at Essen as well as to the Steel Company of Scotland, the West Cumberland Iron and Steel Company and the Dowlais Iron and Steel

[29] A.L. Holley, Private Publication, 2nd Series, No. IX: "Open Hearth Steel Manufacture in Europe", pp. 20-25.

Company in Britain and also to the Bofors Works in Sweden. The compositions of the three grades of "acier sans soufflures" made this way at Terrenoire in 1878 are quoted as follows:

	Carbon	Silicon	Manganese
Very hard steel	0.70-1.20	0.50-0.60	0.70-1.60
Hard steel	0.35-0.60	0.20-0.40	0.50-1.30
Soft steel	0.20-0.30	0.15-0.30	0.40-1.20

This work at Terrenoire, in addition to enabling the production of sound steel castings, was of immense importance for the future of steelmaking in general, since it brought about a major improvement in the general physical properties of bulk steel, particularly in the application of these cheaper materials in more heavily stressed engineering applications. Up to this time, any large forging ingot had been produced by the cumbersome operation of multiple teeming of the contents of innumerable crucibles, as evidenced by operations of Krupp's works in Essen as well as in the major Sheffield steelworks; no steel produced by any other process was considered sound enough for such purposes. Whitworth, indeed, had taken the idea of stopping off the top end of the ingot to prevent the issue of gas to the extreme, by actually compressing the fluid metal immediately after casting into the mould.[30] He found that a pressure of 6 tons per square inch applied to the liquid prevented the formation of blowholes and enhanced the properties of the resulting metal. With the better understanding of the mechanism by which gas was evolved during the solidification of the ingot, building on the work done at Terrenoire, Open Hearth steel began to be accepted for engineering purposes, firstly by the Admiralty for shipbuilding and then for ordnance. By 1880 John Brown was using Whitworth's press, not for compressing liquid steel, but for forging solid steel ingots from the Open Hearth furnaces; in less than ten years, Firths, Vickers and Jessops had all done likewise and the casting of large ingots in crucible steel was a matter of history – except, rather surprisingly, at Essen, where the Krupp practice of multiple pouring of crucible steel was to continue until after 1900.

As a final footnote to this section, which, after all, has mainly been a discussion on the deoxidation of steel, it was as early as 1885 that the final shot in the locker was brought out – the use of aluminium in steelmaking. This was a relatively scarce and expensive metal at the time but its affinity

[30] J. Whitworth, British Patent 3062, 7th September 1874.

for oxygen – or, in other works, the stability of its oxide – was such that the addition of the order of 0.05% or less, dependent on the carbon content, was found to suffice to "kill" a normal steel completely.[31]

Open Hearth Steelmaking at Le Creusot

It is not clear when Open Hearth steelmaking was introduced at Le Creusot; when Holley visited the plant in 1875[32] there were six 10 ton furnaces in an extension of the Bessemer building. These were arranged in two lines, four to one side and two to the other, each furnace with its own reheating furnace and with the casting facilities in the central part of the building between the furnaces. The furnaces were a little unusual in that the roofs were flat and the distance between the roof and the surface of the hearth somewhat less than normal in English furnaces, the regenerators being accommodated beneath the hearth. Silica bricks were employed for the roof and the upper walls; their expansion was contained by the very strong furnace shell so that the roof was allowed to rise. This tended to shorten the life of the roof and ports – they only endured about sixty heats. The end walls at the rear of the ports had large openings filled with loose bricks which were readily knocked out for work on the ports themselves. Steam jets were blown under the furnace bottom to induce air currents which kept the brickwork cool.

For the production of rails, twelve 10 ton charges were normally made each melting week; the actual melting time was of the order of nine hours per charge of 10 metric tons. The charge was composed of Bessemer pig, produced in their own blast furnaces from Algerian and Spanish ores with the following general analysis:

C	Si	Mn	S	P
3.5-4.1	1.5	3.0	0.03	0.03

The later additions consisted of Bessemer or Open Hearth scrap and rails-ends. Rails at Le Creusot were clearly produced from both Bessemer and Open Hearth steel; tyres, however, were produced exclusively from Open Hearth steel. For high grade merchant steel and boiler plate from the Open Hearth furnace, however, Bessemer pig was puddled and the blooms rolled to 75 mm. x 15 mm. bars[33]; this material, cut to metre lengths, was used in place of the scrap and rail ends. An interesting comment is that the

[31] T. Nordenfelt, British Patent 8269, 8th July 1885, a communication from Carl Wittenstrom of Stockholm.

[32] A.L. Holley, Private Publication No. 8: "Steel Manufacture in France and Belgium". New York, May 1876, pp 15-18.

[33] This is very reminiscent of the size of Swedish bar iron used in the making of blister steel.

steel for boiler plate was cast into rectangular ingots, 500 mm. x 225 mm. x one metre long (approximately 20" x 9" x 40" long).

The Open Hearth steel was said to be no better than the Bessemer steel when similar stock was used; clearly, however, it was considered prudent to use the Open Hearth process, with better quality raw materials, when steel of better quality was specified. It appears that the pig and ore process had been tried at Le Creusot but had not found favour.

When Holley next visited the plant, in the autumn in 1878,[34] he noted that the second line of furnaces had been completed; there now was a total of eight furnaces. In addition, the furnaces had been modified by lengthening the hearths, so as to make them nominally of 15 ton capacity (although he indicates that 20 ton charges were not uncommon); the width had remained the same, but the roof design, on which he had commented somewhat unfavourably on his earlier visit, had been completely changed to introduce the "V" form, that is, one with a downward slope from port to centre of hearth. This had become standard practice in England and elsewhere; with the flame well directed downwards on to the bath by improved port design, such a roof reflected the heat better and was found to give a good life and faster furnace working.

There were some individual aspects of the Open Hearth steelmaking at Le Creusot at this time. For the most specialised steels the charge was only 10% of what was referred to as "special" Bessemer pig;[35] the balance consisted of puddled blooms together with scrap arising from previous special heats. The pig and half the blooms, all preheated, were charged together, followed by the remainder of the preheated blooms in batches as usual. The spiegeleisen or ferromanganese was also preheated. It was noted that, whilst the pig and ore process was not used, it was a common practice to charge quantities of from 50 to 100 lb. of red hot ore from time to time to increase the heat of the bath. A further novel feature was the skimming off of the bulk of the slag some half hour or so before tapping; again the object was to utilise the heat of the flame as much as possible, by cutting down the insulation effect of the thicker slag layer. The general arrangements covered the making of only two heats per furnace in twenty four hours.

Interestingly, a "recipe" for cheap steel was quoted as being 20% pig and 80% scrap, with a final addition of 6-7% spiegel; steel had, indeed, been made using scrap alone, the time per heat and the quality of the steel being much as usual, but the damage to the furnace was quite severe. Another innovation had been the addition of nearly blown liquid metal from a Bessemer converter into a molten bath of Bessemer pig in the Open Hearth

[34] Holley, 2nd Series, No IX, pp. 27-33.

[35] It is conceivable that this could well be to a similar analysis to that quoted by Holley in his earlier report: C 4.37%; Si 2.19%; Mn 5.99%; S 0.04%; P 0.055%.

furnace, followed by the melting in of the remaining scrap; this had been done at a time when scrap was scarce, the proportions being 20% Bessemer pig, 40% molten Bessemer metal and 40% scrap. There had been occasional difficulties due to the tendency of the Bessemer metal to set during transfer and the process had finally been discontinued.

The casting of larger ingots at Le Creusot had been well organised. Even with the original smaller furnaces, a system of casting ingots up to 28 tons had been devised using the contents of four furnaces. The ladle used had a capacity of twenty tons; 7 tons were tapped from one furnace and the ladle positioned over the mould. Whilst the metal was running into the mould a further ladle with 7 tons was brought over the first ladle and the metal run into the ladle, and so on with two further ladles, the nozzles in these subsidiary ladles obviously being larger than the casting nozzle. Everything was said to go "like clockwork". With the enlarged furnaces, Holley saw a 60 ton ingot cast in 1878 from four furnaces; this was a rectangular one intended for the production of an armour plate; in the event it was slightly short of metal, so a ladle with some Bessemer steel was brought over to complete the filling of the mould; the comment was that:

> as the Bessemer metal was all in the rising head, the casting came out alright.

Other Open Hearth Steelmaking Activity in France

There are two lists of Open Hearth furnaces installed in France; the first gives details in 1875[36] and the second in 1879.[37]

These may be summarised as follows:

	1875	1879
Societé des Forges & Acieries de Firminy	8	7
Bietrix et Compagnie a la Chaleasserie	2	1
MM. Schneider et Compagnie au Creusot	8	4
Compagnie de Terrenoire et Bessèges	5	11
Cie de Chatillon et Commentry a Montluçon	-	4
Societé de Commentry Fourchambault a Guerin	-	2
Usine d'Aubin a Aveyron	-	4
Societé des Aciers Martin a Sireuil	2	2
Compagnie d'Alais a Tamaris	-	1
MM. Charrière et Cie a Allevard	-	1

[36] J.I.S.I., 1875, p. 194
[37] J.I.S.I., 1880, pt 1, p. 425.

Cie. des Forges et Acieries de la Marine et des Chemins de Fer a St. Chamond	-	3
MM. Mariel Freres aux Etaings	-	1
M. Bussery a Lorette	-	1
Societé des Forges de Denain et Anzin	-	2
Total	25	44

The comment is also made that the capability of the 44 furnaces in being in 1879 was of the order of 200,000 tons per annum.

There are some obvious discrepancies here, particularly in regard to the later establishment at Le Creusot; the figure for Terrenoire also includes the five furnaces at Bessèges. The overall picture, however, is one of growth in the use of the Open Hearth process at a time when the use of the Bessemer process was stagnating. No detail of the activities at most of the smaller establishments has been discovered but it seems fair to comment that the process likely to have been operated was according to the Siemens Martin technique.

Open Hearth Steelmaking in Germany

Information of the development of the Open Hearth process in Germany is even more difficult to obtain than that concerning France. There is a statement that there were 11 furnaces operating in 1875, with a further 7 standing idle.[38] A general comment on the situation in 1879 reads as follows:

> The Siemens Martin process has made immense progress in Germany within recent years. There are eight furnaces at Biskupits, fourteen at Dortmund, twelve at Krupp's, three at the Phoenix, three at the Grafenberg Works and several at Haspe. The Bochum Works has constructed a large establishment for this manufacture. In all there must be fifty furnaces, capable of producing 200 tons a day, supposing only one-third of them to be in daily work. All the modifications of the process are employed – the utilisation of the scrap, the simultaneous fusion of the iron and the pig, the direct employment of the ore as at the Landore Siemens works and the employment of the rotator for previous direct extraction of iron from the ore. This last modification – that of employment of reduced spongy iron in place of ore ... – may be seen at Dortmund and other German steelworks.

This represents the sum total which one of the fullest texts on steelmaking can provide.[39] The other relevant information is that in 1878 there were 42 Open Hearth furnaces in Germany (as against "about fifty" above) and that

[38] J.I.S.I. 1875, p. 205.

[39] J.S. Jeans: "Steel: its History, Manufacture, Properties and Uses", London, 1880, p. 178.

the total production from them in the year was 65,000 tons.[40] There is some small amount of further information available concerning operations at Essen and Bochum.[41]

In 1878 there were twelve 6 ton furnaces at Krupp's Works at Essen, six on each side of a railway. The moulds, standing on cars, were brought down the line under the ladles for casting. An interesting procedure was noted in operation here. Six tons of pig melted in a cupola were charged into the Pernot puddling furnace with an excess of iron ore; this desiliconised and dephosphorised the metal in five to eight minutes and the product was tapped into a ladle and charged to the Open Hearth furnace. There scrap would be fed to produce the desired composition and in this way, not only were four heats obtained in twenty four hours instead of the normal three but good quality steel could be produced from pig relatively contaminated with phosphorus.

The Open Hearth plant at Bochum had ten 6 ton furnaces in a line. It was noted that the older furnaces had flat roofs but the newer one had the "V" roof. They operated the pig and scrap process exclusively. Special comment was made on the casting technique, using what was termed a "pouring funnel", nine inches in internal diameter at the top, nine inches deep, lined with a sand-clay mixture and with a two inch nozzle set in the base. This was preheated to a good red and then lifted off with a forked handle which was suspended from a light crane on the ladle. Manoeuvring this device exactly over the centre of the mould, by levering lightly on the support, the stream was neatly prevented from hitting the side of the mould. Opening up the ladle nozzle, which was one and a half inches in diameter, the metal stream splattered out into the funnel, which then concentrated the stream into the mould, thus avoiding many of the defects caused by the extreme pressure from the ladle and the splayed stream which caused splash on the ingot mould walls.

Open Hearth Steelmaking in Austria

Austria had 15 Open Hearth furnaces in 1878, producing an annual total of about 25,000 ingot tons.[42] Not much more is known, except that there were two Pernot furnaces making steel in Vienna, as indicated above.

There is, however, some specific information available on the operations at what appears to have been the major Open Hearth steelworks at this date, that belonging to the Southern Austrian Railway Company in Graz.[43] Here there were two 10 ton furnaces, with a port design which ensured that the

[40] J.I.S.I., 1880, Pt. 1, p. 425.
[41] Holley, 2nd Series, No. IX, pp. 33-36.
[42] J.I.S.I., 1880, Pt. 1, p. 425.
[43] Jeans, op. cit., pp. 247-249.

flame was directed down into the hearth; it was stated that the redesign of the furnace had extended the life of the brickwork from 30-40 up to 150 or even 200 heats.

All the materials for the charges were preheated. Over a campaign of 28 working days on the two furnaces the following information was collected:

Number of melts		96
Metal charged:	Pig iron	256.4 tons
	Scrap & rail ends	706.6 tons
	Spiegeleisen	35.2 tons
	TOTAL	998.2 tons
Product: 3364 ingots		955.4 tons = 95.7%
Scrap		2.0 tons
Coal for furnace	553 tons	= 1,296 lb per ingot ton
Coal for preheater	287	= 673 lb per ingot ton
TOTAL COAL	840 tons	= 1,969 lb per ingot ton

Working on the previous furnaces, with 5.5 ton charges, had given only a 91.7% yield with a coal consumption of 3,068 lb per ingot ton.

The output of the two current furnaces was put at 10,000 to 12,000 ingot tons per annum. By far the larger part of the output was destined for rail making, the analysis being 0.30-0.40% carbon, 0.01-0.02% silicon, 0.10-0.25% manganese and 0.08-0.12% phosphorus. Soft iron with 0.18-0.22% carbon was produced without problems, provided the furnace was run as hot as possible. Castings were also made – and the influence of Terrenoire shows through:

> an addition of silicoferous pig iron is found to be very beneficial.

The steel so produced, typically 0.65% carbon, 0.15% silicon and 0.35% manganese, was said to yield perfectly sound and dense castings. An interesting point was made that the course of the melting operation was checked by reference to the Eggertz colour method for carbon; in addition, an estimate of both phosphorus and manganese could be carried out in about half an hour; this is a very early example of what came to be known as "pass test" analysis.

Open Hearth Steelmaking in Sweden

Steelmaking in Sweden has always been conditioned by two overriding factors: the virtual absence of coal as an indigenous fuel and the necessity of selling good quality iron abroad as part of the Swedish economy. The Bessemer process was attractive to the Swedish metallurgists as soon as they heard the phrase "without fuel". The Open Hearth process had its attractions since it appeared to give better control than the Bessemer process and could be operated with a gas generator. This was something which had been pioneered in Sweden, using scrap wood, sawdust, peat and the like, and Lundin had already heated iron for rolling and, indeed, had designed a gas fired cementation furnace for the production of blister steel at Munkfors in 1862.[44]

In order to investigate the process, Jernkontoret commissioned its Director, L. Rinman, who was attending the 1867 Paris Exhibition, to visit Sireuil and discuss with the Martin Brothers the possibilities of the process. This was followed by some experimental work in Sweden and the building at Kilafors of an experimental 500 lb. furnace with a Lundin gas producer, so as to produce some trial material by the new process. When it was seen that this was a viable route for the production of high quality material, they experimented further with variations in the charge and also with ore additions, eventually producing first class steel with 60% pig iron and 40% wrought iron charges and it is stated that the establishment of the Swedish Open Hearth practice was almost entirely due to the work of Rinman and the proprietor of the iron works at Kilafors, C.A. Rettig.[45] The fuel consumption for this small furnace was, of course, excessive – from 370 to 560 lb per 100 lb steel. In 1869 a one ton furnace was built, with a consumption of about 275 lb fuel per 100 lb metal. By the end of 1878, there were seven Open Hearth steelworks in Sweden, operating eleven furnaces, the largest of them being 4 ton units,with a fuel consumption of around 200 lb per 100 lb of steel. It was in 1878 that Bofors commenced the manufacture of solid steel castings, using the Terrenoire patent, and from that time the acceptance of the Open Hearth process grew and larger furnaces were built. It was also about this time that the improvements in communications world wide made it possible for Sweden to import coal at an economic price and about 1880 coal fired producers were established to fuel 10 ton furnaces.

It was soon decided that only quality steels should be made in the Open Hearth furnace, of a quality equivalent to that of crucible steel, and this pattern, which was to persist until about 1940, had been fixed by 1880. The plan was to produce only three kinds of Open Hearth steel: the softest steel, containing 0.15% or less carbon, tool steel, with at least 0.45% carbon, and castings. The procedures these three were as follows:

1. **Soft Steel.** The pig iron was selected to have as low a sulphur content as possible, 0.015% maximum. The pig and scrap were charged together and, three quarters of an hour after melt down, the feeding of ore was commenced; this ore was of the richest kind, with not less than 64% iron. The process was operated so that at the end of the melt the proportions of

[44] J.F. Lundin: "Branstålsugn for Sagspan", Jernkontorets Annaler, 1862, Pt. 2, p. 115.

[45] The basis for the information in this section is a paper by E.G. Odelstjerna: "The Manufacture of Open Hearth Steel in Sweden": presented to the American Institute of Mining Engineers in February 1894. See Trans, Amer. Inst. Mining Eng. Vol. XXIV (1894) pp. 288-315.

both sulphur and oxygen would be low and this was more easily achieved with wood fired producers than with coal fired producers. The absolute minimum addition of manganese consistent with obviating red shortness was added.

2. **Tool Steel.** The pig iron here had to be as free as possible from phosphorus and also from sulphur; 0.020% maximum of each was the aim. The steel melting was carried out at a good steady heat, to obtain a melt free from gas at the time of tapping without too high a percentage of silicon, so that the ingots might be solid and free from blowholes. A more liberal attitude existed to the presence of manganese in these steels. There is an interesting note in connection with this steel; if it was feared that it was coming from the furnace too hot, it was either held in the ladle to cool to the proper temperature, or the "so-called De Laval heat rings" were used. These were cast iron rings, lined with fire proof material and heated to a high temperature. Immediately before casting, they were placed on top of the moulds; on casting, they were filled with metal which remained liquid and sank until the ingot below the ring was entirely solid. This is one of the earliest references to the use of what we would now refer to as a "hot top" or a "feeder head", apart from the use by Mushet of his patented "dozzle" at his "Secret" steelworks in the Forest of Dean in 1861.[46]

3. **Steel Castings.** It seems that the castings made at Bofors followed the Terrenoire procedure quite faithfully. They were of excellent quality.

Open Hearth Steelmaking in Russia

There is a tradition that the establishment of the Open Hearth process in Russia was the work of one John Hughes, born in Merthyr Tydvil in 1814, who trained as an engineer and eventually became a consultant in London. Around 1870 he became involved in the formation of the New Russia Iron Company, which was set up on a green field site, with coal and iron mines, blast furnaces, puddling furnaces and Open Hearth steel furnaces, as well as the dwellings for his work force. This activity brought about the founding of the town of Hughesovska, which at the time of his death in 1889, had a population of nearly 100,000.[47] In 1882 there were eight furnaces with an output of about 16,000 tons of Open Hearth steel.[48]

A year later, a report indicates the Open Hearth furnaces handled charges of 12.9 tons each, these being 48.6% pig iron, 30.4% puddled bar, 17.0%

[46] R.F. Mushet, British Patent 1310, 23 May 1861. It is possible that the "rising head" referred to in connection with the casting of the 60 ton ingot at Le Creusot was a similar device; this is not clear from the text.

[47] Obituary Notice of John Hughes, J.I.S.I. 1890, Pt 1, pp. 177-178.

[48] J.I.S.I. 1883, pt, 1, p. 472.

steel scrap, 2.4% ore and 1.6% ferromanganese, the daily production being 23.3 tons per furnace, with an annual production of 40,000 ingot tons, which provided 32,000 tons of rails.[49]

In 1878, Russia was reported to have produced some 22,000 ingot tons of Open Hearth steel from 17 furnaces; there were 30 furnaces by 1882.[50] Some of these furnaces can be located.

Siemens Martin steel was produced at the Aboukoffsky (or Obouchoff) Steel Works in Alexandrovsk; the product was largely railway material and the scrap used in the furnaces was also mainly supplied by the railway companies.[51] Plates for the Russian Admiralty were also produced from these furnaces,[52] the average composition being quoted as:

C	Si	Mn	S	P
0.17-0.24	0.01	0.20-0.25	trace	0.01

There is, however, more information about these activities.[53] Described as "the greatest steelworks in Russia", there were two 10 ton Siemens furnaces, without any preheating facilities. The gas producers were so arranged as to be fired either by wood or coal. The charges were all introduced at the beginning of the process, often including scrap pieces up to two tons in weight. The melt took up to 4 hours, when 100 to 150 lb. of ferromanganese would be added. If tests showed the metal not to have reached the desired degree of softness, either puddled blooms or, preferably, good iron ore would be added. When the conditions were judged correct, the appropriate amount of ferromanganese would be added and the metal tapped. In a separate paper[54] the same author gave details of charges for three types of steel.

Hard Steel:	Charge:	Commenced 4.30 am, finished 6.0 am: 120 cwt steel scrap 17 cwt ladle scrap 17 cwt pig iron (12% Mn)
	Melted 8.15 am; sample taken; sample hard 17 cwt puddled iron bloom charged Sample taken 9.0 am; sample hammered well; bent double 15 cwt. pig iron (9% Mn) charged 10.20 am.	

[49] J.I.S.I., 1884, Pt. 1. p. 296.
[50] J.I.S.I., 1880, Pt. 1. p. 425.
[51] Jeans, op. cit., pp. 255-256.
[52] J.I.S.I., 1883, Pt. 1, p. 464.
[53] S. Kern: "On the Manufacture of Steel and Steel Plates in Russia", J.I.S.I., 1881, Pt. 1, pp. 80-83.
[54] S. Kern: "The Manufacture of Siemens Martin Steel in Russia", Chemical News, No. 41, p. 278, abstracted in J.I.S.I., 1880, Pt. 2, p. 777.

	Furnace tapped 11.00 am.	
	Analysis of steel: C 0.67%; Mn 0.40%	
Medium Steel:	Charge:	Commenced 1.0 pm, finished 2.45 pm:
		135 cwt steel scrap
		23 cwt ladle scrap
		10 cwt pig iron (12%)Mn)
	Melted 7.40 pm. 0.5. cwt ferrmanganese added	
	Sample taken: steel too hard	
	2.5 cwt magnetic iron ore added 7.55 pm.	
	8.30 pm: further sample taken; bent nearly double, slight crack	
	0.5 cwt ferromanganese added	
	Furnace tapped 9.15 pm.	
	Analysis of steel: C 0.35%; Mn 0.18%	
Soft Steel:	Charge:	commenced 11.00 am, finished 12.30 pm
		113 cwt Steel scrap
		26 cwt ladle scrap
		6.5 cwt pig iron (12% Mn)
	Melted and first test 6 pm	
	2 cwt ferromanganese added 6.30 pm	
	Second test bent double after hardening	
	0.67 cwt ferromanganese added 7.00 pm.	
	Furnace tapped 7.15 pm	
	Analysis of steel: C 0.16%; Mn 0.15%	

Analyses of steel scrap and cast iron showed sulphur and phosphorus contents generally 0.02% maximum; it was noted that the high manganese pig iron originated in Sweden.

As regards other establishments in Russia, Poutiloff's Works in St. Petersburg had eight 8-ton Siemens Open Hearth furnaces in 1882 and one Pernot rotating furnace; the Alexander Works in the same city had seven Open Hearth furnaces.[55] There were six Siemens Open hearth furnaces at the Briansk Works and two at the Nijni-Tagilsk Works in the Urals.[56]

[55] J.I.S.I., 1883, Pt. 1, p. 471.
[56] J.I.S.I., 1883, Pt. 1, p. 472.

10 Open Hearth Steel Manufacture in the United States of America to 1880

The activities of Siemens did not go unnoticed in America and he had thought fit to obtain patents in that country, as indeed did the Martin brothers. This, however, did not prevent regenerative furnaces being built without any licence. A Pittsburgh copper works set up a melting and refining furnace, built solely on the basis of what one of their representatives had seen in Birmingham, together with the drawings included in Siemens' own advertising literature; this was in 1863 and is said to have worked well. The next year Park, Brothers and Company erected a billet reheating furnace and Lyon and Company a glass melting furnace, both in Pittsburgh; neither apparently were a success. The publicity given by the Paris Exhibition in 1867 led to further interest being taken and in the autumn of that year licences were given by Siemens for three projects: billet reheating furnaces for Griswold and Company at Troy and for the Nashua Iron and Steel Company of New Hampshire and a crucible steel melting furnace for Anderson and Woods of Pittsburgh.[1]

One of the American commissioners at the Paris Exhibition was Abram S. Hewitt, co-owner with Peter Cooper of the New Jersey Iron and Steel Company at Trenton. He had experimented with the Bessemer process in the early days but had reverted to a form of crucible steel melting. His journey to Paris gave him the opportunity of seeing the Sireuil exhibit and led to a visit both to Sireuil and to Le Creusot to see the process in operation. Sensing that here could be a source of cheaper steel than he was producing of a quality better than Bessemer steel, he negotiated the use of the Martin patents in America. In this way, he felt that Trenton could have a major advantage, since he now held the control over the alternative process to Bessemer by which cheap steel could be made in bulk, with a plant which was cheaper to install and a process which, whilst being capable of absorbing the increasing amounts of cheap scrap – something quite foreign to the Bessemer process – was more easily controlled. He ran into delays, how-

[1] J. M. Swank: The History and Manufacture of Iron in All Ages, 2nd. edition, Philadelphia, 1892, pp. 421-422.

ever, in connection with the fact that Siemens' American patent rights were held by a firm of agents in Boston; negotiations were prolonged by Hewitt's insistence on making the Martin patent secure against attack in the American courts. Hewitt had previously taken on Holley's brother-in-law, Frederick Slade, as his personal assistant and, whilst the legal matters were being sorted, sent him to Sireuil to observe the process in action. He then returned to supervise the erection of the furnace, which had a capacity of four to five tons, together with the associated plant; this was started in the spring of 1868 and by December of that year was ready to operate.[2]

There were considerable initial problems; as it so happened, Siemens was employing S. T. Wellman, a man destined to be to the Open Hearth furnace in America what Holley had been to the Bessemer process, as assistant to the engineer sent over to supervise erection of the regenerative crucible furnace at the Anderson and Woods premises in Pittsburgh. Wellman was despatched to Trenton to endeavour to sort things out and he later set down what he referred to as "vivid recollections" of his experiences:[3]

> Mr. Slade was the only man connected with the works who had ever seen a heat of Open Hearth steel melted and, as he was not a practical man, he probably overlooked some of the important points while watching the manufacture in France. At any rate, everything bad that could happen in the manufacture of Open Hearth steel seemed to fall to their lot. The best man among the workmen was the melter, who had been an old puddler or heater, but of course he was unaccustomed to such temperatures as were necessary to melt mild steel. The first trouble encountered was with the making of good gas. This arose from a too literal interpretation of the printed instructions for running the producers which had been sent them by Messrs. Siemens' agents.
>
> When I took charge of the producers I found they were full to the very top, the coal being some six or seven feet deep, so thick a fire that about all the heat in the bottom of the producers could do was to drive out the moisture from the upper layers. Water and thin tar were running out of the joints through the whole length of the gas flue, or cooling tube, which was of wrought iron plates, placed above ground. We ran the producers for two days without charging any coal. After about twenty four hours work the producers began to do good service and we had no further trouble with them. The only difficulty was to get men to attend faithfully to the making of gas, a very disagreeable job and one that demands constant watchfulness to get the best results. Before trying to describe the melting of the first heats, a word or two about the furnaces and the method of casting. The furnace was a Siemens' regenerative furnace, I suppose copied from the one in use at Messrs. Martins' works near Paris. It had a very flat shallow bottom but there did not seem to

[2] J. McHugh: "Alexander Holley and the Makers of Steel", Baltimore, 1980, pp. 286-290.

[3] S. T. Wellman, Presidential Address, 1901: "The Early History of Open Hearth Steel Manufacture in the United States", Trans. Amer. Soc. Mech. Eng., Vol. xxiii, pp. 80-82.

be much trouble getting it up to a fair melting heat. The regenerators seemed to be of ample size and the furnace fairly well designed for its work, except as to the shape and depth of the bottom, which was entirely too shallow for the quantity of steel it had to hold. On the casting side of the furnace, in front of the tapping hole, was fastened a kind of fore hearth, called by the workmen, from its shape, "a shoe". In the bottom of this, at the outer end, was placed a firebrick nozzle in which was fitted a fire clay stopper, by which the flow was regulated. Running parallel with the front of the furnace was a narrow gauge railroad, on which were placed the cars for the ingot moulds, which were arranged for group casting. These cars were moved by a chain windlass placed at one end of the track. In all the heats I saw made, Franklinite pig iron was used for the bath and the balance of the heat made up from puddled bars and scrap steel. This mixture should have made good steel, but unfortunately they had no ferromanganese, none being made at that time in this country [that is, U.S.A.] and but little, if any, in Europe.

For recarbonization, they used a small quantity of Franklinite pig; according to my recollection, not more than 1.0 to 1.5 per cent of the weight of the charge. As the Franklinite pig contained only about 10 per cent of manganese, this gave them a steel very low in that element, with at most not more than 0.15 per cent and often only a mere trace. The result was a steel which was difficult to roll, except at a high temperature. It would stand a welding heat and a dull red heat, but was yellow short and any attempt to hammer or roll it at that temperature resulted in the steel going to pieces. . . .[4]

In the first heat at Trenton, when the melter thought, from the small tests taken, that the metal was ready to cast, a small quantity of Spiegeleisen was added and, when this was melted, a bent bar was driven through the sand wall between the "shoe" and the furnace. When this bar was withdrawn, it allowed the steel to run out and fill the shoe to the level of the steel in the furnace. The stopper being opened, the steel ran out through the nozzle into the moulds. When the first group was filled, the stopper was closed and the car was moved along until the next group came under the nozzle. In this manner, about one half of the moulds were filled when the nozzle in the bottom of the shoe began to close up by the metal chilling. It soon closed up entirely and the car was then moved on until the next group of moulds was in place under the nozzle, when the whole nozzle was lifted out of its place and the steel then came with a rush for a short time. The operation was repeated until all the steel that was liquid had drained out of the furnace. Only a small proportion made ingots which could be rolled; the larger portion was scrap.

The great difficulty in these pioneer days of steel making was to teach the melters to maintain the steel at the proper temperature. The margin between

[4] This is an unusual but very descriptive way of referring to a phenomenon well known in the iron trade. The welding heat was the temperature normally employed for forge welding together the bars of wrought iron which constituted the "ball" or "faggot" or "pile" and usually represented a temperature somewhat in excess of 1300°C; a yellow heat was of the order of 1150-1250°C; a bright red heat would be of the order of 1000 to 1100°C; a dull red heat was below this temperature.

the proper casting temperature and the melting point of the furnace was very narrow . . . as at that time no silica brick was available and we were obliged to use clay firebrick for the roof and sides of the furnace. Shortly after the heat which I have mentioned, I saw an attempt made to cast a charge which had chilled in the nozzle so that it could not be started again. By the time the shoe could be taken off the steel had chilled solid, clear back into the furnace, which was then shut down half full of steel. The next day the furnace was started up again. A quantity of pig iron was added and the steel finally all melted, but it was so slow coming to a steel melting heat that the carbon was reduced and several more additions of pig iron had to be made, until finally the furnace was overfilled, making it all the more difficult to get the proper temperature. It was finally tapped, but it was so cold that it chilled in the nozzle and the forehearth, so that when the latter was taken off the steel ran down over the moulds, cars and tracks, welding the whole together and making a mess that I was told afterwards took two weeks to clean up so that the furnace could again be started. These are only examples of the difficulties and rough experiences which were encountered in these initial trials and it was no wonder that poor Mr. Slade was discouraged.

As is very often the case, this experiment was not very profitable to the pioneers, Messrs. Cooper, Hewitt and Co. The furnace was operated at intervals for a year or two but was finally abandoned and the manufacture stopped, as it was not a commercial success, though those who saw the experiments were able to profit by them and avoid many of the difficulties encountered in them.

The second Open Hearth steel furnace in America, and the first to be a commercial success, was at the Bay State Iron Works of South Boston, Massachusetts.[5] One of the specialities of this establishment had been the so-called "steel headed rail", produced by forge welding a steel plate to a wrought iron bloom and then rolling to a rail section in a mill. This was cheaper than the imported steel rail and more wear resistant than the plain wrought iron rail. The superintendent at the works, Ralph Crooker, had tried puddled steel and also Bessemer steel from Troy for this purpose but both had proved to be too variable in manufacture; the Bessemer steel in particular was very irregular in quality and sometimes impossible to weld to the wrought iron. Crooker visited Trenton and, after seeing a number of heats and observing how the steel performed, became convinced that a properly conducted operation on these lines would provide him with the material he required. He too had difficulties with Siemens' agents. In the event, Wellman, having left Siemens employ some time earlier, agreed to supervise the construction of the works in the summer of 1869 and operations were commenced early in 1870.

[5] Wellman, op. cit., pp. 83-89.

The furnace, of 5 ton capacity, was designed to be very powerful for its size, to be "sharp and quick in its working". The bath was much deeper than the one at Trenton. The tapping arrangement on the furnace was much as at Trenton, but the moulds were carried on a large turntable, rotated by bevel gears driven by a steam engine. The melter, a skilled heater with many years experience, had the advantage of having spent a few weeks at Trenton and was able to start up operations at Bay State with little trouble.

One notable innovation at Bay State was the production of ferromanganese, in house. This was made by melting together black oxide of manganese, charcoal and lime in crucibles; when the initial charge was melted, Spiegeleisen, or Franklinite pig iron with 10% manganese, was added to the crucibles, the whole then melted together and poured into flat moulds. In this way an alloy with up to 40% manganese was available and by its use it was practicable to make steel with 0.10% - 0.15% carbon which was perfectly malleable at all temperatures, a great improvement on the Trenton experience.

The first steel made by Bay State was produced from a mixture of puddled bar from the works, together with some steel scrap, melted into a bath of imported West Cumberland haematite pig. It was rolled into bars about 7" x 2" and these were used for the top quarter of the rail pile, the balance being old iron rails. The whole was reheated and forged together, the weld being as sound with the Open Hearth steel as with iron. This was, however, a transient trade and the "weldless steel rail " was soon to become universally used wherever there were railways; this, of course, had been the motivation behind the rapid expansion of Bessemer steel production in America. Bay State sought out other outlets for their steel. Having considerable experience in producing iron plates, they turned their attention to making steel plates for such applications as boilers, fire boxes and the like. Initial trials with their in-house iron proved this to be too high in phosphorus; using Lake Champlain blooms, produced direct from the ore in Catalan hearths, they managed to produce a soft, ductile material, perfectly malleable both hot and cold and the plates rolled from it never gave any blistering. The single furnace at Bay State was run day and night for several years on flat products which sold at a good price. In this way this small furnace earned a great deal of money.

Wellman's father was superintendent of the Nashua Iron Company. It will be remembered that this company already had experience with a Siemens regenerative mill reheating furnace and in 1871 the decision was made to install an Open Hearth furnace. It was understandable that Wellman Jr. should be put in charge of the installation of the furnace together with a plate and bar mill.[6] The furnace, which came into operation early in 1872,

[6] Wellman, op. cit., pp. 89-92

was a 5 ton unit and was essentially similar to that at Bay State; a pair of preheating chambers were, however, provided on the charging side of the furnace – blooms could be charged into these and then pushed into the furnace when hot. For the best quality steel, English haematite pig was used for the bath to which were added charcoal blooms, as at Bay State; a cheaper grade of steel was also produced from steel and iron scrap and domestic pig. The casting arrangements at Nashua, however, were altered from those at Bay State; a ladle was used and the shoe dispensed with. Wellman's own comments are interesting:

> . . . it had a great advantage . . . as no matter how cold the steel was, it was always hot enough so that it ran "clean" out of the furnace. While some of it might chill in the ladle, the furnace was not interfered with at all, and we were able to get out the next heat without interference from steel left in the furnace.

The Nashua Company produced forgings for railway and marine applications as well as boiler and fire box plate and also provided steel locomotive and wagon wheels. The original 5 ton furnace served for several years, when it was taken down and replaced by a 10 ton unit.

At about the same time that the Nashua furnace was being erected, the second Pittsburgh crucible steel firm to install gas fired crucible furnaces, Singer, Nimick and Company, took the opportunity of their collaboration with Siemens to put in a 5 ton Open Hearth steel melting furnace, to facilitate the production of lower carbon steel for boiler plates, for which their crucible process was not really suited. This was the only furnace in the first decade of Open Hearth steelmaking in America with which Wellman had no connection.

Most certainly, this was not the case in the setting up of the first American establishment for the exclusive production of Open Hearth steel on a green field site. This was the Otis Iron and Steel Company of Cleveland, near the shores of Lake Erie and with a major railroad either side of the site.[7] Work here started in 1873. Wellman brought Holley in as consulting engineer.[8] The steel melting plant was designed with two 7 ton melting furnaces, a regenerative preheating furnace for blooms and a regenerative crucible furnace for the production of ferromanganese. An important part of Holley's design, an off-spin from his Bessemer experience, was the raising of the

[7] Wellman, op. cit., pp. 93-96

[8] Wellman related that he had met Holley seven years earlier, when, wanting to learn the steel business, he applied for a position at Troy, only to be told by Holley, who was in charge there, that there were no vacancies. When he again met Holley in Cleveland, he laughingly reminded Holley by saying: "You could not give me a position; I have at last succeeded in getting you one!" This was the start of a pleasant and fruitful friendship, which lasted until Holley's death in 1882.

charging platform some ten feet above ground level; there were hydraulic lifts for the charge materials. This made possible the use of shallow casting pits and also rendered the furnaces very accessible for repairs.[9] Ladle casting was employed. The first heat was made in October 1874, using charcoal blooms from the Lake Champlain district melted into a bath of Lake Superior charcoal pig iron; the steel was of high quality and was made into plates for fire boxes. Wellman's comments are again interesting:

> The starting of the works was attended with the usual mishaps which always attend the launching of new enterprises, especially in the iron business. The first thing to happen before any metal had been charged into the furnaces was a terrific gas explosion, which cracked the main gas flues, loosened up the gas and air valves, smashed man hole plates and escape valves and disarranged things generally. After our furnaces were started they were occasionally fused down by careless melters and the steel did not always prove to be what we expected, but a careful system of inspecting prevented these mistakes getting to our customers and after a few months' schooling of this kind things began to work smoothly and we had comparatively little trouble in making what we wanted. Of course we had to educate all our skilled workmen; having no works from which to hire them we had no other way. Our head melter, who originally came from New England during the construction of the works, was by trade a bricklayer – a bright and observing fellow, who was anxious to learn the steel melting business. I sent him to Nashua and Boston some two or three months before we started up the works to learn what he could about the operation of the furnaces there. Of course, for a time he needed constant watching but, after a few months work, he became a very expert melter. Nearly all of our skilled men learned their trade after entering the employ of the Otis Company and when I severed my connection with the company, after sixteen years' service, out of about 1,000 employees I was able to learn of only one skilled workman who had not learned his trade in the shops of the company.
>
> Not only men but all of our materials had to be tried and proven. The question of firebrick, as well as of other refractory materials, was an important and troublesome one, and a great deal of experimenting had to be gone through before the best materials could be found. In the meantime we were building up a good reputation for the quality of our product, which was principally plates for boilers and fire boxes. "OTIS STEEL" was for many years a standard brand for that purpose in this country and sold at the very highest price.

In 1878, to supplement the original two 7 ton furnaces, two more furnaces were added; these were of 15 ton capacity, the largest to have been installed up to that date; a further two of the same size were added three years later.

[9] This set a pattern which had a long existence. With the addition of mechanical charging devices (which themselves were also largely due to the work of Wellman) at the rear of the platform, this design was still quite easily recognisable in the last of the Open Hearth plants in Sheffield in the 1960s.

Chronological details of the individual works which took up Open Hearth steelmaking beyond 1874 are not readily accessible, but there is a list of Open Hearth steelworks in the United States at the end of 1877:[10]

Bay State Iron Company, Boston	1 furnace
Beaver Falls Steel Works, Pennsylania	2 furnaces
Blair Iron and Steel Company, Pittsburgh	1 furnace
Canton Steel Works, Ohio	1 furnace
Cleveland Rolling Mills, Ohio	2 furnaces
Midvale Steel Works, Philadelphia	2 furnaces
Nashua Iron and Steel Company, New Hampshire	1 furnace
New Jersey Steel and Iron Company, Trenton	1 furnace
Norway Ironworks, Boston	1 furnace
Otis Iron and Steel Company, Cleveland	2 furnaces
Pennsylvania Steel Company, Philadelphia	2 furnaces
Pittsburgh Steel Casting Company, Pittsburgh	1 furnaces
Rumford Chemical Works, Providence, Rhode Island	1 furnace
Sheffield Steel Works, Pittsburgh	1 furnace
Springfield Iron Company, Springfield, Massachusetts	2 furnaces
Washburn Car Wheel Company, Hartford, Connecticut	1 furnace

This gives 16 firms with a total of 22 furnaces and agrees precisely with a summary given elsewhere;[11] this also indicates that the total installed capacity was 45,000 ingot tons per annum. A similar survey in March 1880 indicated there were 22 established works with 33 furnaces, an installed capacity of 225,000 tons per annum, with three new works being erected to house 6 new furnaces which would increase the available capacity by 50,000 tons.[12] These figures obviously disclose two features: firstly a growing demand for Open Hearth steel and secondly a marked increase in the size of the newer furnaces.[13]

In conclusion, output figures for Open Hearth steel ingots in America can be provided from 1870 onwards; the following table compares these with the American Bessemer production, the figures being in short tons of 2000 lb:[14]

[10] J.S. Jeans: "Steel: Its History, Manufacture, Properties and Uses", London, 1880, p. 148.

[11] J.I.S.I., 1876, p. 249.

[12] J.I.S.I., 1880, Part 1, p. 284.

[13] The annual capacity per Open Hearth furnace works out as follows:

1876		2,045 tons
1880:	existing furnaces	6,818 tons
	new furnaces	8,333 tons

[14] Swank, op. cit., p. 425, p.415

Open Hearth Steel Manufacture in the USA to 1880

Year	Open Hearth ingot tons	Bessemer ingot tons
1870	1,500	42,000
1871	2,000	45,000
1872	3,000	120,108
1873	3,500	170,652
1874	7,000	191,933
1875	9,050	375,517
1876	21,490	525,996
1877	25,031	560,587
1878	36,126	732,226
1879	56,290	928,972
1880	112,953	1,203,173

11 The Solution to the Phosphorus Problem

The presence of significant amounts of phosphorus in the majority of iron ores, and, therefore in the product of their smelting in the blast furnace, coupled with the inability of the processes devised either by Bessemer or Siemens to reduce this contamination, has been stressed on a number of occasions in preceding pages. It was this factor which gave rise to the phenomenal growth at Barrow, situated on the major source of British low-phosphorus ore, the Furness haematites; from a green field site there arose in less than ten years an establishment which for a time could justifiably claim to be the largest steelworks in the world. Conversely, although vast deposits of iron ore had been discovered in Cleveland, the location of bulk steelmaking in the area was initially by virtue of the convenience of Middlesbrough for the importation of low phosphorus ore from Spain and elsewhere. The growth of bulk steelmaking in France and Germany was likewise stultified by lack of domestic supplies of suitable ores and these countries chafed under their dependency on imported ores from Spain, Algeria and even Britain; so much so that they seriously investigated and, to a marked extent, succeeded in producing steels with significant phosphorus contents which could still perform useful functions. Only Sweden, the United States of America and, to a considerable extent, Austria had suitable ores which enabled them to provide their own requirements under the current conditions.

Bessemer originally had been fortunate – or unfortunate, depending on point of view – in his random choice of one of the few available domestic supplies of low phosphorus cast iron for experimental work at Baxter House, St. Pancras. Subsequent to his announcement to the audience at Cheltenham, he invited interested parties to witness the operations and among the audience was one Dr. R. H. Collyer, who described his observations in a letter to the *Times* on 10th September 1856.[1] He was one of about eighty present who observed a blow carried out, in his words,

> with a degree of precision and neatness which would have done honour to a Faraday, a Turner or a Dalton.

[1] W.T. Jeans: "The Creators of the Age of Steel", 2nd.ed., London, 1885, pp.44–46.

He also reported on a bar of malleable iron which was bent and showed great ductility although, on fracture, the grain was not such as would have led the ironmasters present to expect such ductility. It is the next few sentences of his letter, however, which merit full quotation:

> ... the real secret is getting metallic purity. The particles do not require to be interlaced or fibrous to the same extent as when the iron contains even a small proportion of phosphorus and sulphur. The former I consider the most pernicious of all. I would suggest, with due deference, that a stream of finely pulverised anhydrate of lime [quicklime] be forced at a given time with the compressed air into the incandescent mass of iron. The lime, having a greater affinity for silica and phosphorus, would form phosphate and silicate of lime and be thrown off with the slag. By this contrivance I cannot conceive but that the phosphorus would entirely be got rid of.

This quite reasonable suggestion for the solution of a major problem of whose existence, at that very time, even the inventor himself was unaware, was the first of many over the next twenty three years and an enormous effort was expended on the phosphorus problem by many eminent metallurgists. Probably the most eminent among these was Isaac Lowthian Bell, who was looked upon in those days as "the high priest of British metallurgy". He pointed out as early as 1872 that the limit to the production of Bessemer pig was the lack of ores free from phosphorus and that this had put such a premium on its price that the bringing about of a process which would allow the use of the Cleveland iron stone for steelmaking would save something of the order of £4,000,000 per annum.

> Now there is such a margin to stimulate exertion, we may be sure that the minds of properly qualified persons will be directed toward the solution of a question of such national importance.[2]

In the event, it was still a few years before the problem was solved; even then it was not the trained metallurgists who were to be responsible but someone who has been described as:[3]

> one of the last and perhaps the most important of the line of tinkerers that had made the Industrial Revolution – after him, the professionals just about had the field to themselves.

The difference in the behaviour of phosphorus in the puddling furnace, where a marked degree of phosphorus removal took place, and in the Bessemer converter, where none occurred, was noted on several occasions;

[2] I. Lowthian Bell:"The Chemical Phenomena of Iron Smelting", J.I.S.I., 1872, p.85.
[3] D.S. Landes:"The Unbound Prometheus", Cambridge,1970, p.259.

there was, in fact, usually a small phosphorus increment, due to the removal of the silicon, manganese and carbon from the blast furnace metal during the Bessemer blow. Indeed, one suggestion for the solution of the phosphorus problem was to puddle the blast furnace metal to wrought iron, then to remelt the wrought iron in a cupola so as to recarburise it, cast this into slabs and then repuddle, so as to obtain two reductions of the phosphorus content.[4] Gruner, as early as 1859, had pointed out that a slag with less than 40% silica, and preferably less than 30%, was essential for dephosphorisation in the puddling process;[5] it was for this reason, of course, that a number of wrought iron works, as, for example, Low Moor, employed the refinery furnace for removing some of silicon from the pig, so as to obtain a more actively oxidising slag in the puddling furnace. It followed that a lining rich in basic oxides would be required if adequate dephosphorisation was to be achieved. In the puddling furnace hearth the basic oxides capable of absorbing the oxidised phosphorus in the form of phosphates were, of course, those of iron and manganese; such a lining was impracticable in a Bessemer converter. Trials of a bauxite lining were made, already noted, by Siemens at Montluçon in France in 1863[6] but the lining would not hold together. Gruner's further researches, in conjunction with Emile Muller, led to a suggestion that magnesite could prove serviceable and Muller took out a patent for magnesia linings for both Bessemer and Open Hearth furnaces;[7] it seems bricks were actually produced at Ivry. The problems caused by the Franco-Prussian war, however, disrupted further work. George Snelus was working as chemist at Dowlais at that time and, based on laboratory experiments, came to the conclusion in 1870 that hard burned lime was a possible basic furnace lining material. Further work indicated that the use of a calcined magnesian limestone (dolomite), with a small amount of iron oxide as binder, was superior to straight calcium limestone for this purpose and in 1872 he obtained a patent for such a furnace lining.[8] Indeed, he did carry out small scale trials, using Cleveland pig iron, and had some success. With hindsight, it is clear that he was well on the way to making a major contribution to the solution of the phosphorus problem. It was here, however that fate played a part: only a few months after Snelus took out his patent, he accepted the post of Bessemer Department manager at the West Cumberland Iron and Steel Works – one of the few operating directly on

[4] G. Parry, British Patent 2900, 18 November 1861.

[5] L. Gruner:"Notes Diverses concernant l'Acier Puddlé et l'Acier de Forge", Annales des Mines,xv,(1859),pp. 299–301.

[6] W.E. Newton (on a communication from Guillaume Boigues and Paul Rambourg), British Patent 708, 16 March 1863.

[7] E. Muller, French Patent 84,735,1869; a parallel patent was applied for by J.H. Johnson on Muller's behalf(British Patent 908, 25 March 1869) but only the provisional specification was filed.

[8] G. Snelus, British Patent 908, 25 March 1872.

pig iron smelted from the local Furness haematite ores. As he was rapidly promoted to General Manager of the works, he was obviously fully occupied. At the same time, it was clear that any procedure which assisted the production of good steel from phosphoric ores would only weaken the standing of his own organisation, which relied heavily on the local natural advantage. He did, however, confide his results to Edward Cooper, of the American firm of Cooper and Hewitt, who advised him to take out an American patent.[9]

It was about the time that Snelus went to West Cumberland that the "last of the tinkerers" began his "tinkering". Sidney Gilchrist Thomas, born in 1850, had intended going to London University to study medicine but the premature death of his father altered his plans and he entered the Civil Service and studied science in his spare time; in particular, he went to evening classes at Birkbeck College and it was there that he heard George Chaloner, his lecturer in Chemistry, remark on one occasion that[10]

> the man who eliminates phosphorus in the Bessemer converter would one day make his fortune.

By 1873 he had obtained all the qualifications in Chemistry and Metallurgy he could, by evening studies, whilst still retaining his duties as a Police Clerk. He proceeded to collect all the available technical data on the behaviour of metal in the Bessemer converter as well as analytical data on the reactions of phosphorus, coming to the same obvious conclusion as Snelus and others that a strong base was necessary to unite with oxidised phosphorus and that a lining to the convertcr must also have a basic constitution. He accordingly became convinced, without any knowledge of what Snelus had done – since this work remained unpublished – that lime and magnesia offered the best possibilities as lining materials. There were clearly difficulties with the use of these materials as linings. Ganister, as used in the normal converter, was a good, solid material, which would withstand the attack of heat in the condition in which it was mined; it would withstand the attack of the slags normally produced and was impervious to attack by atmospheric moisture. It was, in other words, a good natural refractory. The basic materials, on the other hand, were far more chemically reactive; indeed, the basic oxides of calcium and magnesium – lime and magnesia, as they are usually termed – do not occur free in nature by reason of their reactivity; they are mined as their carbonates. Thc oxides may be obtained by calcining the carbonates; this drives off the carbon dioxide. Nevertheless, the oxides so formed will readily absorb atmospheric moisture

[9] In the event, Snelus deferred doing this, hoping to confirm his results later on full scale trials and only, in fact, took out an American Patent in 1878 when Thomas made his first provisional announcement.

[10] L.G. Thompson:"Sidney Gilchrist Thomas: an Invention and Its Consequences", London, 1940, p.47.

– lime more readily than magnesia – forming the hydroxides.[11] Only by calcining the carbonates at an extremely high temperature – well in excess of 1200°C – can "dead burned" lime or magnesia, resistant to hydration, be obtained and this high temperature treatment brings about considerable shrinkage and gives a product which is extremely difficult to bond into a brick.

Working from his own firm convictions, Thomas set up a miniature converter in his own small laboratory, lining it with various basic mixtures and thereby achieving some encouraging results. He soon discovered the necessity for extended facilities. It is also of interest to note that he spent his summer vacations travelling around Europe, spending as much time visiting various metallurgical works as admiring the scenery. In the autumn of 1876, his cousin, Percy Carlisle Gilchrist, who had been chemist at the Cwm Avon Iron Works in South Wales, moved to a similar position at Blaenavon. The two cousins had been in regular correspondence with each other on the phosphorus problem and now, in a shed on a Welsh hillside, they secretly built a converter which held eight pounds of metal. Thomas went down to Wales every weekend and innumerable trials were made, using hard burned lime bonded with waterglass, magnesia bonded with iron oxide and all manner of materials. The progress was slow, but sure, and they were becoming more and more convinced of their eventual success. In course of time, they were able to remove the bulk of the 1.5% phosphorus present in Northamptonshire pig iron on a regular basis, using bricks made from burned lime bonded with water glass, although they were not able always to obtain fluid metal on the small scale on which they were working. They nevertheless felt that they had "arrived" at the position where they needed to cover themselves with patents. With his sister's help Thomas went through all the five hundred or more patents which could be thought to have a bearing on the subject and, as a result of this search, considered their way to be clear. The first patent was taken out late in 1877 and covered the production of basic bricks from lime, using sodium silicate (water glass) as a binder.[12] At the stage where the financial implications were beginning to appear somewhat daunting, help came from a somewhat unexpected quarter. The manager of the Blaenavon Works, Edward Pritchard Martin, called the two men in to see him just before Christmas 1877, saying that he was aware they had some secret work in hand and suggested they should take him into their confidence. Gilchrist's analyses were submitted and Martin was so impressed with the work that he at once afforded facilities

[11] The product of the lime kiln, in which limestone is burned to "quicklime" by mixing it with coke and so heating it to around 900°C:

$$CaCO_3 = CaO + CO_2$$

will, on addition of water, give off so much heat that it disintegrates to powder and emits steam, the product being "slaked Lime" or calcium hydroxide:

$$CaO + H_2O = Ca(OH)_2$$

[12] S.G. Thomas, British Patent 4422, 23 November 1877.

for further work on a larger scale and they soon had available two converters, one holding four hundredweights and the other ten hundredweights, together with the promise of larger scale trials at Dowlais. Two further patents were obtained at this stage: the first covered the use of the basic materials in either Bessemer or Open Hearth furnaces, whilst the second related to the used of calcined magnesian limestone (dolomite) bonded with a small amount of a suitable clay.[13] The Dowlais trials were carried out in a five ton converter lined with the lime bricks; the results were encouraging as regards the effectiveness of the desphosphorisation was concerned but the lining failed to stand up to the wear and only five heats were obtained from it.[14]

It was at this stage that the first public indication of this work and its results was given. At the Spring Meeting of the Iron and Steel Institute held in March 1878, Lowthian Bell presented a paper on "The Separation of Phosphorus from Pig Iron".[15] This was a sequel to two the previous year on reactions in the puddling furnace and in the Bessemer converter; to the modern reader this third paper indicates that his massive researches were foundering. It is the discussion of this paper which is of historical importance. Professor Williamson enquired whether Mr. Bell had considered that benefit might be obtained by replacing some of the iron in the slag by a non-reducible basic oxide, such as lime or some other base of comparatively little cost. At this, Mr. Snelus informed the meeting that he had obtained a patent covering the use of lime as a steelmaking refractory some years ago and this was still valid. Then Sidney Gilchrist Thomas, one of the youngest members present, rose to his feet and stated that he had succeeded in eliminating phosphorus almost completely from the metal in a Bessemer converter; he believed the practical difficulties in the way had been overcome and that Cleveland pig might be made into good steel without any intermediate process. Lowthian Bell's reply included an acknowledgement of Professor Williamson's comments, indicating that he was, indeed, trying the effects of lime but that his tests were not yet at a sufficiently advanced state to enable him to deal with the matter. He ignored the comment by Snelus; his further comment was reported as follows:

> With regard to what Mr. Thomas hoped to do with the Bessemer converter, he was so much interested in freeing iron, and particularly Cleveland iron, from phosphorus, that he should hail as a public benefactor any gentleman who would come forward and do the work more perfectly and more economically than he had been able to effect this object himself.

[13] S.G. Thomas, British Patents 289, 22 January 1878 and 908, 6 March 1878.

[14] The above history is in the main from Thompson, op.cit.,pp. 54–121 and Jeans, op.cit.,pp. 303–306.

[15] I. Lowthian Bell, "On the Separation of Phosphorus from Pig Iron", J.I.S.I., 1878, Pt. 1, pp. 17–34; discussion pp. 34–47.

After some further work, Thomas prepared a paper for presentation at the Autumn Meeting of the Iron and Steel Institute, to be held in Paris. As things turned out, preference was given to the numerous French contributions and Thomas' paper was excluded by pressure of business. Nevertheless, this turned out to be a blessing in disguise. E. Windsor Richards, manager of Bolckow, Vaughan and Company in Middlesbrough, had read the paper and took the opportunity of discussing it with Thomas during a visit to the works at Le Creusot. The outcome was a visit by Windsor Richards, accompanied by J. E. Stead, to Blaenavon. There they saw three experimental heats in the experimental converter, this being an old cupola, lined with crushed dolomite bonded with water glass, with side tuyeres fixed near the bottom of the cupola. Stead later commented that the three heats were all supposedly blown to give complete dephosphorisation; he pointed out, however, with the small amount of metal and with the side blowing, it was difficult to determine the exact moment when the carbon flame disappeared. A standard addition of spiegeleisen was made and the metal cast into ingots. Stead took samples of metal and slag and analysed them back in Middlesbrough. He reported the results as follows: [16]

	Metal	Steel		
		No. 1	No. 2	No. 3
Carbon	3.33	0.32	0.43	0.13
Silicon	2.26	0.01	0.02	0.02
Manganese	0.29	0.86	1.03	0.40
Sulphur	0.16	0.11	0.10	0.11
Phosphorus	1.44	0.41	0.88	0.11
			Slag	
Protoxide of iron 2		7.13	11.83	27.71
Peroxide of iron		3.00	6.85	3.07
Protoxide of manganese		0.60	0.58	0.61
Alumina		3.49	3.30	2.94
Lime		24.86	29.51	31.92
Magnesia		10.00	9.56	5.76
Silica		19.10	27.10	16.95
Sulphur		0.09	0.09	0.11
Phosphoric acid		11.57	11.19	10.90
Weight of slag per 100 metal		20.0	11.5	23.0
Cwt. Fe as oxide		4.6	1.7	7.9

[16] J.E. Stead, Presidential Address, J.I.S.I., 1920, Pt.1, p.57. The significance of these figures will be discussed later.

Windsor Richards took the opportunity of discussing the experimental work with William Menelaus at Dowlais and returned to Middlesbrough convinced of the potential of the work; he put this to his directors and the outcome was a proposal that Thomas and Gilchrist should transfer their experimental work to Middlesbrough, full facilities to be provided by Bolckow, Vaughan and Company. This was agreed by E. P. Martin and Gilchrist moved to the North East, Thomas retaining his London base but commuting regularly. It was about this time that Thomas took out two further patents, the first covering the production of "dead burned" lime, brought about by heating it to temperatures in excess of 1500°C, the other bringing together the salient features of basic refractories covered in his earlier patents.[17]

There were advantages on both sides from the new arrangement. The two cousins were offered support for further work at a plant where Bessemer steel was already being made from haematite iron, in an area where there was an overriding need for a solution to the problem of phosphorus. They were also made a payment to reimburse them for some of their earlier expenses. At the same time, Bolckow, Vaughan and Company exacted an agreement that, should the process be successful, they could produce basic steel free from royalty to a total of 150,000 tons per annum and, moreover, would be due to one fifth of any royalties accruing from the granting of licences to other steelmakers.[18]

The first task was to produce a satisfactory lining for the converter, the only previous production scale trials, those at Dowlais, having demonstrated a very limited life.[19] The initial trials in their new premises were made by grinding dolomite, moulding into bricks and firing them in a normal brick kiln; the bricks shrank up to 50% of their volume and came out from the kiln as misshapen lumps without any plane faces. Mixing the dolomite with clay was equally disastrous because some of the bricks, on cooling down from the firing, spontaneously disintegrated to powder. Some of the misshapen bricks were ground to coarse powder, the product mixed with silicate of soda as binder and the plastic mixture rammed into position; the water in the silicate solution slaked the lime-magnesia powder and the enormous pressure set up actually burst the outer shell of the converter. In order to avoid water, oil was suggested as a binder by Gilchrist; it seems that Edward Riley, working at Brown, Bayley in Sheffield, had the same idea at about

[17] S.G. Thomas, British Patents 3975, 9 October 1878 and 4063, 14 October 1878.

[18] The details of the developments carried out under the aegis of Bolckow, Vaughan and Company are taken in large measure from an excellent recent account by Dr. J.K. Almond entitled "Making Steel from Phosphoric Iron in the Converter", which appears as Chapter 7, pp. 177–230, in "Cleveland Iron and Steel", published by the British Steel Corporation in 1979 to record the centenary of the successful conclusion of the work by Thomas and Gilchrist.

[19] Stead,op.cit., p.59.

the same time.[20] Stead reports that a "member of the staff"[21] advised using tar. Trials with cold tar gave a mixture which was difficult to ram and fell apart when heated. J. R. Skinner was a young chemist who was given the task of assisting Gilchrist and Thomas; fifty years or more later he recalled:[22]

> the almost heartbreaking knocks Mr. Thomas received with so many failures on account of the shrinkage of the bricks and then to find a binder seemed an almost hopeless task ... Mr Thomas was repeatedly upset by these failures and often came for a sit down in the laboratory and a cup of tea in a room set apart for his materials. At the same time as drinking tea he would watch the blowing of a Bessemer heat.

Thomas later came up with the idea that there could be an advantage in mixing the dolomite with hot tar and then ramming the mixture whilst still hot with heated rammers. This was the breakthrough needed; it worked remarkably well. With the refinement that the tar should be boiled before use, to expel any water it might contain, this was the real foundation of the basic steelmaking processes. Thomas subsequently took out patents covering the use of tar-dolomite.[23]

It was now possible to line a converter and hopes ran high once more. Indeed, there was a paragraph in *The Times* which read.[24]

> We understand that the experiments which have been so long carried on by Messrs. Bolckow, Vaughan and Company of Middlesbrough with a view to find means of making steel from Cleveland ore have been crowned with success. The results are said to excced expectations and it is claimed that this discovery will enable Cleveland, which has long produced the cheapest pig iron in the world, to make steel at prices equally beyond the reach of competition. When it is borne in mind how steel has already supplanted iron for rails, for bridges, even for shipbuilding and for all minor purposes, it will be at once seen what an important bearing this discovery is likely to have upon the trade of the district.

In the event, however, this announcement was premature as there were still to be further "heartbreaking knocks" for Mr. Thomas. A number of blows in the 30 hundredweight converter, rammed up with the tar-dolomite mixture, gave erratic results as far as dephosphorisation was concerned; despite the addition of extra lime, the phosphorus did not invariably transfer itself to the slag as was expected. Stead had studied the evidence from

[20] E. Riley, British Patent 4780, 23 November 1878.

[21] Just one more of the innumerable anonymous contributors to the major advances in the steel industry; Le Play, commenting in 1842 on the Sheffield crucible steel trade, had referred to "those true metallurgists – the simple workmen".

[22] Almond, op.cit., quoting a pencil manuscript believed to have been prepared for the 1937 meeting of the Iron and Steel Institute in Middlesborough.

[23] S.G. Thomas, British Patents 1079, 19 March 1879 and 1313, 2 April 1879.

[24] An article in "The Times", 14 March 1879, quoted by Almond, op.cit., p. 189.

Blaenavon, the variable removal of phosphorus and the best result with the highest amount of iron oxide in the slag, and the lowest recovery of carbon and manganese from the spiegeleisen addition, and had come to the conclusion that the phosphorus would not be fully retained by the slag until the carbon had been completely removed. He therefore suggested that the blow should be continued for a reasonable period of time after the carbon flame had dropped – the so-called "afterblow". His suggestion was not immediately accepted, since it was against established Bessemer traditions to over oxidise the metal; not only did it give higher metal losses but it was considered to be bad practice in that steel in a poor condition had to be rectified by excessive additions of spiegeleisen and the quality of the product was considered suspect. Only after a further ten days or so of frustrating lack of consistency in their results did Stead manage to persuade Windsor Richards to try the "afterblow"; it was tried and was found to be completely successful; the steel was low in phosphorus, the slag was fluid and high in phosphorus and the ingots rolled down well, producing excellent rails. Stead, with characteristic modesty, reported many years later:[25]

> I have explained in detail what may be described as the evolution of the overblow – the real dephosphorising part of the process – because statements have been made and published that it was I who invented the afterblow, whereas I only concluded that it was fundamentally necessary and my conclusions were based on Thomas' and Gilchrist's own experiments.

Thomas put in his patent application covering the afterblow on 2nd. April 1879, just two days before Windsor Richards had invited a party of ironmasters and local businessmen to witness the production of steel by the new method. This was completely successful and in 31 minutes a ton and a half of No. 3 iron was tapped from one of the three blast furnaces, transferred to the special converter, blown down, tests taken for malleability and, when satisfactory, spiegeleisen added and the metal cast into ingots. From one of the ingots a bar was drawn down to 20mm x 16mm and when cool was bent into the form of a horse shoe and then flattened double under a steam hammer without fracture. The whole cycle of operations was then repeated quite successfully.

Five weeks later Thomas and Gilchrist presented their nine month old paper, postponed from the Paris meeting, to the Spring Meeting of the Iron and Steel Institute; they also added a postscript covering the work at Eston.[26] According to Thomas,[27] it was the largest meeting of its kind ever held,

[25] Stead, op.cit., p.58.

[26] S.G. Thomas and P.C. Gilchrist:"On the Elimination of Phosphorus",J.I.S.I., 1879, Pt. 1, pp.120–134.

[27] Thompson, op.cit., p.142.

attended by many continental as well as British members; the paper was very well received and

> we became, pro-tem, junior lions!.

At the same meeting, George Snelus revealed that he felt that he had a solution to the phosphorus problem as long ago as 1872 when he produced steel with only 0.018% phosphorus from a Middlesbrough pig iron with 1.5% phosphorus in an experimental 2 hundredweight converter lined with hard burned lime. He continued:[28]

> The results of the experiments I have shown to many private friends ... I have pleasure to place before the meeting what I believe to be the first sample of Bessemer steel made entirely from Cleveland ore by one operation in which the phosphorus has been reduced to a mere trace. A portion has been forged into a chisel, while a rough portion of the same still has part of the lime lining attached to it. With the samples I have placed the original wrapper bearing the date when the sample was made and also my notebook with the original entries of the details of the analysis.

The next paper read at the meeting is also relevant, since it covered the making of basic refractory materials.[29] This covered the practice at Messrs Brown, Bayley and Dixon in Sheffield and the work of Edward Riley, who pioneered the use of petroleum oil as a binder for calcined lime or dolomite prior to pressing them into bricks and then firing them in ordinary kilns.

In the general discussion which followed, Windsor Richards gave some details of the operations in the 6 ton vessel at Eston. Using an iron with 3.5% carbon, 2.9% silicon and 1.5% phosphorus, the vessel would be turned down, 9 hundredweights of lime added and then 6 tons of liquid metal run in and the converter turned up, blowing with air at about 25 lb. pressure. Taking samples every three minutes, it was shown that the silicon quickly reduced over the first nine minutes, then, oxidising at a slower rate, was completely removed in eighteen minutes. The carbon was untouched for three minutes, when it started to burn off at a more or less constant rate and was removed almost completely in eighteen minutes. Removal of the phosphorus did not commence until after 12 minutes, being reduced by about a third at eighteen minutes. Three minutes "afterblow" then reduced the phosphorus to negligible proportions. The final slag contained 21% silica and 11% phosphoric acid. They had found that more rapid removal during the afterblow required the silica to be lower than this, certainly less

[28] G. Snelus: "On the Removal of Phosphorus and Sulphur during the Bessemer and Siemens Martin Processes of Steel Manufacture", J.I.S.I., 1879, Pt. 1, pp. 135–143.

[29] E. Riley: "On a Ready Means of Moulding Lime and Making Lime of Basic Bricks and Linings for Furnaces and Converters", J.I.S.I., 1879, Pt. 1, pp. 144–149.

than 15% and preferably nearer 10%; this could be achieved by a further addition of 10 hundredweights of lime after six minutes, together with 5 hundredweights of iron ore. This shortened both the carbon blow and the afterblow. Windsor Richards reported that the loss in blowing ordinary haematite iron in the normal Bessemer converter[30] was 15%; with Cleveland iron in the basic converter he was coming to the conclusion that a realistic figure was 17%.

The following week, a party of members of the Institute, including people from Le Creusot, Seraing, Terrenoire and St. Chamond, visited Middlesbrough and saw two blows in the 6 ton vessel; due to a strike of the Durham miners, the blast furnaces were damped down and the metal had to be taken from the cupola. Both heats, however, were satisfactory; after lunch they returned to the works to see the ingots produced during the morning rolled into rails, which were reported to have been the equal of any produced from haematite iron. It was also stated that there had been thirteen heats the previous week in the basic converter and that the lining had stood well. It was reported that the foreign visitors took away pieces of the steel produced as souvenirs. The following day they visited the experimental brickworks at Hunwick, near Bishop Auckland, where the basic bricks for the converters were being made.[31]

One major problem still remained, however. Reference back to the paper presented by Thomas and Gilchrist will reveal a table[32] giving phosphorus contents at various times through a number of blows in the 35 hundredweight converter, together with the final phosphorus content of the steel. In all cases, there is a reversal of around 0.04-0.06% phosphorus after the addition of the spiegeleisen. A typical example is Heat 35, where the figures are:

Minutes from start	12	13	14	15.5	16	Steel
% Phosphorus	0.37	0.43	0.08	0.023	0.01	0.05

There was a violent reaction when the spiegeleisen was added and Windsor Richards recommended the addition of a small amount of siliceous haematite pig prior to the addition of the spiegeleisen; whilst this calmed the reaction, it did not prevent phosphorus reversion to any real extent. The practical answer to the problem came from the pioneers of the basic process

[30] This was beginning to be referred to as the "Acid Bessemer Converter" to distinguish it from the new "Basic Bessemer Converter". The terminology of "the basic process" was, it would seem, proposed by Snelus at the Spring Meeting of the Iron and Steel Institute in 1880 (J.I.S.I., 1880 Pt 1, p.97); the process was invariably referred to on the Continent as the "Thomas" process.

[31] J.I.S.I., 1879, Pt 1, p.315.

[32] Thomas and Gilchrist. op.cit., p.133.

in Germany, the Hoerde Company.[33] The practice they developed was to run off the bulk of the slag at the completion of the afterblow before adding the spiegeleisen. In this way, the phosphorus was mechanically removed so that it could not react with the deoxidising addition. The figures given for Charge 69, for instance, indicate this:

	C	Si	Mn	S	P
Iron from the cupola	2.73	0.72	1.39	0.27	1.04
After 12 min (disappearance of C spectrum lines)	0.08	0.14	0.45	0.18	0.74
After 120 secs. overblow	0.07	0.004	0.19	0.12	0.06
After adding Spiegeleisen	0.24	trace	0.40	0.063	0.06

The final slag composition was low in silica and relatively high in phosphorus.

SiO_2	P_2O_5	Fe_2O_3	FeO	Mn_3O_4	CaO	MgO	$CaSO_4$	Al_2O_3
11.10	12.43	1.75	8.90	4.40	50.21	9.84	1.15	trace

This procedure was seen at Hoerde both by Windsor Richards[34] and by Arthur Cooper of Brown, Bayley and Dixon[35] and was adopted by both of them. By the end of 1880, therefore, a satisfactory standard procedure had been evolved by Bolckow, Vaughan and Company, almost entirely by the team headed by Windsor Richards and Stead, working on the principles evaluated by Thomas and Gilchrist. This procedure had also been adopted by a number of continental workers who were themselves evolving modifications to suit their own situations.

There were two problems which had to be solved for fully reliable operation of the Basic Bessemer Process in the country of its invention. The first concerned converter design to avoid the blocking of the throat of the converter by accretions of slag; the second was the consistent provision of blast furnace metal of the composition best suited for the basic process.

The first problem was solved by taking note of what was done elsewhere and adapting British practice to it. Hoerde was the continental pioneer of the new method. They quickly came up against the problem of the blocking

[33] R. Pink:"On a Dephosphorisation of Iron in the Bessemer Converter", J.I.S.I., 1880, Pt. 1, pp. 57–67.

[34] E. Windsor Richards, Presidential Address to the Cleveland Institute of Engineers, 15 Nov 1880, Proceedings Cleveland Inst. Eng., 1880–81, p.36. The author is indebted to Dr. J.K. Almond for providing a copy of this address.

[35] C.B. Holland and A. Cooper:"On the Manufacture of Bessemer Steel and Ingot Iron from Phosphoric Pig", J.I.S.I., 1880, Pt. 1, pp. 79–89.

of the throat of the converter with an accretion of slag, this being much more noticeable in basic operation than in acid operation. Knocking this off mechanically was generally found to cause damage to the surface of the basic brickwork, as well as weakening its adhesion to the shell. Alterations to the width of the throat gave little assistance and it was eventually found that the building of the upper part of the vessel – the part not normally in contact with slag or metal other than on turning the vessel down – with good quality firebrick, rather than with basic brick, greatly eased the problem since the slag was less firmly adherent and came off quite well. This again was something which was adopted in the early British Basic Bessemer works. A more radical method of dealing with the early problems with the converter was adopted at Eston by redesign of the converter and making the throat concentric rather than the eccentric form which had been current since the commercial introduction of Bessemer steelmaking in the early 1860s. In addition, Windsor Richards aimed to make the converters large enough to contain most of the splashing within the body of the vessel. He therefore installed a pair of 15 ton vessels – 50% larger than any then operating in the country – shaped like "a Scotch haystack". The first of these converters came into operation on 18th October 1880 and was pronounced to have "answered the purpose well".[36]

The second problem was never really solved in Britain. The necessary heat to give the rise in temperature during the process so as to provide fluid metal for casting obviously comes from the oxidation of the "impurities" in the blast furnace metal. In the Acid Bessemer Process, as we have already seen, the major heat production comes from the oxidation of the silicon, with useful additions from the removal of the manganese and the carbon. The products of the combustion are silica, manganese oxides and carbon monoxide. The last, being a gas, issues from the mouth of the converter where it burns to carbon dioxide; the other two combine with the iron oxide present to form a slag and this slag is compatible with the ganister lining, which simply becomes coated with a thin layer of the slag and is then protected against any further attack. In the Basic Bessemer Process matters are different. The silicon present in the metal will react in the same way, providing a source of heat and producing silica. The silica, however, in the absence of any lime addition, would attack the lining; in the presence of lime it forms a stable silicate slag, thus tying up a proportion of the lime and preventing it being effective in absorbing the phosphorus pentoxide when this is produced later in the blow. Indeed, it has already been pointed out that for the dephosphorisation to be effective the silica content of the slag must be held below 15% and preferably significantly lower than this. This implied that a silicon content normally considered to be necessary in the

[36] E. Windsor Richards, op.cit., pp. 37–38.

Acid converter would require a very high slag bulk, which absorbed a considerable amount of heat itself and also gave rise to problems in the converter simply by virtue of its bulk. The alternative was to produce a blast furnace metal low in silicon; this, in general, implied a grey cast iron composition, and also would be difficult to produce with a low sulphur content. Moreover, such a blast furnace metal would not provide the necessary heat in the blow. As a compensation, however, the oxidation of phosphorus provides almost as much heat as the oxidation of silicon, as the following table, showing the heat evolved, in calories, for the oxidation of one kilogram of the various elements present, indicates:[37]

C	Si	Mn	S	P	Fe to FeO	Fe to Fe_2O_3
2,450	6,414	1,654	2,166	5,890	1,232	1,611

The oxidation of the phosphorus, of course, occurs late in the process, giving a boost of heat at the time when it is most required. The real requirement for the Basic Bessemer Process, therefore, was an iron relatively low in silicon and high in phosphorus, preferably with a reasonable manganese content. The bulk of the phosphoric ores in Britain were not as high in phosphorus as the French and German ores and were richer in silicate minerals, with the result that they were really only marginally suited to the purpose; the Cleveland ores, in particular, were high in silicate minerals and it was difficult to produce a really low silicon content from the blast furnace. Windsor Richards reported towards the end of 1880:[38]

> It is not an easy matter to accomplish in a blast furnace the manufacture of a metal low in silicon and, at the same time, low in sulphur. It would, no doubt, very much help to keep the sulphur low if manganese were added in the blast furnace but manganese is a costly metal. At present we have succeeded in making a mottled Cleveland iron with 1.0 per cent silicon and 0.16 sulphur, and white iron, with 0.5 per cent silicon and 0.25 sulphur, which, taken direct from the blast furnace, have both made excellent steel. But we have another method of operating which relieves us from the necessity of making a particular quality of Cleveland pig iron. We call this mode of working the "transfer" system[39] because we transfer the metal from the acid to the basic converter. Thc transfer system enables us to take any grey iron direct from the blast furnace without any consideration as to the percentage of sulphur, which is always low in grey iron. This grey metal is

[37] F.W. Harbord and J.W. Hall:"The Metallurgy of Steel", 3rd. ed., London, 1907, p.104.

[38] E. Windsor Richards, op.cit., pp. 38–39.

[39] This was generally referred to as the "Duplex Process". Whilst it solved certain problems, it generally proved inconvenient and expensive. It is not clear whether it was of any lasting significance at Eston; clearly a plant with existing acid converters alongside the basic vessels, as here, was suitably placed to operate such a process.

> poured into a converter with a siliceous lining and desiliconised, when, after say 12 to 15 minutes' blowing in the ordinary manner, it is poured out of the converter into the ladle, and poured again from that ladle into a converter lined with dolomite, taking care that the highly siliceous slag is prevented from entering the basic lined converter. Then in the second converter it is only necessary to add sufficient lime for the absorption of the phosphorus of the metal and blowing then need not occupy more time than is necessary for the elimination of the phosphorus, say about three minutes. This mode of operation will, no doubt, give the basic lining and bottom a much longer life, but we have not yet been long enough at work to obtain the necessary experience to determine which is the better system of working, but both are good and effective and have given excellent results.

Another point which needs to be underlined is that the phosphorus content of the Cleveland ore was only moderate as far as basic operation was concerned; typical phosphorus values in the blast furnace metal were of the order of 1.6 to 1.8%. There was, however, a ready source of high phosphorus material to hand, in the piles of "puddling cinder" which had been accumulating as waste in the area for a quarter of a century. By judicious incorporation of such material in the blast furnace burden it was found to be of advantage to produce a liquid metal with up to 3% phosphorus which eased the working as far as temperature was concerned and, at the same time, removed the disfiguring heaps of waste around the area. Obviously, this was a "wasting asset" and it eventually ran out.

The difficulties under which the Cleveland steelmakers operated were underlined many years later in a German review of the first thirty years of basic steelmaking. This reiterated the disadvantages in Cleveland as compared with Germany and in particular the problems associated with the necessity of using a high silicon and relatively low phosphorus iron, if sulphur was to be maintained at a reasonably low level. This led to the more rapid erosion of the converter lining, the use of larger quantities of lime in order to obtain a sufficiently basic slag with the attendant difficulties of such an increased slag bulk, and higher losses of iron due to increased afterblow times. The account is useful, however, in that it also details procedures used subsequently in Cleveland to offset these disadvantages. These include the addition of manganiferous ores to the iron ore in the blast furnace, the manganese serving to reduce the sulphur content and thus make possible the production of a low silicon, low sulphur iron. A further improvement was to add just sufficient lime, often admixed with some iron oxide, to the converter before running in the metal and then burning off the silicon, so producing a fluid slag, not sufficiently acid to cut the lining but capable of absorbing the silica produced. This slag was then run off, the extra lime required then added, and the metal blown to remove the phosphorus under almost ideal conditions. This can be considered to be the previously suggested "duplex" process adapted to allow the two stages to be carried out in the one vessel.[40]

Despite these earlier drawbacks, however, confidence in the new method was such that two further 15 ton basic converters were installed at Eston in 1881 and a third pair in 1882. It should be noted that these were all on elevated staging so that the ingot casting could be done more or less at ground level, on the pattern which had been established for Bessemer steelmaking in America under the influence of Holley. With this equipment, together with the earlier acid converters still operating on imported haematite ores, Bolckow, Vaughan and Company could justifiably claim in 1882 to be the largest steelmaker in the world and three fifths of the output was derived from the local Cleveland ore, the total capacity being approximately 225,000 tons of finished product per annum.[41]

In this way, the belief expressed by the young Sidney Gilchrist Thomas at the Iron and Steel Institute meeting in 1878, that Cleveland pig might be made into good steel, was triumphantly vindicated. The advantage conferred by this technological breakthrough, however, was of much more benefit to the Continental steelmakers than it was in Britain, some 38% of the steel made on the Continent in the ten year period, 1881 - 1890, being by the basic process, against only 10.6% of British steel; for the year 1890 the corresponding figures were 64% as against only 14%, as will be observed from the tonnage figures quoted in Table IV:[42]

There is a postscript to be added to the story of the solution of the phosphorus problem. Lowthian Bell, as early as 1872, had suggested, as a bonus to the elimination of this element from steel, that the phosphorus from the current use of the Cleveland ironstone could be worth of £250,000 per annum as fertiliser.[43] Strangely enough, while Thomas fully realised the potential, very few steelmakers in Britain appeared to be interested in this aspect. In Germany, however, as early as 1882, there is a report of trials being made in Westphalia.[44] A slag containing 19.33% P_2O_5 was treated with differing quantities of sulphuric acid (the standard treatment of phosphate rock to render the phosphorus water soluble) with little success. It was found, however, that about 60% of the phosphate present was soluble in ammonium citrate solution, this being an indication of the proportion which would be assimilated by plants from soil treated with the slag, and it was concluded that the use of the finely powdered slag without any prior

[40] J. Massenez, Stahl und Eisen, Vol.xxix, pp. 1465–1477, abstracted in J.I.S.I., 1909, Pt 2, pp. 452–454 under the title:"History of the Basic Process in Germany".

[41] Almond, op.cit., pp. 205–206.

[42] "Thomas and Gilchrist – Bolckow and Vaughan – 1879–1929", a commemorative booklet issued by Bolckow, Vaughan and Company, September 1929, pp. 22–23.

[43] I. Lowthian Bell;"The Chemical Phenomena of Iron Smelting", J.I.S.I., 1872, p. 85.

[44] "Utilisation of Slag from the Basic Steel Process in Agriculture", abstract in J.I.S.I., 1882, Pt 2, pp. 787–788.

TABLE IV THE RISE OF BASIC STEEL PRODUCTION ON THE CONTINENT AND IN BRITAIN FROM 1878 TO 1890

	MAKE OF STEEL ON THE CONTINENT			MAKE OF STEEL IN BRITAIN		
	Basic Steel (tons)	Total Steel (tons)	% Basic Steel	Basic Steel (tons)	Total Steel (tons)	% Basic Steel
1878	Nil	1,349,000	Nil	20	983,000	0.002
1879	50	1,433,050	0.003	1,150	1,009,511	0.11
1880	40,000	1,867,000	2.14	10,000	1,295,382	0.77
1881	289,880	2,215,880	13.08	46,120	1,779,629	2.59
1882	340,636	2,515,636	13.54	109,364	2,106,649	5.20
1883	511,993	2,509,993	20.39	129,380	2,008,880	6.44
1884	685,000	2,323,000	29.48	179,000	1,774,926	10.08
1885	799,610	2,280,610	35.06	145,707	1,888,095	7.72
1886	1,055,165	2,465,165	42.80	258,466	2,264,670	11.41
1887	1,589,025	3,070,025	51.76	435,046	3,045,507	14.28
1888	1,544,640	3,277,640	47.12	408,594	3,305,536	12.36
1889	1,789,633	2,893,633	61.84	493,919	3,569,960	13.84
1890	2,021,903	3,161,903	63.95	503,400	3,579,043	14.06
TOTAL	**10,667,535**	**31,562,535**	**33.80**	**2,720,166**	**28,610,788**	**9.51**

Source: "Thomas and Gilchrist — Bolckow and Vaughan — 1879 - 1929", a commemorative booklet issued by Bolckow, Vaughan and Company, September 1929, pp. 22 - 23. (The percentage figures have been added by the author.)

chemical treatment would be reasonable. A suggestion was made that, to render the sulphides innocuous, the powder should be mixed with stable manure and spread in the autumn, so that oxidation of the sulphides would be complete by the time the seed was sown. Small scale trials with summer rye and vetches on a poor sandy soil showed equally good results from treatment with the slag powder as with a mixture of potash, nitrate of soda and guano-superphosphate or with superphosphate prepared from the slag.

Certainly, by 1885, the use of powdered basic slag was widespread in Germany and it gives some idea of the difference in attitude in Britain and Germany when it is realised that from 1885 to 1895 the bulk of the output of slag in Cleveland was being shipped to Germany for grinding and use there, the remainder actually being dumped as waste into the North Sea! There was a report indicating that "a new industry" was being introduced in Staffordshire in 1886, in connection with the Basic Bessemer operations at the Staffordshire Steel and Iron Company at Bilston.[45] A "slag-house", some 140 feet by 50 feet, had been built for the accommodation of the grinding equipment and for storage. The slag was first fed into crushers, which reduced the raw slag to three sixteenths of an inch pieces and the major pieces of steel trapped in the slag were extracted. The material then went on to a runner mill followed by a magnetic separator to take out the remaining metal. The last stage completely pulverised the slag, "since the value of the material as a fertiliser depends on its impalpability".

From the rough slag to the finished powder the process was continuous and the feed automatic. The ground slag contained about 40% lime and from 15 to 20% phosphorus pentoxide. Some slag was ground in the North East from 1887 onwards; the slag was roughly crushed and magnetically separated to remove the bits of steel entrained in it and then ground in edge runner mills. Considerable dust was produced and, indeed, health problems, known as "slag cough", caused the Inspector of Factories to insist that the workmen should wear respirators. The powdered slag here was said to contain 16-20% phosphoric acid, together with around 50% lime and was sold at 27s. per ton.[46] It was to obviate the dust problem that the European works installed ball mills for the grinding of the slag[47] and this practice was also taken up in America at the one Basic Bessemer plant to be established there in the late 1880s.[48]

[45] "Grinding Basic Slag", Iron and Coal Trades Review, Vol. xxxiv, p. 479.

[46] Almond, op.cit., pp. 224–225. By 1913, some 5 million tons of basic slag were ground as fertiliser per annum, 3.5 million tons of this in Germany and Luxemburg, Britain contibuting only 300,000 tons per annum. This gave credit of around five shillings per ingot ton.

[47] H. Wedding:"Progress of German Practice in Mattallurgy", J.I.S.I., 1890, Pt. 2, p. 535.

[48] J. Hartshorne:"The Basic Bessemer Plant of the Pottstown Iron Company". Trans. Amer. Inst. Mining Engs., Vol xxi (1892–3), pp. 743–744.

12 The Basic Bessemer or Thomas Process

Sidney Gilchrist Thomas was careful to take out patents covering the basic steelmaking process in Germany, France and America. Such was the appreciation of the great potential of the new process that, as Windsor Richards puts it:[1]

> The news of this success spread far and wide and Middlesbrough was soon besieged by the combined forces of Belgium, France, Prussia, Austria and America. ... Directly the meeting [of the Iron and Steel Institute in London] was over, Middlesbrough was again besieged by a large array of Continental metallurgists and a few hundredweights of samples of basic bricks, molten metal used and steel produced were taken away for searching analysis at home. Our Continental friends were of an inquisitive turn of mind and, like many other practical men who saw the process in operation, only believed in what they saw with their own eyes and felt with their own hands, and were not quite sure even then!

Thomas was by no means so affected by patent problems as had been Bessemer or Siemens. His approach was more reasonable; he accepted that both Snelus and Riley had some claims with regard to the production of suitable refractory linings and, to safeguard his own position and also to prevent any tedious litigation, he negotiated the formation of "The Dephosphorising and Basic Patents Company Limited"; the original members were Sidney Gilchrist Thomas, Percy Carlisle Gilchrist, George J. Snelus and Edward Riley, together with Bolckow, Vaughan and Company Limited.[2] The only real problem arose in Germany. The original rights for Germany had been negotiated with Thomas by G. Pastor, the representative of the Rheinische Stahlwerke at Ruhrort, much to the disappointment of the Hoerde works; these two organisations eventually joined in the patent rights for Germany and Luxembourg whilst Hoerde took up the rights for Austria

[1] E. Windsor Richards, Presidential Address to the Cleveland Institute of Engineers, 15 Nov 1880, Proc. Cleveland Inst. Eng., 1880-81, pp. 34-35.

[2] "Thomas and Gilchrist - Bolckow and Vaughan - 1879-1929", a commemorative booklet issued by Bolckow, Vaughan and Company, September 1929, pp. 11-12.

and Hungary.[3] Krupp of Essen and the Bochumer Verein, however, challenged the validity of Thomas' German patents, mainly on the grounds of irregularities in the procuring of the patents. In 1880, however, two cases in this connection were tried in Berlin, before the head of the Imperial Patent Office, with Professor Wedding, Bruno Kerl, Dr. Hoffman, Dr. Siemens and others as expert technical witnesses. In both cases the court held that the validity of the patents had been thoroughly established and considered substantial novelty and the great value of the invention to have been proved and to be such as to override any minor technicalities in the procurement.[4]

It is generally accepted that the parallel development at the two German Works of Rheinische Stalwerke and Hoerde led to the first basic steelmaking outside Britain. It is quite clear that these two firms organised their activities so that the first blows were made on the same date, 22nd September 1879, and that the first rails in the basic steel were rolled on the same day at Hoerde, eminently satisfactorily.[5] There is, however, a strong claim that the first basic steel of the Continent was made at Kladno, near Prague, Thomas having granted a licence direct to the Prager Eisen Industrie Gesellschaft and, indeed, having supervised operations there between 11 and 13 May 1879, since he was anxious to establish his claim for patent rights in the Austro-Hungarian empire. In addition, the manager of the works at Witkowitz (Vitkovice), Paul Kupelweiser, visited Middlesbrough in the company of Gilchrist in June 1879, to examine the process and to study the manufacture of the necessary refractory materials. On his return, he negotiated a patent with Massenez of Hoerde and successful trials were made at Witkowitz in August 1879, again prior to the German work at Hoerde and Ruhrort in September.[6] Hoerde, in fact, was more or less used as an experimental steelmaking station for the whole of Germany and the activities there became of interest throughout Europe. It was there that the idea of running off the bulk of the slag before the addition of the spiegeleisen was developed; this was noted by Windsor Richards on his visit to Hoerde on 23 August 1880.[7] It was also there where the lining of the throat of the converter with firebrick, to obviate the build up of basic slag in that area, was developed; this was noted

[3] J. Massenez, Stahl und Eisen, Vol. xxix, pp. 1465-1477, abstracted in J.I.S.I., 1909, Pt. 2, p.452.

[4] W. T. Jeans: "Creators of the Age of Steel", 2nd. ed., London, 1885, p. 314. It is significant that Thomas taught himself German so as to be able to participate in these court cases. This was typical of the active manner in which he prosecuted the extension of the basic processes. He literally wore himself out by constant travel and involvement and his constitution, never robust with a proneness to chest infections, could not stand the strain. He died in Paris on 1st February 1885, two months short of this 35th Birthday. Thereafter the affairs of the Patents Company were handled by Percy Carlisle Gilchrist.

[5] J. Massenez, loc. cit., pp. 452-453.

[6] I. Krulis-Randa: "The Introduction of the British Metallurgical Techniques into Austria", J.I.S.I., Vol. 194 (1960), pp. 43-45.

[7] E. Windsor Richards, loc. cit., p. 36.

during a visit paid by representatives of Brown, Bayley and Dixon to Hoerde as early as November 1879.[8]

A report of the early operations at Hoerde is available.[9] This pointed out that the cost of producing a good grey foundry iron at Hoerde was almost as high as that for Bessemer pig and that the aim at Hoerde was to use the basic process for the production "of a reliable and cheap steel from the worst classes of pig iron" and this they had achieved. Since good reliable rails had been produced by the acid process with 0.08% phosphorus, it was considered that rail steel by the new process with 0.06% or so of phosphorus would be satisfactory but that for axles, tyres, plates and wire a lower phosphorus content would be desirable. On 22 September 1879 the first heat was blown from a 3 ton converter lined with dolomite; the phosphorus was only reduced to 0.12%, but a rail rolled from one of the ingots "stood exceedingly well under the tup" – this was the first rail manufactured under the Thomas patent in Germany. By the third blow, a phosphorus content of 0.06% was obtained and, at the time this paper was submitted".[10]

> with but one exception, no difficulty has been found in keeping the phosphorus within the limits of 0.08 per cent to 0.04 per cent.

Initial practice was based on that at Eston, with pig iron containing around 1.5% silicon. Trials with various types of pig iron – the metal remelted in a cupola – had indicated the benefit of a lower silicon content, of around 0.75%, and eventually cheap white forge pig iron was used, with about 1% manganese to assist in removal of sulphur and with phosphorus up to 2%. Such a metal had been shown to give a fluid slag and sufficient heat in the product for the steel to be "cast ascensionally" – that is, bottom poured or uphill cast in clusters.[11] The lime was shown to be best used fresh and, if possible, added hot to the empty vessel before running in the metal; this was used in the proportion of 15% minimum of the metal weight and up to 20% if the metal contained over 2% phosphorus. The method of operation was given as follows:

[8] C. B. Holland and A. Cooper: "On the Manufacture of Bessemer Steel and Ingot Iron from Phosphoric Pig", J.I.S.I., 1880, Pt. 1, p.80.

[9] R. Pink: "On the Dephosphorisation of Iron in the Bessemer Converter", J.I.S.I., 1880, Pt. 1, pp. 57-67.

[10] Certainly a period of at least four months.

[11] There are three British Patents, taken out by W. and J. Ireland, on communications from R. Pink, covering "uphill casting of ingots in groups around a central git". No. 2548, 27 September 1871, gave provisional protection only; No. 429, 10 February 1872 was a substantive cover for casting in groups of six ingots or more; No. 263, 23 January 1873 covered the casting in groups of solid steel ingots, using closed or "bottle top" moulds for the making of railway tyres. Pink clearly was a pioneer in this method of casting.

> After heating up the converter [with coke] the lime, mixed with a little small coal is added and, after blowing through for a few minutes, the metal is run in. In the three ton converters the time occupied in blowing a charge of from 3 tons 10 cwts to 4 tons up to the vanishing point of the carbon lines of the spectrum varies from 9 to 13 minutes and the afterblow from 115 to 200 seconds. A test piece is taken shortly before the charge is considered as finished; this, when forged, is cooled in water and then broken, after which the blowing is continued as considered necessary. A second test is now rarely taken, as this first is so timed that a further half or three-quarter turn up of the converter suffices to finish the charge. The slag is then run off and the spiegeleisen added. This running off of the slag prevents in great measure the reduction of phosphorus from the slag by means of the final addition.

Details of the working of three charges are given in Table V.[12]

It was pointed out that there were no problems in producing harder steel, although most of the requirements to date had been for the medium strength materials or for soft steel for plates and wire and it was commented that for these it was quite usual to find reductions of area of 70-75% with maximum stresses of around 23 to 25 tons/sq.in.

The lining of the top part of the converter vessel with good quality firebrick, referred to in Chapter 11, was developed at Hoerde and is fully described in this paper. As regards furnace life, the vessels themselves, with patching, were found to last from 90 to 130 charges, but the removable bottoms had a life of only nine charges on average.

It is interesting to note that Herr Pink clearly indicates that the plant of Rheinische Stahlwerke at Ruhrort was working the process even more successfully that he was at Hoerde, that there were eight other German works which had been arranged for working basic converters, together with Maximilianhutte in Bavaria and Witkowitz, Teplitz and Kladno in Austria. When it is remembered that this was within a year of the first successful trials at Eston, the enormous impact made by the "Thomas Process" will be appreciated.

In England, the only firm working the process, other than Bolckow, Vaughan and Company, was Brown, Bayley and Dixon in Sheffield.[13] Here extensive trials had been carried out with various types of pig iron and note that been taken of the processes both at Eston and Hoerde, adopting, as has been outlined, the firebrick lining of the upper part of the vessel to avoid the accretions of the basic slag blocking the throat. They eventually standardised on the following metal composition:

C	Si	Mn	S	P
4.0	0.5	1.0	0.20	1.4

12 Some detail from Charge 69 has already been quoted in connection with the question of rephosphorisation in Chapter 11.

13 C. B. Holland and A. Cooper: "On the Manufacture of Bessemer Steel and Ingot Iron from Phosphoric Pig", J.I.S.I., 1880, Pt 1, pp. 79-86.

TABLE V DETAILS OF EARLY BASIC BESSEMER OPERATIONS AT HOERDE, GERMANY

	Charge 67	Charge 68	Charge 69
Raw Material:			
Foundry No. 3 Pg, lb.	2,400	2,400	1,800
White Forge Pig, lb.	3,000	3,000	3,600
Steel Scrap, lb.	1,000	1,200	1,200
Spiegeleisen, lb.	420	420	420
Total, lb.	**6,820**	**7,020**	**7,020**
Ingots Produced, lb.	**6,074**	**6,250**	**6,302**
% Yield	89.06	89.03	89.77
Final Slags:			
Silica	9.50	13.81	11.10
Phosphoric Acid	9.76	7.38	12.43
Iron Sesquioxide	1.56	1.30	1.75
Iron Protoxide	7.24	6.70	8.90
Mangano-manganic oxide	6.16	6.20	4.40
Lime	59.35	57.85	50.21
Magnesia	5.01	6.93	9.84
Sulphate of lime	0.87	0.22	1.15
Alumina	tr	tr	tr
Tensile tests on bars:			
Maximum stress, tons/sq.in.	31.8	34.0	34.5
Elongation, %	20.6	22.0	19.0
Reduction of area, %	44.8	47.0	42.0

Metal Anaylses:	C	Si	Mn	S	P
Charge 67					
Cupola Metal	2.58	1.08	1.35	0.22	1.04
End of "blow", 11'45"	0.08	0.09	0.39	0.19	0.82
after 100" overblow	0.06	0.007	0.39	0.15	0.08
After 115" overblow	0.04	0.005	0.37	0.14	0.05
Final Steel	**0.28**	**0.002**	**0.46**	**0.067**	**0.06**
Charge 68					
Cupola Metal	2.82	0.45	1.04	0.16	0.96
End of "blow", 14'00"	0.09	0.02	0.40	0.16	0.70
After 110" overblow	0.085	0.003	0.33	0.15	0.09
After 125" overblow	0.08	tr.	0.29	0.09	0.05
Final Steel	**0.26**	**tr.**	**0.31**	**0.055**	**0.06**
Charge 69					
Cupola Metal	2.73	0.72	1.39	0.27	1.04
End of "blow", 12'00"	0.08	0.14	0.45	0.18	0.74
After 125" overblow	0.07	0.004	0.19	0.12	0.06
Final Steel	**0.24**	**tr.**	**0.40**	**0.063**	**0.06**

(End of "blow" = disappearance of carbon lines)

Source: R. Pink: "On the Dephosphorisation of Iron in the Bessemer Converter", J.I.S.I. 1880, Pt. 1, pp. 57 – 67.

With a vessel which gave a mean output of 7.25 tons per blow, they had been able to produce 3,380 tons in an eight week period, mainly on what was termed ingot iron (soft steel). Attempts to produce harder steels were not always successful but some progress was being made in this direction. As regards vessel life, one vessel had to have the fireclay nose replaced after the 50th blow, equivalent to 360 tons steel; after the 87th blow, equivalent to a total of 630 tons, the front or blowing side of the lining was largely replaced and a third fireclay nose put in, whereupon a further 37 blows were made in the same vessel, totalling 124 blows, equivalent to 900 tons of steel from one vessel before complete relining. Bottoms only averaged eight blows.

The Autumn Meeting of the Iron and Steel Institute in 1880 was held in Dusseldorf and it obviously became the forum for discussion of the state of the art as regards the basic Bessemer process. There was a further contribution on activities at Hoerde[14] in which it was pointed out that five German works had adopted the Thomas process: the Rhenish Steelworks at Ruhrort, the Hoerde Iron and Mining Company, the Bochum Company, the Rothe Erde Company at Aachen and the Steel Works at Kaiserslautern. The first two were also making their own basic bricks. There were enormous masses of good quality dolomite available near Hoerde, as well as plentiful supplies of limestones which were virtually free from silica. The paper gives the following details of five charges blown consecutively on 30 June 1880; to follow the progress of a typical operation, analyses of metal and slag samples taken at intervals during the second of these heats may be found in Table VI.

HEAT NUMBER	1	2	3	4	5
Hoerde White Iron, kg	902	870	765	775	815
Hoerde Light Grey Iron, kg	804	780	790	780	815
Metz White Iron, kg	962	597	620	880	605
Wendel White Iron, kg	917	655	820	900	885
Steel Scrap, kg	...	595	520	500	510
Spiegeleisen, kg	260	260	260	260	260
Ferromanganese, kg	60	60	60	60	60
Total Metal, kg	**3,905**	**3,817**	**3,835**	**4,155**	**3,950**
Lime, kg	700	700	700	650	650
Slag, kg	795	595	745	905	755

[14] J. Massenez: "Dephosphorising in the Converter", J.I.S.I., 1880, Pt. 2, pp. 475-487.

HEAT NUMBER	1	2	3	4	5
Steel, kg	**3,285**	**3,070**	**3,354**	**3,271**	**3,029**
Yield, % (Mean 81.52%)	84.12	80.42	87.46	78.72	76.68
Analysis of Steel:					
Carbon	0.21	0.24	0.20	0.25	0.27
Silicon	tr	tr	tr	tr	tr
Manganese	0.50	0.46	0.51	0.58	0.56
Sulphur	0.075	0.090	0.070	0.065	0.067
Phosphorus	0.027	0.020	0.027	0.025	0.009
Time after Start of Blow:					
To fall of carbon lines:	14′30"	9′30"	9′00"	8′30"	9′30"
Afterblow: 1st Test	17′30"	10′45"	12′00"	11′10"	12′20"
2nd Test	17′48"	11′45"	....	11′55"	12′05"
3rd Test	....	11′55"	....	....	....
Spiegeleisen addition	17′54"	12′05"	12′08"	11′30"	12′43"

On the basis of this further work at Hoerde, the following features were considered to be of importance:

1. Hot melting in the cupola.

2. The use of fresh, good, calcined lime, free from water and used as hot as practicable.

3. Quick manipulation of the process itself, the quicker the charges were produced the better the product. The use of larger tuyeres and greater wind pressure was desirable.

During the discussion, the author pointed out that whilst the main demand was for relatively soft steel, there was no problem in making the harder varieties and that within a few months of starting the process they had produced steel for tyres and axles quite satisfactorily.

At the Autumn meeting of the Institute a year later, in the introduction to a paper reporting on the work carried out at the Witkowitz Works in Austria, it was recorded that the following thirty Continental works had taken out licences under the Thomas patents and the greater number of them were already at work, whilst others were adapting old plant or building new works for its introduction:[15]

> Angleur Steelworks, Ars-sur Moselle Ironworks, Athus Ironworks, Bohrbach Company, Bochumer Verein, Chatillon et Commentry, Denain Works, Dietrich of Niederbron, Dillingen Company, Dortmund Works, Gutehoffnungshutte, Hoerde Verein, Ilsede Works, Kladno Works. Longwy Works,

[15] P. Kupelweiser: "On the Basic Bessemer Process", J.I.S.I. 1881, Pt. 2, pp. 397-406.

TABLE VI ANALYSIS OF METAL AND SLAG SAMPLES TAKEN DURING A BASIC BESSEMER BLOW

Samples of metal and slag were taken at intervals during the course of the second blow on 30 June 1880 at the Hoerde Works in Germany.
Chemical analysis of these samples provided the following information:

	Metal					Slag								
	C	Si	Mn	S	P	SiO_2	P_2O_5	Al_2O_3	Fe_2O_3	FeO	MnO	CaO	MgO	CaS
Metal from Cupola	2.83	0.66	0.52	0.29	1.28									
After 2 mins. blow	2.72	tr	0.43	0.26	1.32	18.47	1.09	0.42	0.80	4.21	1.91	67.81	4.94	0.47
After 4 mins. blow	2.48	tr	0.42	0.27	1.29	18.05	1.81	0.38	0.60	3.69	2.80	68.02	4.37	0.63
After 6 mins. blow	1.70	tr	....	....	1.25	17.16	2.37	0.82	1.18	3.24	2.43	67.19	4.60	0.76
After 8 mins. blow	0.70	tr	0.30	0.29	1.22	21.25	3.46	0.72	1.81	2.90	2.98	61.82	4.64	0.90
After 9 mins. 15 secs. blow	0.16	tr	0.25	0.33	1.18	14.85	5.55	0.49	2.54	5.42	2.08	64.00	3.66	0.92
After 10 mins. 45secs. blow	0.15	tr	0.19	0.27	0.48	11.32	12.41	0.39	0.57	4.45	1.96	63.32	4.37	0.83
After 11 mins. 45 secs. blow	0.10	tr	0.17	0.20	0.07	10.90	13.68	1.83	2.95	11.21	2.15	51.01	5.29	1.65
After 11 mins. 55 secs. blow	0.09	tr	0.12	0.16	0.04	9.85	12.88	1.68	4.94	12.27	2.07	49.55	5.08	1.98
After 12 mins. 5 secs. blow (finished metal)	0.24	tr	0.46	0.09	0.02	9.72	10.88	2.21	3.81	8.58	5.93	49.75	6.42	2.26
Spiegeleisen added	4.01	0.67	11.25	tr	0.21									

Source: J. Massenez: "Dephosphorising in the Converter", J.I.S.I., 1880, Pt. 2, pp. 475 –487.

> Maximilanshutte, Societe de Montataire, Ougree Ironworks, Phoenix Works of Ruhrort, Rothe Erde Works of Aachen, Ruhrort Steelworks, Schneider of le Creusot, Schneider and de Wendel of Joeuf, Stumm of Neunkirchen, St. Chamond Company, Societe de Nord et l'Est, Teplitz Works, Warsaw Works, De Wendel, Witkowitz Works.

This impressive list tends to confirm the somewhat surprising statement made by Thomas a year earlier[16] that, only just over a year after it was substantiated on a full production basis in Cleveland, his process was in daily operation at ten of the leading continental works and that there were then in course of actual erection, under his licence, new works with an aggregate capacity of 600,000 tons per annum.

The manager of the Austrian works at Witkowitz, Paul Kupelweiser, stated that the progress of the basic process over the past year had been one of regular development without any remarkable novelty. He was still using the refractory materials proposed by Thomas. Using a dolomite with very little silica, however, he had been able to provide the works with a superior brick which was used for most of the lining. Ground brick mixed with 10% tar was used elsewhere and also for repairs. The changeable Holley bottom was greatly preferable; bottoms could be produced by ramming round series of iron pins or by inserting fireclay tuyeres and ramming or bricking basic material round them. Magnesite tuyeres were being made for trial purposes. An important point was made that, in the basic works, it was essential to have more clear space around the vessels for charging the lime and disposing of the slag; it was important, therefore, in the construction of new works or the adaption of old ones to arrange the casting facilities in separate pits, preferably at ground level, whence the ladle could be transferred for casting. As regards the economics of the process, the cost of conversion was always likely to be higher in the basic process than in the acid, but this was offset by the lower cost of the metal charged to the furnace. On the other hand, Kupelweiser was quite clear that:

> ... the basic process, as regards the quality of its products, is not only completely equal to the acid process but even, in my opinion, superior to the latter.

Basic Bessemer steel had been produced at Witkowitz for rails, tyres, railway sleepers, boiler plates, ship plates, wire, bolts, rolled tubes and forgings of every description. It was pointed out that the scrap from the purified metal was an excellent material for use in the Siemens-Martin process. Here again, the same comment is made that the demand for soft steel, or ingot iron, had recently been so great, and the basic product had been found to be so satisfactory, that it had formed the bulk of the basic material so far produced. So much so had this been the case that rail steel had originally been confined to the acid furnaces on this

[16] S.G. Thomas, letter to The Times, 2 September 1880.

type of production. This had given rise to the suggestion that the basic process was only suited to the production of soft steel and Kupelweiser made a point of refuting this; they had produced rail steel of requisite hardness and purity with perfect facility and when the complete installation of the four converters specially designed to work the Thomas process[17] was available:

> it is intended to make none but basic steel.

To show the capabilities of the basic process he provided the following table:

	1. Moderately hard Rail steel	2. Harder "Ingot Iron" for plates, angles & rivets	3. Soft "Ingot Iron" for wire & stampings
Max. Stress kg/sq.mm.	58 to 63	45 to 50	36 to 39
Reduction in area, %	51 to 37	64 to 55	77 to 72
Elongation, %	20	25 to 20	37 to 33
Chemical Composition:			
Carbon	0.45	0.19	0.06
Silicon	tr	tr	tr
Manganese	....	0.34	0.30
Sulphur	0.06	0.04	0.03
Phosphorus	0.04	0.04	0.02
Chemical Composition of Pig Used:			
Silicon	0.54	0.11	0.62
Manganese	1.00	1.16	1.38
Sulphur	0.23	0.09	0.08
Phosphorus	1.95	3.46	2.00

What seems to have been an interesting innovation was employed at Witkowitz: the basic slag was recharged to the blast furnace in order to raise the phosphorus content of the metal and thus the heat evolved during the afterblow, providing a capability of incorporation of some steel scrap and yet have steel which would cast satisfactorily. The composition of the Basic

[17] These converters were of the concentric type, as favoured by Windsor Richards, with moveable cowls over them to take the fumes and flame outside the building. They were built on stages, with pits either side to accomodate slag vessels and with long casting pits at right angles at ground level, in accordance with the strategy of giving plenty of working space.

Bessemer slag was reported to render it an excellent flux for the gangue in the ores which were being smelted in the blast furnace.

A final comment highlighted one of the difficulties encountered at Witkowitz:

> I must here mention a circumstance which renders the production of the softest material somewhat difficult and comparatively expensive. It is the unquiet casting of the softest qualities of ingot metal. The lively evolution of gas from such soft metal, while cooling in the mould, causes at present, even with the most careful pouring, considerable loss in bad heads. This scrap, although it forms an excellent material for the Siemens-Martin process, nevertheless considerably enhances the loss in the manufacture of the softest qualities. Lately,we have succeeded in overcoming this drawback to a considerable degree but our experiments have not yet proceeded far enough to allow me to place the results before you at present.[18]

About the time that this paper was presented, a second Basic Bessemer plant was being installed in Cleveland. This was the North Eastern Steel Company, incorporated 9 July 1881, with Thomas and Gilchrist among the founders.[19] Shortly afterwards Arthur Cooper, who had been involved with basic Bessemer production at Brown, Bayley and Dixon in Sheffield, was persuaded to become manager of the new enterprise. He drew up the plans for the entire installation of steelmaking plant and rolling mills; it was designed without blast furnaces; arrangements for the supply of pig iron from the neighbouring works had been made in advance. The steel plant consisted of four 10 ton converters built in line on staging some 22 feet above ground level. The vessels were to the latest Holley design, with the shells in three sections which were removable from the trunnion ring to facilitate relining. Seven cupolas were available close to the vessels for remelting the pig and spiegeleisen. The materials were taken to the vessels by hydraulic lifts and chutes were being provided to enable the lime and the steel scrap to be charged to the vessels in the upright blowing position. The designed output of the works was 2,000 tons per week or 100,000 tons per annum.

The first blow was made on 31 May 1883 in the presence of Gilchrist; a charge of 5 tons of metal containing between 3 and 4% phosphorus was

[18] One wonders what the experiments were. One feature which immediately strikes the modern steelmaker is the complete absence of any silicon in any of the early analyses of these basic steels. One wonders also whether the work on solid castings at Terrenoire was as widely known as would be assumed from the publicity given it in the English speaking metallurgical circles by Holley. There certainly appears to have been a prejudice in the 1870s and early 1880s against the use of silicon, it being considered to cause brittleness in steel.

[19] J. K. Almond: "Making Steel from Phosphoric Iron in the Converter", Chapter 7 in "Cleveland Iron and Steel", published by the British Steel Corporation in 1979 to record the centenary of the successful conclusion of the work of Thomas and Gilchrist, pp. 207-211.

successfully blown; the final phosphorus, allegedly, was only 0.02%. After three months, output was over 1,000 tons per week; during the first six months 24,600 ingot tons were made, giving 19,200 tons of saleable product and the rated output was apparently reached within a year of commissioning the plant.

Reference to the layout of new Bessemer plant always seems to bring in the name Holley. The interesting feature in this present context is that Holley had already appreciated the potential of the basic process in certain American locations and had drawn up his own proposals as to how the requirements of the new process could best be met. At a meeting in Philadelphia in 1880 Holley had outlined these.[20] The suggestions were taken up by Mons. Greiner, manager at the works at Seraing and presented at a meeting in Liege in November 1881.[21] The principle was to divide the operations into a converter shop, a casting shop and a repair shop. The converters were erected on staging, with casting carried out at ground level. The essential features of the arrangement for basic operation, with its requirement for handling the extra raw materials and the larger bulk of slag, were detailed by Holley as follows:

1. Two ladle cranes were provided, the first to receive metal from the converter and the second to transfer it to the main crane in the casting house, in the centre of a shallow pit served by four lateral ingot cranes. The workmen here were separated from the heat and the other inconveniences of the blowing of the vessels and there was space provided for the installation of an overhead rail system for transferring the liquid metal from the cupolas directly into the converters. Whilst the first crane was receiving steel from the previous blow; the time required for this operation was roughly equal to that needed for removing the slag, testing a sample, adding the spiegel and pouring into the ladle. As soon as the ladle was empty, it would be set down on a support by one of the lateral cranes, examined and, if in good condition, taken up by the receiving crane for further use. If it required repair it would be placed on a wagon and taken to the repairing shop by a locomotive, another previously heated ladle being substituted for immediate use. The transfer from one crane to another was facilitated by a special hydraulic system which brought the heads of the cranes into close proximity.

2. The bottom of the pit was arranged to be on the general level of the casting house, so that slag from converters and ladles could be emptied directly into special wagons to be removed by locomotives.

3. The converters were so arranged that the slag and metal projected from them fell down outside the building.

[20] A. L. Holley: "An Adaptation of Bessemer Plant to the Basic Process", Trans. Amer. Soc. Mech. Eng., vol. 1 (1880), pp. 130-132

[21] A. Greiner: "Les Installations Bessemer Americaines Adaptées au Procédé Basiqueàpres les Plans de Monsieur Holley", Revue Universelle, Tome ix, 2me Serie, 1882, pp. 1–14.

4. The lime, brought directly from the calcining kiln, was charged into the converter during the blow.

5. The converter bottoms could be lowered on to special wagons by a hydraulic lift system and, whenever one had to be removed to the repairing shop, another was ready heated to be fitted to the converter.

6. The converter shell itself was exchangeable. As soon as the lining required more repair than could be executed during the time at disposal when a new bottom was being fitted, the shell would be detached from the trunnion belt and another put in its place. For this purpose, a specially constructed wagon was placed on the hydraulic lift stationed directly under the converter. The latter was then turned mouth downwards and the wagon raised to receive it. A dozen strong bolts, by which the shell was fastened within the belt, were then withdrawn, the wagon lowered and, with its load, removed to the repairing shop, the belt with the trunnions remaining in place. The replacement of a relined and heated shell was effected by the reverse operation.

7. The repair shop, separated from the converting department, would provide ample space and ventilation for the men employed there, an important condition for ensuring good workmanship.

8. An important feature of the repairing shop was an immense central turntable, sufficiently large to receive, at the same time, a locomotive and the wagon holding the converter shell, or any other part requiring attention. Since there were no cranes in this building, each special wagon had to be fitted with the necessary means of giving access to the pieces under repair.

9. The ladles and bottoms were heated by a gas blast, used with marked success in American steelworks.

10. The converters were easily repaired on their seat on the wagon, whatever the method of relining or material used; a gas supply was also at hand for any heating required.

In comparing the cost of such an installation, using only one pair of converters, with that of the normal European plant, which required two pairs of converters for the same output, the newer design required three supplementary shells and their wagons, two powerful hydraulic lifts and two large turntables. To offset this, the older plant would have two extra converters, complete with their belts, supports and turning gear, as well as the complete additional pit with cranes and foundations and the extra buildings. Holley is said to have estimated that the new design would give something of the order of a 20% saving on the initial capital cost; in addition, the better working conditions and the uninterrupted regularity of the working were calculated to go a long way to ensuring the uniformity and good quality of the product. Despite this elegant design for the application of basic operation to the Bessemer process, it was never seriously adopted in the

United States. Possibly the early death of Holley at the age of fifty in 1882[22] had some bearing on this. Only one basic Bessemer plant, reference to which is made below, was established in America; it was the Open Hearth plant, in its basic form, which gave serious competition to the acid Bessemer process in America.

There is a report of trials conducted in June 1881 by Alfred Hickman at the Patent Shaft and Axletree Company in Wednesbury.[23] Part of the pig contained 3% phosphorus; the balance was the most phosphoric pig available with up to 4% phosphorus and between 1% and 2% each of silicon and manganese. The trials were carried out under the most disadvantageous conditions; the pit was a very old, unused one, without any special appliances for the working of the basic process. Nevertheless the trials, with from 70 to 80 tons of pig, fully established the following conclusions, which will be seen to be in accordance of what has previously been derived from German and Austrian sources:

1. Cinder pig[24] containing a very high percentage of phosphorus could be treated by the basic Bessemer process for the manufacture of either moderately hard steel or ultra soft steel (ingot iron) without any drawback other than the use of a greater amount of lime.

2. The steel so produced was well adapted for the manufacture of tin plates, sheet, wire, tubes and plate, as well as for rails, axles and tyres.

3. The extreme heat of the operation when blowing such iron, owing to the large amount of phosphorus available for combustion, enabled upwards of 10% cold scrap to be added to the vessel without any risk of excessive chilling.

4. On the other hand, whilst up to 4.5% phosphorus was permissible in the pig, a lower proportion was preferable.

5. The waste on conversion of Staffordshire cinder pig would probably not exceed that on the Cleveland pig, probably due to the higher manganese content, which not only thinned the slag, thus preventing the retention of metallic shots, but also retarded the oxidation of the iron.

As a result of these trials it was considered that a basic steelworks should be set up in South Staffordshire; this indeed was the case; the Staffordshire

[22] Holley did not live long enough to receive the Bessemer Gold Medal which had been arranged to be presented to him by the Iron and Steel Institute at its Spring Meeting in 1882. In 1890 a bronze bust to his memory was erected in Washington Square, New York, carrying the inscription:

> "In honour of Alexander Lyman Holley, foremost among those whose genius and energy established in America and improved throughout the world the manufacture of Bessemer steel, this memorial is erected by engineers of two hemispheres."

[23] J.I.S.I., 1882, Pt. 1, pp.195-196.

[24] "Cinder Pig" was the material smelted in the blast furnace from a mixture of puddling cinder with a relatively minor proportion of iron ore; it was, therefore, a cheap material but with excessive phosphorus content for normal forge purposes.

Steel and Iron Company was formed and the erection of the Bilston steelworks adjacent to the furnaces of Sir Alfred Hickman was commenced in 1882. The plant was fully operational early in 1884.[25] These activities obviously relied heavily on the local availability of "puddler's tap cinder", which could be obtained more or less for the cost of carting and which originally provided from 50% up to 80% of the blast furnace burden. The balance was made up of Northamptonshire ore, Oxfordshire ore and some of the local clayband ore. Eventually, however, as the Basic Bessemer process replaced the puddling furnace, mild steel superseding wrought iron, the stocks of cinder became depleted and the raw material costs, based on the replacement of cinder with ore in the blast furnace, made the operations in this locality much less attractive – this, however, was beyond the period under consideration in this survey.

The earliest statistical summary of the activity in Basic Bessemer steel-making which has been found covers operations during the month of October 1882 and the details are as follows:[26]

Country	Firm	Number & Capacity of Converters	Ingot Tons during Oct. 1882
England	Bolckow, Vaughan, Middlesbrough	4 @ 12T	9,578
France	Schneider, Le Creusot	2 @ 7T [27]	678
Belgium	Acieries d'Angleur	2 @ 6T	1,687
Germany	Aachen Hutten Verein, Rothe Erde	3 @ 5T	3,900
do.	Bochumer Verein	3 @ 4.5T	2,835
do.	Gutehoffnungshutte, Oberhausen	2 @ 6T	1,335
do.	Hoerde Verein	3 @ 10T	4,100
do.	Peine Walzwerke	3 @ 10T [28]	418
do.	Rheinische Stahlwerke, Ruhrort	2 @ 6T	3,500

[25] F. W. Harbord: "The Thomas Gilchrist Basic Process, 1879-1937", J.I.S.I., 1937, Vol. II, p.87P.

[26] From a report on "Basic Process Progress" by Gilchrist abstracted in J.I.S.I., 1882, Pt. 2, p. 684. The absence of Brown, Bayley and Dixon from this list will be noted. It appears that the departure of Arthur Cooper to take control of the development of the North Eastern Steel Company in Cleveland coincided with a decision to abandon work on the process at the works in Sheffield.

[27] In addition to the Basic Bessemer operations, Le Creusot had an experimental 10T Open Hearth furnace with a basic lining, which provided 562 ingot tons during this month.

[28] It was pointed out that the works at Peine was not fully commissioned; only one converter was operational in October, the first trial blows having been made at the end of the previous month.

Country	Firm	Number & Capacity of Converters	Ingot Tons during Oct. 1882
Germany	Dortmunder Union	2 @ 9T	7,000
do.	MM. de Wendel, Hayange	4 @ 8T	3,000
Austria	Prager Eisen Gesellschaft	3 @ 5T	1,854
do.	Teplitzer Walzwerke	3 @ 7T	3,000
do.	Witkowitzer Bergbau	2 @ 8T	3,000
Russia	Societe des Acieries de Varsova	....	1,270

The summary to this table, which was taken to represent an annual output of around 560,000 tons – not far short of the 600,000 tons postulated by Thomas two years earlier – reads as follows:

Total Output for Month:	England	9,578
(ingot tons)	Continent	37,577
	TOTAL	47,155

A useful survey of the means available for lining the basic converter was provided by Percy Gilchrist at about this time.[29] Dolomite had been ground up, pressed into bricks and then fired in the original trials at Eston, the misshapen bricks then being ground up; this hard burned dolomite was subsequently mixed with tar, hot rammed into wrought iron brick moulds and these bricks then fired. Sometimes it was found convenient to ram the mixture into moulds built up from wrought iron plates held by cotter pins so as to give shapes which would fit the contour of the vessel; these would be released from the moulds and were often used "green" (that is, without prior firing). At Le Creusot, the magnesian limestone was gas fired; elsewhere the limestone was fired with coke in a cupola and shrunk material raked out every two hours or so. Fired magnesian limestone obtained by any of these means would be ground up and mixed with hot tar, the proportion of tar being adjusted so that the material would ram solid using red hot rammers. Alternatively, it would be made into a "slurry" by using more tar. Basic converters could then be lined in one of three ways:

1. Simply ramming the hot tar-dolomite mixture round a wooden or iron core and drying and setting the lining with a coke fire after the core had been removed.

[29] This comes from a paper presented to the South Staffordshire Mill and Forge Managers Association in November 1882, abstracted J.I.S.I., 1882, Pt. 2, pp. 685-686.

2. Building with the fired dolomite bricks, using ground up waste bricks mixed with tar for the joints.

3. Building with unfired moulded bricks or shapes, regular in section, thus building "dry" without jointing.

Renovating the worn lining could be done using a cast iron core, heated to redness and inserted into the still hot body of the vessel, the nose having been removed. The slurry was then thrown into the gap between the core and the old lining; the heat caused the slurry to run liquid and then to dry up hard, following the contour of the former.

The German works at Peine was brought into operation in the autumn of 1882 with three purpose built 10 ton converters.[30] Supplies of dolomite were available locally; this was broken into fist size pieces and burnt in a cupola, the burnt stone being ground in a mortar mill and incorporated with hot tar in a mixer. The resulting compound served for ramming the vessels and the bottoms and also for moulding into bricks. The vessel linings were said to withstand 80 to 95 charges and the bottoms averaged about 18 charges. During the first six months of operation a thousand charges were blown. The pig iron all came from Ilsede and was melted in cupolas in charges of nine tons, from where it was run into one of the three converters which had previously been charged with lime to the extent of 15% of the metal weight; the lime was drawn hot from the lime kiln and charged into the vessel by a chute. Typical compositions of the pig iron and the remelted metal from the cupola were given as follows:

	Ilsede Pig Iron	Pig Remelted in Cupola
Carbon	2.7 to 3.0	2.0 to 2.5
Silicon	Up to 1.1	Up to 1.1
Manganese	2.5 to 3.0	0.7 to 1.2
Sulphur	Up to 0.1	Up to 0.1
Phosphorus	2.5 to 3.1	2.6 to 3.2

With the low silicon content of the metal, decarburisation began almost as soon as the blast was applied and was finished within 8 to 9 minutes. The afterblow lasted 3 to 4 minutes, after which the slag was poured off and a sample taken, which almost invariably proved satisfactory. The finishing additions, of spiegel with 12% manganese and of ferromanganese with 54 to 60% manganese, were added to the converter after being heated to

30 J.I.S.I., 1883, Pt. 2, pp. 822-823.

redness. The slag bogey was removed to a mill where the slag was eventually pulverised for use as fertiliser; the metal ladle was taken to the casting house and the casting completed within ten minutes. The average tap to tap time was 45 minutes. The ingots were, in general, rolled down to billets and sections. Here again, the bulk of the product seems to have been in softer steel, with around 0.10 to 0.15% carbon, although some higher carbon steel was produced. From the few analyses quoted, typical sulphur and phosphorus contents were from 0.03 to 0.06%, which appear to have been somewhat better than the average continental material at this time.

The various reports so far quoted give a good impression of the European approach to the "Thomas" process. In 1883 a new Bessemer plant was installed in America – the South Chicago Plant of the North Chicago Rolling Mill Company.[31] This was built on a green field site on the lake shore and it was thought fit to follow the scheme laid down by Holley for the provision of all the necessary facilities for basic steelmaking. The plant and its initial operations were described by the plant manager at the beginning of 1884 and it was clear that it had been utilised only for the manufacture of acid Bessemer steel to the pattern of earlier American activities; in the discussion which followed a number of contributors expressed the hope that the possible virtues of the new basic process would soon be evaluated in America.

There is a comment that the first experimental heat of Basic Bessemer steel in America was made at the Pennsylvania Steel Company on 24 March 1884[32] but no other detail is available; there is, however, an entry for a furnace at Harrisburg in a list provided by Dr. Wedding at the end of 1884, as will be observed below; these two may refer to the same operation. The first, and only, substantive account of basic Bessemer steelmaking equipment in America, however, comes eight years later, in a description of operations at the Pottstown Iron Company,[33] situated much nearer Philadelphia than to Harrisburg; it must be concluded this was a different establishment. It had two 10 ton vessels and one 13 ton vessel, all being of the eccentric type with removable bottom sections, the layout being a typical American-Holley one, but obviously taking advantage of all the European experience in practice; Ernst Bertrand, formerly at Teplitz and currently the manager at Kladno, had been employed by them as their consultant engineer. The lining material, for instance, was prepared according to a process developed by Bertrand of Kladno; using a limestone low in silica, magnesia, iron oxide and alumina, this was crushed quite fine and then 2% ground basic Bessemer slag incorporated. When thoroughly mixed with just sufficient water to make it mouldable, it was fed to a brick making

[31] R. Forsyth: "The Bessemer Plant at South Chicago", Trans. Amer. Inst. Min. Engs, Vol xii (1883-4), pp. 254-274.

[32] J. McHugh: "Alexander Holley and the Makers of Steel", Baltimore, 1980, p.360.

[33] J. Hartshorne: "The Basic Bessemer Plant of the Pottstown Iron Company", Trans. Amer. Inst. Min. Engs., Vol xxi (1892-3), pp. 743-765.

machine; the highly compacted shapes were then hard-fired in tunnel kilns. The bricks were then ground, the powder separated into coarse and fine by sieving and the correct proportion of the two grades hand mixed together with 10-12% hot tar, before being used for the ramming operation. The average lining life was given as 150 heats. The rate of production was high by normal standards; from the pair of 10 ton converters, 25 heats in 12 hours was not unusual, giving 50 minutes cycle on each converter. The average ingot size was large by European standards, weighing two and a half tons, 17.5" x 20.5" at the bottom and 16" x 20" at the top; the smallest was 12" square.

The pig used contained less than 0.5% silicon, around 0.8% manganese, 0.02-0.04% sulphur 2.5-3.0% phosphorus. 11-13% of pure calcined lime was charged cold to the vessel, followed by the metal, as hot as possible from the cupola. The blowing engines were controlled to 50 rpm with a blast at about 30 lb. per square inch; revolution counters were installed so that the number of revolutions for each part of the blow could be recorded. During the blow, scrap could be added through the lime chutes to control the temperature. The "change" was determined by means of a spectroscope. The blow averaged about 16 minutes: 11 minutes of "foreblow"and 5 minutes of "afterblow". As the vessel was turned down, 0.6-0.8% of an 80% ferromanganese was added through the lime chute, passing immediately through the slag and about two thirds of the metal then run into the ladle. When slag just began to run over the slag spout (a shallow bricked channel cut into the top level of the ladle) the pouring was stopped and from 1.0-1.5% of a 10% spiegel thrown into the ladle, the pouring then being completed (the slag, presumably, being allowed to run over into a slag pan below). The casting was done slowly, through a 2" nozzle; every care was taken to make the ingots solid, the mould being filled up, the metal allowed to settle and the mould again filled up. With the very low carbon steel made, it was general for the metal to sink a little in the mould when cast this way; a cast iron plug was placed in the top of each mould after casting. The paper provides a log of complete shift (12 hours) work, listing 25 heats; a partial transcript is as follows:

Blow Number	**6266**	**6267**	**6268**	**6269**	**6270**	**6271**	**6272**	**6273**
Vessel Number	3	1	3	1	3	1	3	1
Heat on vessel	98	80	99	81	100	82	101	83
Heat on bottom	98	71	99	72	100	73	101	74
Turned down	8′43"	9′12"	9′37"	10′05"	10′30"	10′55"	11′20"	11′45"
Revs: Blow	840	900	868	840	880	760	840	868
Afterblow	464	464	464	464	464	464	464	464
Length of blow	13′	14′	13′	13′	13′	12′	13′	14′
Av. Pressure, psi.	32	32	32	32	33	33	32	33
Total heat time	28′	29′	25′	28′	25′	25′	25′	25′

Analysis:	**6266**	**6267**	**6268**	**6269**	**6270**	**6271**	**6272**	**6273**
Carbon	0.13	0.09	0.09	0.11	0.10	0.09	0.09	0.10
Manganese	0.35	0.31	0.29	0.50	0.38	0.39	0.48	0.42
Sulphur	0.044	—	—	0.048	—	—	0.050	—
Phosphorus	0.035	0.040	0.030	0.035	0.040	0.045	0.045	0.040

It is to be noted that the slag was ground at the plant in ball mills, on the European pattern, with an output of around 10 tons in 10 hours; 95-98% of the product passed a 100 mesh sieve whilst 70-75% would pass a 150 mesh sieve; this is exceedingly fine. The product, sold as fertiliser, ran from 18-24% P_2O_5, of which 14% was "reverted".[34]

This is the one example of the impact of the full American Bessemer engineering expertise on the method evolved by the tireless experiment by an English police clerk and his analyst cousin, only ten years previously, enhanced by European ingenuity which the Americans had been ready to accept, as they had employed a European consultant; here the operation of two 10 ton converters in a well organised environment was capable of producing the astonishing amount of over 100,000 tons of steel in a year.

Going back to the time when American interest was just being shown in the process, there are isolated statistics to be found. In the year 1883, for example, of the British total of 1,553,580 tons of Bessemer steel, the proportion of basic steel produced was only 120,000 tons. This came from 13 converters, although it was noted that a further 11 converters were being constructed or modified for basic operation. During the same year, the output of Thomas steel in Germany alone had reached 419,000 tons, with an estimated 100,000 tons elsewhere in Europe – a sum total of over 600,000 tons.[35]

At the end of 1884, Dr. Hermann Wedding produced "a work on the basic process" from which the following list of works operating with basic converters was quoted:[36]

[34] This implies the percentage of P_2O_5 which is soluble in dilute citirc acid. The comment is also made that 600,000 tons of basic slag were used as fertiliser in Europe in the previous year.

[35] J.I.S.I., 1884, Pt. 2, p. 588.

[36] G. W. Maynard: "A Biographical Note on Sidney Gilchrist Thomas", Trans. Amer. Inst. Min. Engs., Vol xiii (1884-5), pp. 789-790.

Germany:	Hoerde	3	@ 10T
	Rheinische Stahlwerke	2	@ 6.5T
	Bochumer Verein	3	@ 4.5T
	Oberhausen	2	@ 6T
	Dortmunder Union	2	@ 9.5T
	Hoesch (under construction)	3	@ 10T
	Phoenix	3	@ 10T
	Rothe Erde	3	@ 10T
	Peine	4	@ 10T
	Neunkirchen	2	@ 10T
	Hayingen (de Wendel)	4	@ 8T
	Dudelingen (Luxembourg)	4	@ 10T
	Friedenshutte (Upper Silesia)	3	@ 10T
	Konigshutte (Upper Silesia)	1	@ 7.5T
	do. (under construction)	2	@ 10T
Austria:	Kladno	3	@ 5T
	Teplitz	2	@ 6.5T
	Witkowitz	2	@ 8T
Russia:	Warsaw	2	@ 10T
Belgium:	Angleur	2	@ 6T
	Athus	2	@ 10T
France:	Le Creusot	2	@ 7T
	Joeuf	4	@ 10T
	Commentry	2	@ 10T
	Longwy	3	@ 10T
	Valenciennes	2	@ 10T
Britain:	Eston	6	@ 15T
	North Eastern Steel Co	4	@ 10T
	Sellerhall (Middlesbrough)	1	@ 4T
	Staffordshire Steel & Iron	3	@ 5T
	Glengarnock	2	@ 9T
	Glasgow Iron Works	3	@ 7T
America:	Harrisburg	2	@ 10T

These figures may be summarised:

	No. of Works	No. of Converters	Total tons Capacity
Germany	13	41	367
Austria	3	7	44
Russia	1	2	20
Belgium	2	4	32
France	5	13	124
Britain	6	19	188
America	1	2	20
TOTAL	**31**	**85**	**795**

In the twelve months to 30 September 1885, and the corresponding period to 30 September 1886 the following European output of Basic Bessemer steel was recorded:[37]

	1884-5	1885-6
United Kingdom	145,707 tons	258,466 tons
Germany & Austria	617,514 tons	883,859 tons
France	130,582 tons	122,711 tons
Other Countries	51,514 tons	48,595 tons
TOTAL	**945,317 tons**	**1,313,631 tons**

Similar figures for the calendar years of 1888 and 1889 show a continuing rise in output;[38] these figures also show the proportion of soft steel produced:

	1888		1889	
	Total	Under 0.17%C	Total	Under 0.17%C
United Kingdom	408,594 T	276,476 T	493,919 T	348,828 T
Germany & Austria	1,276,070	1,026,033	1,481,642	1,185,323
France	223,333	158,223	222,392	159,271
Belgium & others	46,237	32,390	76,599	71,217
TOTAL	**1,953,234**	**1,493,032**	**2,274,552**	**1,764,639**

[37] J.I.S.I., 1885, Pt. 2, p. 570; 1886, Pt. 2, p. 1028.
[38] J.I.S.I., 1890, Pt. 1, p. 406.

The total European output for 1890 had risen to 2,603,083 tons with 72% being soft steel with under 0.17% carbon. It was in this year that Hermann Wedding gave a review of German practice.[39] He pointed out the two stages in the revolution in German steelmaking:

1. The adoption of American practice of virtually continuous melting, permitted by the improvements in cupola melting and, above all, by the Holley movable converter bottom.

2. Even more importantly for Germany, the introduction of the Thomas dephosphorising process, which opened up the vast German deposits of iron ore for steelmaking.

Again, as in the consideration of the Basic Bessemer plant at Pottstown, here we have a tacit acknowledgment of the universality of technology developed in part by Britain, by Germany and by America.

As regards the various features of the process, after many trials of other ways, there had been a universal return to the use of the tar-dolomite mix originally suggested by Thomas for lining the converters; bottoms were almost always made using the same ramming mixture, often around basic tuyers. The liquid pig iron from the blast furnace was used wherever practicable; at some locations, as at Hoerde, transfer of the liquid metal over distances of up to 2 kilometres had not proved unfavourable. The most suitable pig iron was one containing 2-3% phosphorus; in Upper Silesia, where the ores carried too little phosphorus, this had been rectified by charging basic slag, with its phosphorus, back to the blast furnace. The shorter period of blow in the basic furnace, rarely over a quarter of an hour, had made it possible to produce more heats per shift than with the acid process. The quantity of blast for the overblow was now universally regulated by engine measurement, considering the number of revolutions in combination with monitoring the blast pressure. A higher blast pressure, of up to 40 to 45 lb. per square inch, could be an advantage. The liquid metal, after deoxidation and recarburisation, was invariably tapped into a ladle, which then generally was taken to a separate shop for casting. The elimination of sulphur in the converter had proved to be a problem and steps were generally taken to eliminate it in the blast furnace, notably by additions of manganese which, although unnecessary in the basic converter, effectively trapped the sulphur in the blast furnace slag. As regards the basic converter slag, where not required as a blast furnace addition, this was invariably ground to a fine powder for use as fertiliser. The grinding in open mills had proved injurious to health and the use of ball mills was becoming standard practice.

[39] H. Wedding: "Progress of German Practice in Metallurgy", J.I.S.I., 1890, Pt. 2. pp. 533-544 refer to Basic Bessemer Practice.

The same paper also gives detail of more recent working at Hoerde and one interesting feature was the division of slag into two portions according to what was termed Scheibler's method. About two thirds of the necessary lime addition was made at the start of the blow and, about three minutes after dephosphorisation had set in, the metal bath then containing about 0.3% phosphorus, the slag was quite fluid and contained 20-30% phosphoric acid and very little iron oxide. The slag was poured off separately and then the other third of lime added; the resulting slag inevitably contained any oxidised iron as well as the remaining phosphorus and, if this was subsequently charged to the blast furnace, both the iron and the phosphorus would be recovered. The initial slag, being freer from iron than a normal basic slag, was even better as a fertiliser. An interesting table brings together metal, slag and gas analyses taken at regular intervals during one blow:

	Pig	1	2	3	4	5	Steel
Metal:							
Carbon	3.60	2.81	2.02	1.33	0.71	0.10	0.14
Silicon	0.58	0.28	0.07	0.07	0.06	0.02	0.02
Manganese	1.57	—	0.30	0.34	0.26	0.21	0.55
Sulphur	0.079	0.079	0.073	0.074	0.079	0.054	0.046
Phosphorus	2.75	2.57	2.08	2.25	1.55	0.061	0.084
Slag:							
Silica		9.20	9.50	9.30	10.28	6.99	4.79
Iron Protoxide		0.55	0.73	2.30	7.13	11.98	26.03
Manganese Protoxide		8.87	9.70	8.42	9.45	5.40	4.62
Lime		76.10	71.40	66.17	50.71	46.84	42.05
Magnesia		4.86	5.33	6.47	6.90	4.09	6.83
Phosphorus Pentoxide		2.94	6.90	7.32	15.87	24.73	16.33
Calcium Sulphide		0.63	0.34	0.31	0.27	0.52	0.65
Gas:							
Carbon Dioxide		3.5	3.0	3.0	1.8	1.2	
Oxygen		8.1	3.0	0.3	0.2	0.8	
Carbon Monoxide		2.0	10.6	28.3	29.8	1.6	
Methane		0.9	1.0	1.6	1.8	0.9	
Nitrogen		85.0	81.4	66.6	65.0	95.6	

Some comparison of the utilisation of the Acid and Basic Bessemer processes in 1890 can be made. In Britain, for instance, the official statistics[40] indicate 1,612,730 ingot tons of Acid Bessemer steel as against 402,113 ingot tons Basic Bessemer steel, giving a proportion of just 20% Basic material. In Austria, by 1890, the output of Basic Bessemer steel had almost equalled that from the Acid converters;[41] the figures are in metric ingot tons:

	Acid Bessemer	Basic Bessemer	Basic as % of Total
1870	25,491	....	....
1875	94,705	....	....
1880	87,881	17,385	16.5
1885	149,557	76,821	33.9
1890	149,660	138,021	48.0

The position in Germany is not clear; there is a specific figure for the production by the Basic Bessemer process of 1,493,157 ingot tons for 1890[42] but the Acid Bessemer output is not separately provided. It is clear that the production by the Acid Bessemer process would have fallen with the advent of the Basic process since a number of converters were adapted for the new method with its more accessible iron supplies. It is known that the Acid Bessemer output in 1880 was 742,000 tons and that in 1900 was 223,063 tons. It would seem likely, therefore, that in 1890 the proportion of Basic Bessemer steel in the total was something of the order of 75%.

These figures underline the great benefit that the invention of Sidney Gilchrist Thomas conferred on the continental steelmakers. In Britain there were problems. The available ores were generally only moderately high in their phosphorus content whilst being rather too high in silicate minerals and, for these reasons, the production of a suitable metal from the blast furnace was difficult. In Britain, therefore, there was a predilection for the Acid process right up to the First World War. In America, on the other hand, the Acid Bessemer proved eminently satisfactory on the bulk of the available ores. So it was that the paradox arose that, based on one British invention, the United States was able to displace Britain as the leading steelmaking nation by about 1885 and that Germany, based on another British invention, was to push Britain into third place within the next ten years.

40 British Iron and Steel Federation, Official Statistics.
41 J.I.S.I., 1892, Pt. 1, p. 504.
42 J.I.S.I., 1912, Pt. 1. p. 48.

13 The Basic Open Hearth Process

The application of basic linings to the Open Hearth furnace was an obvious extension of the Thomas and Gilchrist inventions. There was some hesitation, however; whilst the basic Bessemer furnace was in service in Europe within a few weeks of the successful trials at Eston, it was at least three years before a Basic Open Hearth furnace was reported to be in operation. There is some doubt as to who was the pioneer in this field. According to one authority, it was actually in Russia that Basic Open Hearth steel originated:[1]

> ... the first dephosphorisation in the Open Hearth furnace that was carried out at Alexandrovsky, near St. Petersburg, was not made the subject of an application for patent, but rendered public property by publication in the technical press.

This work was being actively carried out early in 1882 and there is the following information available:[2]

> The steel melting furnaces used are the same that are employed in the ordinary process, except that the hearth and walls are made of a mixture of dolomite and from 17 to 18 per cent of coal tar. In rebuilding a furnace, the gas and air flues are built up in the usual manner with Dinas bricks, and upon them the bottom is rammed in the form of a 7.2 inch layer of dolomite. Then wooden walls are erected parallel to the side walls of the furnace, leaving a space of 18 inches below and 12 inches inches above. This space is filled with the basic mixture in 2 inch layers, but in order to keep the Dinas bricks and the basic mixture from running together, the joint of the two is made of a 6 inch layer of a mixture of chrome ore and coal tar. The roof is borne simply by the furnace casing. It is only touched by the side walls and is made of Dinas bricks as usual. When the furnace is finished, rail ends are laid as closely as possible over the whole of the bottom and the rammed walls are protected by thin sheet iron, held in place by struts of rail ends. By this means the air is kept from the tar, so that it does not burn when the furnace is dried out, before the dolomite is well sintered. It also prevents swelling and foliating. The crop ends are allowed to remain in place as long as possible,

[1] J Massenez: "History of the Basic Process in Germany", J.I.S.I., 1909, Pt. 2 p. 454.

[2] O. Tellander in Jernkontorets Annaler, abstracted as "The Basic Open Hearth Steel Process at the Alexandrovsky Works", J.I.S.I., 1883, Pt. 1, pp. 465-467.

until they show an inclination to weld together, when they are taken away. Firing is continued until the well-sintered mass is hard, which takes a day and a half with gas and two and half days with wood.

Charging is then commenced.

In the beginning of the work a series of trial charges were made. Now the work appears to have settled down to the following as an average:

35% Cleveland Pig (3-4% C, 0.7-2.2% Si, 1.2-2.0% P, 0.06-0.15% S)
10% cast scrap (3.5-4.0% C, 0.3-0.1% Si, 0.4-1.2% P, trace to 0.05% S)
10% iron rails (0.25-0.65% P, 0.04-0.20% S)
30% miscellaneous wrought scrap (0.20-0.50% P)
4% own scrap (0.30-0.45% C, 0.03-0.10% P, 0.01-0.05% S, 0.4-0.8% Mn)
4.5% haematite pig (3-4% C, 0.7-1.2% Si, 0.03-0.05% P, 0.06-0.18% S)
6% Spiegel (3.5-5% C, 0.03-0.05% P, 10-20% Mn)
0.5% Ferromanganese (5.5-7.5% C, 0.08-0.15% Si, 0.08-0.15% P, 50-80% Mn)

In order to get cinder which would be sufficiently basic, lime to the extent of about 6 per cent of the total charge is added from the start. Toward the end of the charge, bricks pressed from a mixture of slack lime and roll cinder are added, according to requirements.

The phosphoric pig is charged first, followed by a layer of lime and as much scrap as the furnace will hold. The first lot generally melts in four hours and the rest of the scrap is then put in as occasion requires, in lots of 320 kilograms. The level of the melted metal is kept pretty high, so that, in the beginning, when the cinder is very fluid, a part will flow out of the working door when the charge is boiling. When all the scrap is melted down, and after the cinder has been drawn off as thoroughly as possible, the bath is stirred and a sample taken, which is hammered and broken. If the fracture shows any traces of crystalline structure, 120 pounds of basic bricks, consisting of a mixture of 5 parts by volume of lime and 4 parts of roll cinder, are added and this is repeated until the desired quality is reached. The metal shows a tendency to rise in the moulds.

Repairs of the furnace after every heat, heating up and putting in the cold charge take 2 hours 20 minutes, while the time for completing the charge is 9 hours 35 minutes,thus making it possible to make two casts in 24 hours. The basic hearth will stand from 150 to 200 charges, while the roof, walls and flues must be repaired after from 30 to 40 heats. The time for working in a small 8 ton furnace is the same as for a large 10 ton furnace; the average product of the works is not therefore as great as it would be if all the furnaces were of the larger size. During the first five months of 1882 six furnaces, working 428 days and making 854 charges, produced 6,081 metric tons of ingots, using 1,970 tons of Cleveland pig, 670 tons cast scrap, 1,409 tons old rails, 2,076 tons miscellaneous scrap, 303 tons haematite pig, 577 tons spiegeleisen and 35 tons ferromanganese, a total of 7,040 tons. The consumption of coal was 5,503 tons and 453 tons limestone and 120 tons lime-cinder bricks were used besides. The yield in ingots was 86.4 per cent and scrap 2.6 per cent, so that the waste was 11 per cent.

The following table gives the chemical composition of the various grades of steel produced and their physical properties, as revealed by mechanical tests:

Chemical Analyses

Grade	C	Si	Mn	S	P
Rail Steel	0.30-0.45	trace	0.40-0.80	0.01-0.03	0.03-0.07
Hard Steel	0.20-0.25	trace	0.30-0.50	do.	0.03-0.05
Soft Steel	0.10-0.15	trace	0.20-0.35	do.	0.02-0.03
Extra Soft	0.06-0.10	trace	0.20-0.30	do.	do.

Mechanical Tests

Grade	Tensile Strength, psi.	[tsi]	Elongation, %
Rail Steel	71,120-85,344	[32-38]	15-25
Hard Steel	56,896-71,120	[26-32]	20-25
Soft Steel	49,784-56,896	[22-26]	25-30
Extra Soft	42,672-49,784	[19-22]	Over 30

The average percentage of phosphorus in the raw materials of the charge was 0.50%, thus showing an elimination to the extent of 90 per cent. The most striking fact in the series of analyses is the almost perfect elimination of silicon, so that in this respect the Basic Open Hearth process does not differ from the Basic Bessemer process.

The Alexandrovsky works have four cupolas [for the burning of the dolomite] each with a sheet iron shell 11 feet 6 inches high and 5 feet 9 inches diameter. At first the shells were lined with the basic material used for the hearth of the steel furnace but it was found this lining would not last. After some experimenting, the best was found to be a lining of chrome ore from the Ural Mountains, containing 8 per cent silica, 13.5 per cent oxide of iron, 13 per cent of alumina, 51 per cent chromic oxide, 1.5 per cent lime and 13 per cent magnesia. This cupola lining lasts from three to four months, as against fifteen days for the dolomite lining. In burning the dolomite the charges are 560 pounds of dolomite to 260 to 300 of coke, the yield of a cupola being 6.4 metric tons of burnt dolomite, or 51 per cent of the raw rock charged. The burnt dolomite is sorted into three grades, the first being highly burnt or dense. The third grade, in quantity about 25 per cent of the whole, is porous and light and is used only for patching, without admixture or tar.

The other claimant for priority in the manufacture of steels from a basic lined Open Hearth furnace is Le Creusot; there was a notice of such operations in early 1882. As to which country has the most valid claim, the evidence is not sufficiently precise. More importantly, however, there is interesting information also available from the French records:[3]

[3] There is a reference to a paper by Monsieur d'Elaford as appearing in Annales des Mines, 1882, Vol. ii, p. 366 (J.I.S.I. 1882, pt. 1, p.337); this is further elaborated by a lengthy abstract in J.I.S.I., 1882, Pt. 2, pp. 773-774, from which the details quoted here are taken.

The furnace employed at Creusot is the same as that used in the manufacture of Siemens Martin steel, the only difference being that the bed is made of a mixture of magnesian lime of the same kind as that employed in the Bessemer converter. The roof is of silica bricks, a layer of bauxite being placed at the junction of the magnesian lime and the bricks. The furnace is heated in the ordinary manner, by the combustion of gas from the Siemens generators. The charge of phosphoric pig is run in[4] and common iron dissolved gradually in the bath. The refining is produced under the influence of the current of gas, as in an ordinary operation. Lime is added, at three or four different times, in order to make the slag very basic. From time to time this slag is removed by means of a rake. Owing to this excess of base, silicon disappears completely and phosphorus is almost totally eliminated. The different bodies disappear in the same order as in the Bessemer converter, so that the operation consists of the following successive periods: scorification, decarburisation and afterblow. The moment of stopping the refining and proceeding to recarburise by an addition of spiegel is found by means of trial tests, as in the converter. One operation lasts about twelve hours and about fifteen tons are obtained.

The advantages which the reverberatory furnace presents over the converter are:

1. The construction and repair of the basic hearth is much more easy.

2. The temperature of the bath is obtained independently of the combustion of silicon, carbon, phosphorus, etc, so that there is no necessity for pre-treating a pig containing a high proportion of these bodies.

3. The slag is more easily removed by means of a rake,it being easy to remove it at any time during the operation, while in the converter it can only be removed when it has become fluid. The elimination of phosphorus is also more assured and the return of this body into the steel on the addition of spiegel is less to be feared.

4. The duration of the refining is longer and the trial tests more easily taken, while the operation is more easily regulated. The manufacture of basic steel thus being easier in the reverberatory furnace than in the converter, Messrs. Schneider have been quite successful from the very commencement in making this metal in the the Siemens Martin furnace.

The discussion then becomes somewhat philosophical; whilst the argument might well have appeared perfectly valid to the Continental steelmaker, after limited experience with the processes and a domestic shortage of raw materials suited to the acid processes, it has to be said that the steelmakers in Britain took a somewhat different view. The argument is, nevertheless, worthy of recording here:

[4] A curious statement, presumably implying the charging of molten metal to the furnace.

> Metallurgists are now in possession of two different methods of manufacturing steel, both in the converter and in the reverberatory furnace, the one consisting in treating pure products in an apparatus furnished with a siliceous lining, the other in treating impure products in presence of a basic lining. The following question, therefore, seems to present itself. Since, other things being equal, linings of magnesian limestone offer a more complete purification than siliceous ones, why not treat all pig metal, even that resulting from pure ores, in a basic apparatus? So far as concerns the reverberatory furnace, it appears probable that in the majority of cases the siliceous hearth will be replaced by one of magnesian lime, the carrying out of a basic operation presenting, as explained above, no difficulties. But with the converter the case is different; it would not be found advantageous to treat in a basic lined converter metal such as that used in the acid process, since this metal is high in silicon and this circumstance presents a grave obstacle. On the other hand, if the blast furnace charges were so regulated that the resulting pig was but slightly siliceous, there would probably not be sufficient heat generated by the molecular combustion to assure the liquidity of the bath and slag.

There was also an interesting early production of basic steel using the Pernot furnace at St. Chamond; here were no problems in using a basic hearth and a silica roof since the two were separate. A hearth, already carrying a charge of limestone and iron ore, had pig and scrap charged on top; the metal on melting down dropped on to the oxide and reacted to give a disengagement of carbon monoxide; the rotation of the hearth assisted and the slag gradually absorbed the phosphorus. It seems, however, that the degree of dephosphorisation was limited; it was said that a charge containing 2.5% silicon and 0.5% phosphorus provided a steel with around 0.1% phosphorus which was "a quantity not objectionable in rails".[5]

As far as the introduction of the Basic Open Hearth process in Great Britain is concerned, there is again conflicting evidence as to priority. The generally accepted story is that it was started at the Brymbo Steelworks which were built by J. H. Darby in 1883, commencing operations in 1884.[6] There is, however, a record of earlier work by the Farnley Iron Company;[7] this was put in hand as early as May 1882. The trials were carried out in a modified 55 cwt furnace, originally acid lined. Brackets were provided for the support of the upper walls above the hearth lining level and two slag tapping holes were provided either side of the normal tapping hole but at a higher level. The hearth was built up with hot tar-burnt dolomite mixture, around wooden plugs for the three tap holes, shaping the mixture up to the wall brackets. On top of the rammed wall was laid a course of hard fired

[5] Comments on "The Progress of Dephosphorising", J.I.S.I., 1882, Pt. 1, p. 331, quoting a paper in Memoires de la Societe des Ingenieurs Civiles, 17 February, 1882.

[6] J. E. Stead, Presidential Address, J.I.S.I., 1920, Pt. 1, p.63; also A. Jackson: "The Rise and Fall of the Open Hearth", Ironmaking and Steelmaking, Vol. 3, 1976, p.1.

[7] T. Gillott: "The Basic Open Hearth Steel Process", Proc. Inst. Civ. Engs, Vol 77, (1883), pp. 297-310.

magnesite bricks, with a course of bauxite bricks on these, before building the rest of the upper wall in silica bricks; the ports and roof were made in silica bricks as usual. After drying out, the wood plugs in the tap holes were burned out and the holes closed using a mixture of burnt lime mixed with powdered coke. A layer of well burnt and ground magnesian limestone, mixed with a few per cent of ground firebrick, was packed on the hearth and up the sides over the rammed material. The furnace was then slowly fired, so as just to fuse this as protective cover. After about seventy two hours firing, the furnace was ready for operation.

Initially, freshly calcined lime, to the extent to a tenth to a twentieth the weight of the subsequent charge, was spread on the hearth, the pig then placed over the bottom and the scrap placed on the pig, all being charged cold. After four hours firing the pig would have melted and a fluid slag formed; this being somewhat siliceous (due to any sand on the pig plus the silica formed by oxidation of the silicon in the pig), as much as possible was run off through one of the slag notches and some further lime added if the slag was considered too fluid. After about six hours the whole would be melted, more slag could be run off and more lime added, together with some ore, if the proportion of pig to scrap was high, and the boil allowed to proceed. When the boil had slackened and the metal well stirred, a sample would be taken, flattened under a hammer, quenched and broken cold. If the sample was considered tough enough, the slag would be run off or, as usually was the case, would have thickened so that it had to be raked off through the central door, placed a few inches lower than the other two. Then haematite pig would be added to the extent of about 5% of the weight of the charge; this caused a heavy boil which would last about fifteen minutes and functioned as very efficient stirring mechanism for the whole of the bath. After a further fifteen to thirty minutes rest, all excess slag, other than that required to provide a sufficient cover for the metal when tapped into the ladle, would be removed and the metal treated with spiegeleisen and/or ferromanganese and tapped into the ladle. After casting the metal and raking out the remains of the slag, the furnace bottom and sides were "fettled" using well burnt and ground magnesian limestone, the heat being turned on to frit the repaired surfaces. This obviously was a well organised procedure, as the following details of a cast made 29 June 1882 will confirm:

			T.	C.	qr.	ib.
Cast No. 229. Charged	6.30 am:	Farnley No. 3 pig	1	8	0	0
		Steel Scrap		8	2	0
	2.45 pm:	Haematite No. 3 pig		2	0	0
	3.10 pm:	18% Mn Spiegel		1	1	0
	3.25 pm:	53% Mn Ferromanganese			1	5
		TOTAL	2	0	0	5

Procedure:

10.30 am.	2 cwt slag run off; 4 cwt 2 qr Sommorostro ore and 1 cwt 3 qr lime added
2.40 pm.	Sample taken and flattened, quenched, broke tough, no crystal visible in fracture; 3 cwt 2 qr slag run off
3.28 pm.	Metal cast; one 15 in. sq and one 12 in. sq full ingots.

Metal analyses:	C	Si	Mn	S	P
Farnley Pig	3.64	1.24	1.19	0.013	0.60
Haematite Pig	4.21	2.29	0.20	0.004	0.058
Finished Steel	0.24	trace	0.53	0.060	0.067

Final Slag:[8]	P_2O_5	SiO_2	Fe_2O_3	FeO	Al_2O_3	CaO	MgO	MnO
	1.97	13.64	1.37	18.57	2.24	39.70	11.75	8.76

Mechanical tests on 0.466" plate rolled from 12" ingot:

Max stress, tsi	Elongation, %			Reduction of Area, %
	On 10"	On 8"	On 2"	
25.89	25.0	27.5	53.1	55.8

These results certainly indicate that a good quality material was being produced. A series of phosphorus results from some 22 casts indicate a phosphorus range of 0.050-0.097%, with a mean of 0.073%; the six latest cast numbers, however, showed a mean figure of 0.060%.

As far a yields were concerned, figures covering Casts 302 to 408 were quoted:

[8] The low figure quoted for phosphorus pentoxide content is for the final slag, two previous portions of the slag having been run off; one of these earlier slags was analysed and gave a figure of 5.09%

	T.	c.	qr.	lb.	
Farnley Pig	90	8	0	0	= 38.5%
Haematite Pig	9	13	2	0	= 4.1%
Iron and Steel Scrap	130	5	0	16	= 55.4%
Ferromanganese	4	14	2	14	= 2.0%
TOTAL	**235**	**1**	**1**	**2**	
Ingots produced	219	6	3	326	= 93.3%

The overall raw dolomite usage was calculated to be of the order of 9 cwt. per ingot ton.

The original work at Brymbo Steel Company started towards the end of 1883 with the setting up of what was known as a "Batho" furnace, with an almost circular hearth; this had a capacity of about 5 tons.[9] This furnace gave considerable trouble, being short of heat and with refractory problems due to its design, but the work done with it indicated that phosphoric pig produced on site could be made into good soft steel. On the basis of this work, four conventional 12-ton furnaces and, subsequently, a further two 20-ton furnaces, all said to be specially designed for basic work, were installed. These were arranged in line, with a charging platform in front of them and a casting pit below so that the metal could be run into a ladle on a carriage. The ladle was then run over to the moulds for casting. The lining of the furnace was done very much as has previously been described, the rammed basic hearth being separated from the silica walls by a rammed layer of chrome ore-tar in this case.

The established melting practice was to use 80% pig and 20% scrap. The pig had a typical composition of 3% each carbon and phosphorus, with 1.75% manganese, 0.4% silicon and 0.04% sulphur, the scrap being either steel or wrought iron, the percentage of phosphorus in the wrought iron being immaterial. Limestone was charged in a quantity sufficient to produce a slag which was basic from the start. The charge of pig iron and scrap was not preheated. When melted and sufficiently hot, feeds of iron ore mixed with limestone began to be made and would continue for about five hours. It was pointed out that the quality of ore, as regards its freedom from phosphorus, was unimportant and the usual ore at Brymbo was "pottery mine", the same as used in the blast furnace. When it was judged that sufficient ore had been added and the boil was beginning to slacken, additions of lime only would be made. Samples were taken from time to

[9] J.H. Darby: "The Manufacture of Open Hearth Steel", J.I.S.I., 1889, Pt. 1, pp. 79-83.

time and their fractures noted; these would continue to improve with time and, when refining was complete, the sample would be capable of being bent into four thicknesses after flattening without any sign of cracking. Ferromanganese was then added, part in the furnace and part in the ladle, and the charge teemed. It was said to run quietly and solidly, such that the total mill loss in rolling to slab or billet and including the fire loss, was only 8% of the ingot weight.

The 20-ton furnaces were each capable of producing 180-200 tons of ingots per week – an average of 23.3 cwts per hour. The ingot yield was given as 93.5% of the metallic charge. Usage of coal averaged 11.07 cwt per ingot ton. A typical analysis of the soft steel generally produced was

C	Si	Mn	S	P
0.12	nil	0.40	0.018	0.030

This gave a maximum stress of about 24.5 tsi. and an elongation of 31-33% on 8 inches. No examples of "red shortness" had been found on the production of over 60,000 tons of Basic Open Hearth steel over a five year period; all the metal worked well and welded freely.

As far as furnace life was concerned, one furnace which was in service at the time the paper was written had produced 3,600 tons of dead soft steel without relining. It seems clear from this account, together with the previous evidence, that the Basic Open Hearth process was capable of providing a good product from a relatively poor quality raw material with a minimum of trouble.

It was at this stage of development that the steel industry in the Scunthorpe area was founded. The oolitic ores of Lincolnshire and Northamptonshire were vast resources but with a relatively low iron content – around 25-30% only – and these had received little attention until about 1860 when the Dawes family thought fit to institute trials with them in their blast furnace at Elsecar, north of Sheffield. As a result, they founded the Trent Iron Company in Scunthorpe, with two blast furnaces, in 1862, to be followed by the Frodingham Iron Company in 1866. By 1876 there were five companies operating in the region of Scunthorpe, with nineteen blast furnaces, two more being under construction for the Appleby Iron Company. Meanwhile, supplies of the ore were being mined and shipped to Staveley, Parkgate and Sheffield. The local make of pig iron had risen to 110,000 tons per annum and some 627,000 tons of ore were mined in 1875. The pig iron typically contained 0.9-1.4% phosphorus.[10]

The Cliff family, operating the Frodingham Iron Company, decided in 1888 to commence making steel; accordingly they consulted Gilchrist as to the appropriate process. His recommendation of the Basic Open Hearth process was, perhaps, to be expected; the blast furnace metal contained too much phosphorus

[10] G. Dove: "The Frodingham Iron Field", J.I.S.I., 1876, pp. 318-326.

for an acid process but insufficient phosphorus to give the necessary heat in a Basic Bessemer converter. Gilchrist also recommended that Maximilian Mannaberg, an Austrian engineer who had overseen the introduction of the Basic Bessemer process at the Wishaw plant of the Glasgow Iron and Steel Company, would be the appropriate person to install and operate the plant for them. Accordingly, Mannaberg came to Frodingham in 1888; not only did he pioneer the process for them but stayed in charge of the operations until his retirement in 1920.[11]

The original plan was for an output of 400 tons of steel per week. Accordingly two 15-ton furnaces were installed, together with a cogging mill and a billet mill. The furnaces were housed in a building some 500 ft. long, 80 ft. wide and 45 ft. high, the mill building being of similar dimensions and centrally at right angles to it.[12] The first steel was tapped on 21st March 1890; it is not clear whether the mills were ready at that date, since there had been considerable difficulties with the civil engineering on the site.

The furnaces were hand charged, with raw materials being brought in by rail to the back of the charging floor. There was, it seems, no overhead crane in the building until much later. The Wilson type gas producers, steam jet blown, were in line behind the railway. The steel was tapped into ladles carried on bogies; these were transferred to the middle of the shop, where a casting crane rotated around the periphery of a circular casting pit – reminiscent of the earlier Bessemer shops. The pit was 6 ft. deep with the moulds set on bottom plates curved to the outer radius of the pit. The same crane stripped the moulds off the ingots and lifted out the red-hot ingots with dogs; these then went straight into the mill furnaces to be reheated for rolling. The original ingots weighed 35 cwt. and this method of handling served well until larger ingots became necessary.[13]

The origins of the Basic Open Hearth operations in Germany are not well documented. It was obviously well established by 1890 and was gaining in

[11] The details of Frodingham Iron and Steel Company operations come entirely from G.R. Walshaw and C.A.J. Behrendt: "The History of Appleby Frodingham", published privately by the United Steel Companies in 1950, pp. 63-68. It is surprising, particularly in view of the rapidly growing importance of Scunthorpe as a steelmaking centre in the period 1890-1920, that there is a complete absence of any record in J.I.S.I. of the Frodginham operations.

[12] These buildings were, it seems, originally designed for the Antwerp Exhibition and were purchased second hand for erection at Frodingham. It is related that, when they were dismantled in 1949, the wrought iron framework, which had never been painted, was found to show no signs of corrosion.

[13] The method was abandoned in 1908 due to the excessive labour and heat involved with increased production; it was replaced by straight pit casting which, in turn , in 1917, was superseded by car-casting. For details of the growth of steelmaking at Frodingham up to 1901, please refer to the section on the Talbot process in Chapter 14.

favour as against the Basic Bessemer process. Wedding, in a review of the progress of metallurgy in Germany in 1890 wrote:[14]

> Basic Open Hearth metal is generally regarded as an excellent material, for the reason that it is practicable to make accurate tests during the manufacture and, since not more than three or four heats can ordinarily be run in twenty-four hours, there is sufficient time for both chemical and mechanical examinations. A by no means unessential cause of the common preference for Open Hearth steel over converter metal lies in the better opportunity offered by the Open Hearth for a uniform distribution of the manganese introduced as ferro or spiegel. For the lack of such opportunity, much difficulty has heretofore been encountered in producing high carbon metal in the basic converter.

By 1890, the works at Hoerde, in addition to four Basic Bessemer converters, had two 7-ton and seven 15-ton Basic Open Hearth furnaces; it was said that six furnaces were in operation at any one time. The hearths were of dolomite and the walls and roof were in silica brick which was manufactured at the works. The larger furnaces gave 300 to 360 heats without relining. The general process was to run with 20% pig and the balance scrap, making up to six heats in twenty four hours. Similar operations were being carried out at Phoenix Ruhrort and at several other German works, but details do not appear to be available.

It was not until 1886 that the United States established its first Basic Open Hearth plant; this was at the Otis Iron and Steel Company in Cleveland. Strangely enough, only about a thousand tons of basic steel was made here before the bottom was taken out and replaced by an acid lining. It was another two years before the process was again introduced in an American location; this was at the Homestead Works of Andrew Carnegie.[15] It was a slow starter but was to have phenomenal growth.

[14] H. Wedding: "Progress of German Practice in Metallurgy", J.I.S.I., 1890, Pt. 2, pp.544-548.

[15] S.T. Wellman: "The Early History of Open Hearth Steel Manufacture in the United States", Trans. Amer. Soc. Mech. Engs, Vol. xxiii, pp. 96-98

14. Steelmaking from 1890 to 1900

By 1890 the stages which between them had contributed to one of the most significant changes in technological history – the availability of steel in bulk – had been brought together to give viable production methods on a worldwide scale and, indeed, many aspects of the technology developed over a century ago still have relevance today. In considering the situation at the beginning of the period under survey here, the 1843 figures given by Le Play for European production of steel were quoted, with an overall total of 54,250 tons.[1] At the same time, production of steel in the United States was quoted as being 6,078 tons,[2] giving an estimate of world wide production of the order of 60,000 tons in the mid 1840s. Between 1856 and 1879 had come the momentous work of Bessemer, Siemens and Thomas; the first two had established the two procedures by which steel could be produced in bulk, but only from selected raw materials whilst the third, building on these two technologies, had modified the technique whereby the whole of the world's supplies of iron ore could be utilised for the making of steel. By 1890, when the techniques had been established and there were available the Acid and Basic variants of both the Bessemer converter and the Open Hearth furnace and proved steelmaking procedures for all four types of plant, the estimate for world production of steel in that year stands at over 12 million tons, a two-hundredfold increase in a mere forty five years! Moreover, over the next twenty years output was to double every eight years or so – the figure for 1900 was 28 million tons and that for 1910 some 58 million tons. Britain, which had led the world in the iron and steel industry since the early days of the Industrial Revolution, found herself overtaken as a steel producer by the United States in the mid 1880s and by Germany as well in the mid 1890s – it is not always an advantage in the long run to perform the role of a pioneer, since prior capital investment and obsolete plant eventually has to be dealt with, one way or another.

[1] F. le Play: "Memoire sur la Fabrication de l'Acier en Yorkshire", Annales des Mines, 4me Serie, Tome III (1843), p. 692.

[2] A report of the Philadelphia Convention of Ironmasters in 1847, quoted by J.M. Swank: "History of the Manufacture of Iron in All Ages", 2nd. ed., Philadelphia, 1892, p. 385.

This discussion, however, is a little too simplistic. Prior to the revolution in the trade brought about by the Bessemer converter, the general engineering material had been wrought iron, a soft, malleable, weldable material with a reasonable strength, and some two million tons or more of this were made every year in Britain, with probably the same amount elsewhere in Europe and America. Admittedly, this was produced in parcels of up to 100 pounds in weight, by a most arduous procedure known as "puddling" and any larger pieces required had to be welded together, rather painstakingly, under a forging hammer, but it was available for most engineering applications – railway axles, ships shafts, crank shafts and flywheels, as well as boilers and ship plate and railway rails were all produced as required in wrought iron. It was only the difficult pieces, such as those which required a cutting edge or higher resilience, that were even more painstakingly made from steel. The advantages of the Bessemer and the Open Hearth processes, in the main, were that much larger amounts of "ferrous metal" could be more easily brought together – ingots could be made and forged, quite straightforwardly, without having to stick together numerous small pieces before the real task of forging the required shape could begin; alternatively, shapes could be cast in moulds using the sample supplies of molten metal which had never before been readily available. In doing this, however, the distinction between "iron" and "steel" became blurred. Before 1856, the steel in Britain – which, by and large, was the "crucible steel" produced in Sheffield – rarely contained less than 0.5% carbon. By 1890, very little of the steel produced by the Bessemer and Open Hearth processes contained as much carbon as this; the demand for the low carbon "soft steel" from the basic processes which have been followed in the last two chapters patently underlines this. It could, indeed, quite reasonably be argued that the crucible furnaces in Sheffield in 1890 were still producing much the same material in much the same way, probably in somewhat larger quantities, as they had been at the time of the visit of Le Play; again, this is a little too simplistic an argument, since these Sheffield steelmakers had also moved with the times and were producing more advanced materials; there was a growing demand for these special varieties of metal which, at least up to the end of the century, only the crucible furnaces could provide. This demand arose mainly for the provision of metal cutting tools; although to some extent the steel produced by the Bessemer converter and the Open Hearth procedures replaced wrought iron, the total overall production, of steel and wrought iron, was steadily rising and engineering requirements were becoming ever more involved.[3]

The uses to which steel was being put also changed with the times. By coincidence, the early years of the Bessemer production were also years in which railways were spreading far and wide across the globe. When the superior properties of the new steel rails over the previously appropriately named "iron way", with its wrought iron rails, became clear, the railway companies all the world over became interested. It was not long before the inherent economy of this replacement produced an enormous market and the accent on steel production through to the time of the fulfilment of the work of Thomas and Gilchrist was to produce rails. Thus the Acid Bessemer and the Acid Open Hearth furnaces were called upon to provide materials with a sufficient degree of wear and resistance and spent most of their time producing steels with around 0.4% carbon from low phosphorus raw materials – except in Central Europe where, in the absence of phosphorus-free ores, they replaced half the carbon by a similar amount of phosphorus for this purpose. Just by coincidence, about the time that the Thomas invention became a commercial possibility, the demand for rails had dropped but there was an ever increasing need for constructional materials – plates, tubes, girders and the like – and the production of the soft steel became the lifeblood of the new basic processes in Europe.

This, then, was the situation in 1890. An overall picture of the world production of steel at this date can be seen in Table VII.[4] In this final chapter, the story is not of any further world-shaking developments, but rather of consolidation and the introduction of modifications in procedure, aimed at

[3] Figures for the production of wrought iron in Britain are not easily found. The earliest relevant estimate is 2.6 million tons in 1871 (L. Beck: "Die Geschichte des Eisens", Vol. 5 Brunswick, 1903, p. 937). It is estimated that it exceeded 3 million tons in the peak year of 1873 (J.C. Carr and W. Taplin: "History of the British Steel Industry", Oxford, 1962, p.93). Beck subsequently gives figures from 1881 to 1890 (in metric tons) and Carr and Taplin from 1886 to 1896 (in avoirdupois tons). Converting to the latter, we get the following combined figures in thousands of tons:

	1881	*1882*	*1883*	*1884*	*1885*	*1886*	*1887*	*1888*
Wrought Iron	2681	2841	2703	2240	1910	1616	1701	2032
Steel	1779	2210	2009	1862	1988	2264	3044	3304
TOTAL	4460	5051	4712	4102	3898	3880	4745	5336

	1889	*1890*	*1891*	*1892*	*1893*	*1894*	*1895*	*1896*
Wrought Iron	2254	1923	1734	1561	1364	1339	1148	1214
Steel	3520	3579	3157	2920	2956	3111	3260	4132
TOTAL	5774	5502	4891	4481	4320	4289	4408	5346

[4] A. Cooper, Presidential Address, J.I.S.I., 1912, Pt. 1, p.48

TABLE VII WORLD OUTPUT OF STEEL INGOTS IN 1890

	United States	Great Britain	Germany	Austrian Empire	France	Russia	Belguim	Sweden	Italy	TOTAL
Acid Bessemer	3,611,091	1,612,730		149,660						
Basic Bessemer	77,780	402,113	1,493,157	138,021						
Bessemer Total	**3,688,871**	**2,014,843**	**1,493,157**	**287,681**	**444,801**			**92,247**		**7,929,353**
Acid Open Hearth	423,232	1,462,963		33,904						
Basic Open Hearth	90,000	101,287		178,015						
Total Open Hearth	**513,232**	**1,564,250**		**211,919**	**244,190**			**67,779**		
Unclassified			668,660							
TOTAL	**4,202,103**	**3,579,043**	**2,161,817**	**499,600**	**688,991**	**372,625**	**236,226**	**160,026**	**157,899**	**12,058,330**

Source: A. Cooper, Presidential Address, J.I.S.I., 1912, Pt. 1, p. 48.

improving the flexibility of operation and range of capabilities of already well-established techniques. It is virtually impossible to present these wide ranging topics in a systematic manner; the technological modifications appeared at an ever-increasing rate. It seems appropriate, therefore, to select the more important topics and attempt to set them in context.

The Introduction of the Mixer Furnace

The figures quoted in Table VII are derived from a survey of steelmaking over the period 1880 to 1910 presented by one of the pioneers of basic steelmaking, Arthur Cooper. He had originally planned that the North Eastern Steelworks in Middlesbrough – he took responsibility at the green-field site stage of their development in 1882 – should take the molten metal direct from the blast furnaces. In the event, the lack of regularity in composition and temperature of the metal coming to the converters made it necessary to revert to cupola melting. Pig iron from six or seven blast furnaces was regularly used and, by careful selection and mixing of the various grades, fairly uniform results were obtained, although there was still something to be desired in regularity of operation.[5] It was towards the end of 1888 that attention was drawn to the provision at the Edgar Thomson Works in America of a large holding furnace into which metal from the blast furnace could be poured and metal for the converters taken, acting as a large reservoir and thus evening out any variations in the metal composition; this was one of the improvements due to Captain "Bill" Jones[6] and became known as the "Jones Mixer". This was seen in operation by representatives of a number of foreign establishments, among them Massenez, of the Hoerde Works in Germany, and David Evans, manager at the Barrow Works in Furness, who each took out licences and installed "mixers" around 1890. Massenez reported that, provided the iron contained upwards of one per cent of manganese, the period of storage in the mixer was capable of bringing about a large reduction of the sulphur content of the metal, most desirable in the basic steelworks. Evans had, in the meantime, moved from Barrow to Bolckow, Vaughan and Company, in Middlesbrough, and it was on his recommendation that Arthur Cooper at the neighbouring works of the North Eastern Steel Company had thought fit to install a 150 ton mixer;

[5] A. Cooper, "Metal Mixers as Used at the Works of the North Eastern Steel Company", J.I.S.I., 1895, Pt. 1, pp. 66-70.

[6] William Richard Jones was a legend in the American Steel Industry. Having served in the Civil War, he took up work as assistant to the Manager at the Cambria Works, eventually putting the knowledge he gained in Bessemer steelmaking to good use as Superintendent of the new Edgar Thomson Works, which became renowned for its high output and super-efficient working. Jones was clearly involved in all improvements in plant and technology and, indeed, met an untimely death in a works accident whilst endeavouring to clear a problem at a blast furnace at the works in September 1889; he was only 50 years old.

this was at the end of 1892. The vessel was built on the lines of that at Hoerde, but the hydraulic cylinder fitted above the vessel instead of beneath it, a wise precaution in case of any metal breakout. The layout can be seen in Fig. 26. Also included were weighing machines to measure the quantity of blast furnace metal introduced and the metal taken to the converter.

Originally some 2,000 tons of blast furnace metal per week were passed through the mixer and treated separately in the converters (without any admixture of cupola iron); it soon became clear that there was benefit in regularity of working to be gained by adding the metal from the cupolas to the blast furnace metal in the mixer, particularly since the 2,000 tons per week of blast furnace metal was barely sufficient to maintain the temperature of the metal so that the slag remained free-running. Under these conditions the working was so successful that a second mixer was installed to give continuity of working when the first vessel required lining. The original lining had been of firebrick; when this had worn too thin – in about six to eight weeks – it was replaced with a new lining; it had later been found eminently satisfactory to repair a worn lining with ganister, as with acid converter repairs, reducing costs and downtime. Normal practice was to start charging the mixer at midnight on Sunday, so that it was full by start of work at 6 am on Monday. It was then drawn on at such a rate as to maintain a minimum of 80 tons in the vessel until Saturday morning, when it would be emptied. If no repairs were necessary, the mouth would then be bricked up and the charging hole sealed with fireclay and, in this way, it would be kept red hot until charging recommenced. One point which was stressed was that the air should be excluded as much as possible; there was a close fitting sliding door, lined with firebrick, at the mouth and a plate to cover the charging hole. In this way, it had proved possible to pass all the 2,000 tons of blast furnace metal and from 1,800 to 2,000 tons of cupola melted iron through one or other of the mixers for the two year period from May 1893, with results said to have fully justified the Company's expectations.

The discussion which followed the presentation of these details proved to be particularly interesting in that Andrew Carnegie was in the audience and he intimated that the use of the original 200 ton mixer at the Edgar Thomson Works had proved most satisfactory, but it was now considered too small and they were now preparing to put in one of 600 tons capacity with a greater depth which should have some advantage in heat retention. Herr Massenez reported that five works on the continent were using mixers and all were giving great satisfaction; David Evans confirmed that he had built two at Eston of a similar size to that described by Arthur Cooper, one operating on haematite iron for the acid converters and the other on basic iron; the uniformity of composition of the metal, and the knowledge of the exact weight of metal, led to a marked reduction in the number of defective ingots. In addition, particularly with the basic furnaces, there was a welcome

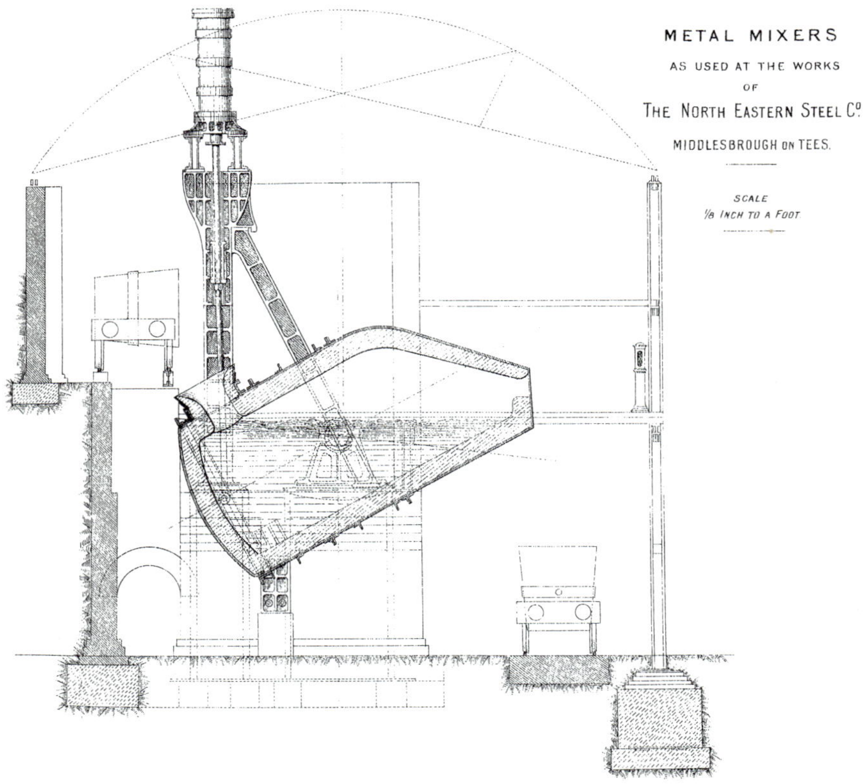

Fig 26. The Metal Mixer as installed at the Works at the North Eastern Steel Company in Middlesborough. The illustration comes from the paper by A. Cooper, J.I.S.I., 1895, Pt. 1, and is typical of those installed in British works towards the end of the nineteenth century. Its capacity was 150 tons of liquid blast furnace metal.

prolongation of the life of the furnace lining and particularly of the bottom. The effect of the mixer on regularity of composition was demonstrated by Windsor Richards, referring to operations at Eston on haematite iron:[7]

Metal charged		Metal withdrawn	
1.25-2.00% Si	24.26%	Below 1.50%	Nil
2.01-3.00% Si	51.90%	1.51-2.00%	3.64%
3.01-4.00% Si	20.80%	2.01-2.75%	96.03%
4.01-5.00% Si	3.04%	2.76-3.00%	0.33%

The introduction of the mixer into virtually all the large Bessemer steelworks was certainly a feature in this last decade of the twentieth century.

The Tilting Open Hearth Furnace

The major advance in the Open Hearth practice in this period was a mechanical one, concerning the furnace structure. One advantage of the Bessemer converter was the capability of moving the metal within the furnace or, putting it a different way, the capability of rotating the furnace structure about one axis. There had been the Pernot furnace, where the hearth had slowly rotated on an inclined plane; apart from the somewhat unsatisfactory trials with a basic hearth at St. Chamond, there is virtually no mention of this furnace in the literature after about 1880 and it must be concluded that it went out of use. The development of Tilting Open Hearth furnace was something completely different and occurred in America; the furnace shell, hearth and roof were so built as to be capable of rotating about its longitudinal axis, through an angle of up to 30°. The first furnace of this type was built at the Steelton Works of the Pennsylvania Steel Company in 1889 to the design of their Engineer, H. H. Campbell. Ten years later there were ten furnaces of the same type operating there: six of 45 Ton capacity,[8], two of 18 Ton capacity and two (presumably the early experimental ones) of 4.5 Ton capacity.[8] In this design, illustrated in Figure 27, the furnace body revolved around its axis on four rings of live rollers running in circular paths, operated by a hydraulic cylinder. The gas and air ports were a fixed block at each end and when the furnace was rotated, these were only partially closed with a gap of about half an inch to allow free motion of the body;

[7] E. Windsor Richards in a paper presented to the West of Scotland Iron and Steel Institute in 1899, quoted by F. W. Harbord and J. W. Hall: "The Metallurgy of Steel", 3rd. ed, London, 1907, p. 19.

[8] A. P. Head: "Tilting Open Hearth Furnaces", J.I.S.I., 1899, Pt. 1, pp. 69-75.

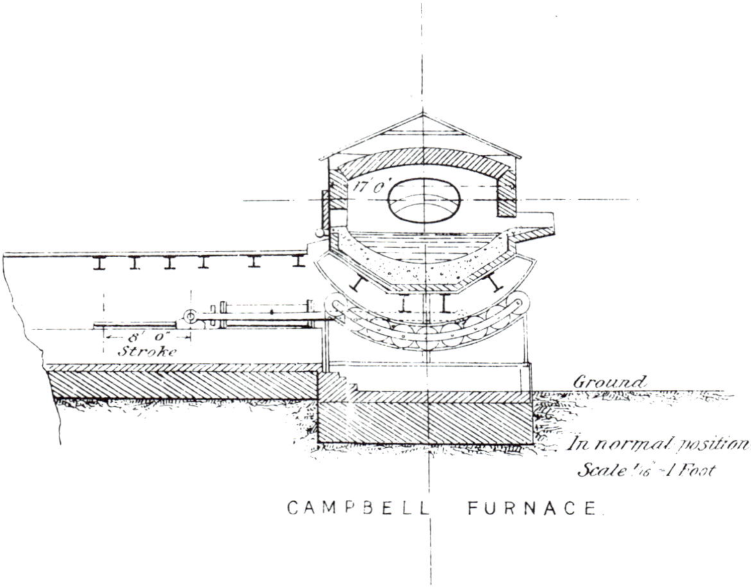

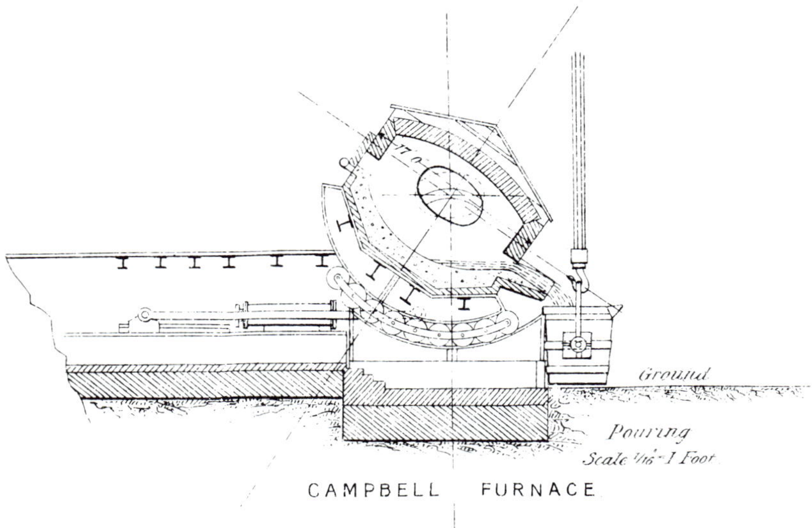

Fig. 27. The Campbell Tilting Open Hearth Furnace. This was the earliest design of tilting Open Hearth furnace, in which the furnace body could be rotated from the horizontal on rollers set in the arc of a circle, so enabling slag to be run off or metal to be tapped by mechanical means. This illustration (and the following one) are derived from the paper by A. Head, J.I.S.I., 1899, Pt. 1.

some adventitious air leakage clearly occurred. The ports and furnace ends were both faced with water-cooled cast iron plates. The charge was normally 80% liquid metal, taken direct from the blast furnaces, the balance being scrap, charged through chutes with the furnace tilted backwards towards the charging platform. It was usual to expect 14 or 15 heats per week with these furnaces.

The other type of tilting furnace in operation in America by 1900 was the design of S. T. Wellman, who had introduced Basic Open Hearth steelmaking to America at the Otis Works in Cleveland in 1886. This furnace had on its underside two circular steel castings which rolled on, and were supported by, strong steel supports with horizontal upper surfaces. The following surfaces were provided with rackwork which kept the furnace parallel without supporting any weight. The tilting was effected by two hydraulic cylinders, mounted on trunnions at their lower ends, the upper ends of the pinions being attached directly to the body of the furnace on the tapping side. To tilt the furnace, water would be admitted to the top end of each cylinder; in case of hydraulic failure, the weight of the furnace would return it to the horizontal position. In 1899 there were at least eleven of these furnaces in operation in America; six of these were at the South Chicago Works, four of 45 Ton capacity, the other two 27 Ton furnaces. In addition, there was under construction at this time a completely new works for the Alabama Steel and Shipbuilding Company at Ensley, near Birmingham, Alabama. This installation exemplified the growing philosophy in the United States that the Open Hearth process, more capable of mechanical improvement than the Bessemer converter, was the method for the future; foremost among these improvements were mechanical charging, which had been introduced by Wellman in 1896[9] and the tilting furnace, now considered to be a working proposition. The situation in Alabama was rather unusual; it was to work the cheaper pig produced from phosphoric ores in the Southern United States, rather than that from the low phosphorus Lake Superior ores which had fed the acid Bessemer converters for the previous twenty years and had made the United States the foremost steel producer in the world. An impression of the installation in Alabama can be gained from Figure 28, bearing in mind that there were to be ten such furnaces in a row,

[9] The mechanical charging device was an electrically driven trolley with an arm which could pick up an already charged pan, containing up to a ton of material, from a rack in front of the furnace. It could then be raised to the level of the charging door of the furnace,push the pan through the open door, rotate it on the arm through 180° and deposit its contents into the furnace. It would then return, replace the empty pan on the rack, pick up the next charged pan and repeat the process. The time saved in this way, compared with manual charging, obviously reduced the down time during which the furnace was not being actively used for melting.

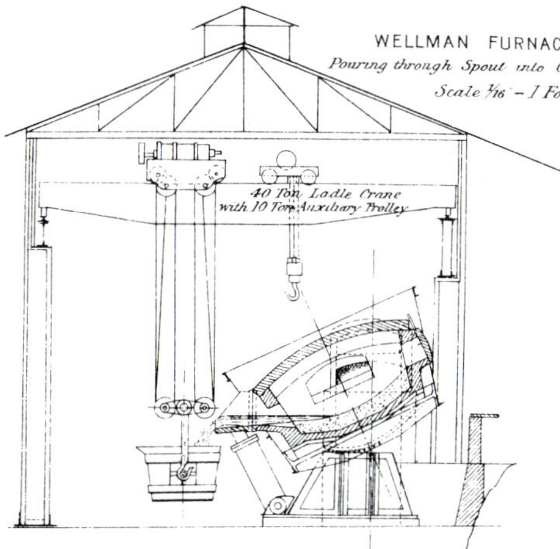

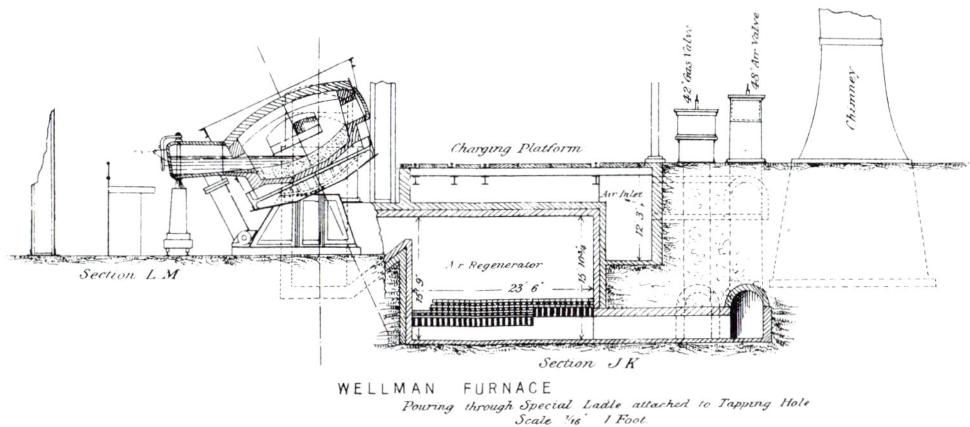

Fig. 28. The Wellman Tilting Open Hearth Furnace. This design of furnace relied on a hydraulic ram to actuate the tilting of the furnace body. These tilting furnaces were installed in increasing numbers in the early years of the twentieth century and, fed from with liquid metal from mixers and with mechanical charging devices for scrap loading, ensured that the Basic Open Hearth furnace became the major producer of steel for the next half century, rendering the Bessemer process obsolescent.

all of 50 Ton capacity, all with basic linings.[10] The angle of tilt for teeming was 25° with the horizontal, regulated by stops which could, however, be overridden manually for giving extra tilt to drain the furnace slag after tapping. The furnace doors were operated by pneumatic cylinders through wire ropes; on tilting they remained firmly closed. The only disadvantage of the tilting furnace seems to have been its obviously higher installation cost; against this were set the following advantages:

1. The slag which, especially in the basic process, is somewhat abundant and troublesome can be poured off at intervals during the melting process. If allowed to remain, it forms a more or less non-conducting covering to the bath, preventing the access of heat.

2. As the pouring hole of the furnace is above the level of the bath in the normal position, it is never closed up, but only loosely covered to exclude air. Consequently the time taken to open the tap hole before pouring, and to repair and close it after pouring, which sometimes amounts to one hour in an ordinary fixed furnace, is saved, as well as the labour connected therewith. The moment after the furnace is emptied the charging of materials can begin.

3. Since no injury is done to the pouring hole by opening and closing, the life of the furnace bottom is much prolonged. The occasional trimming required by the hole can be done at any convenient time.

4. The cold air which enters at the end ports when the furnace is tilted is an advantage, in that it chills a layer of slag on the surface of the metal, which is as effectively prevented from boiling or spurting as if it were in a ladle.

5. In every fixed furnace small inequalities in the bed must exist in which pools of metal lodge, and can with difficulty be removed. This leads not only to a waste of metal but to a diminution of the capacity and often to a distortion of the furnace bottom. In the tilting furnace every particle of metal and slag can be removed after each charge. A saving of metal of at least 2 per cent has been effected by the tilting furnace as compared with the fixed furnace, which is ascribed to this cause.

6. The tapping of the charge can take place at the exact moment when the metal is of the desired composition. As no opening of the tapping hole is required, no time need be lost in that operation. The advantage of this is specially felt when steel of exact composition is required for special purposes.

7. In case of any hitch or accident during pouring, the furnace can instantly be tilted back and pouring cease.

8. The tilting furnace lends itself readily to the transfer of metal from an acid to a basic furnace or vice versa.

[10] There seems to be a discrepancy here in that the crane shown in the drawing is designated as being of 40 tons capacity. The size of furnace was such that, if used with an acid lining and no extra capacity allowed for the bulky slag as in the basic process, it would have been rated at 60 ton. The integral tapping ladle shown here attached to the furnace should be noted. It is not clear, however, whether such an apparatus was actually used.

9. The whole body of the furnace is easy of access for repairs or examination. Should a breakout occur, the regenerative chambers will not suffer, not being underneath the body, as in the case of the older form of Siemens furnace.

10. The furnace bodies do not become deformed, as in the case of fixed furnaces, since as structures they are much stronger.

Desulphurisation in the Basic Furnace

As has already been pointed out, one of the chief problems in Basic Bessemer operation using the Cleveland ores was the provision of metal from the blast furnace which was sufficiently low in silicon as to produce a satisfactory slag without the use of excess lime, to have sufficient phosphorus to provide the necessary heat to render the metal fluid and to have a sufficiently low sulphur content. The problem of the phosphorus content was temporarily solved by the addition of puddling cinder to the blast furnace charge; this contained significant amounts of phosphorus removed from the Cleveland pig iron during the production of wrought iron and there were vast dumps available. The problems of silicon and sulphur, however, were inter-related. With an ore relatively rich in silicates and also containing significant amounts of sulphur containing minerals, the running of the blast furnace to provide a white cast iron, low in silicon, would automatically restrict the removal of sulphur; the production of grey cast iron, with free graphite, would give a material lower in sulphur, but the silicon content would be excessive. Windsor Richards in 1880 stated that his furnace-men had succeeded in producing a mottled Cleveland iron with 1% silicon and 0.16% sulphur and a white iron with 0.5% silicon but with 0.25% sulphur. He also suggested that both materials would produce excellent steel when taken in liquid form to the Basic Bessemer converter.[11] These problems persisted, however, and despite flirtations with duplex acid-basic processing and the use of active mixers to modify the blast furnace metal, the production of Basic Bessemer steel in the North East was never entirely satisfactory and by 1920 had ceased. It was, however, different with the more flexible operations in the Basic Open Hearth furnace, which became the major steelmaking operation world wide for the first half of the twentieth century. Even so, it had its problems, particularly with regard to sulphur removal; indeed it was stated in 1890 that virtually no sulphur was removed in the Basic Open Hearth furnace as normally worked and that, if the ore fed contained any substantial amounts of sulphur, almost all of this could be retained in the metal, giving an overall increment of sulphur during melting.[12]

[11] E. Windsor Richards: Presidential Address, Proceedings Cleveland Institute of Engineers, 1880-81, p. 38.

[12] H. Wedding: "Progress of German Practice in Metallurgy", J.I.S.I., 1890, Pt. 2, p. 547.

It was a metallurgical chemist, trained by Stead in Middlesbrough and having found his first employment at the North Eastern Steelworks, who applied himself to this problem around 1890 and whose process was still considered worthy of special mention in Stead's Presidential Address to the Iron and Steel Institute some thirty years later.[13] The chemist in question was Ernest H. Saniter, later to be Chief Metallurgist to Steel, Peach and Tozer Limited and a prolific contributor to the discussions at the Iron and Steel Institute meetings over quarter of a century. In 1890 he left the North Eastern Steelworks to become Chief Chemist to the Wigan Coal and Iron Company, one of the firms which had adopted the Basic Open Hearth process for steelmaking. He commenced a series of experiments in October 1890:[14]

> ... in recognition of the fact that sulphur was the worst enemy, and the only one over which the iron and steel manufacturer had not as yet obtained complete control.

The initial trials were made by running liquid blast furnace metal into a fused mixture of equal proportions of lime and calcium chloride, using 50 lb. of the mixture for each ton of metal. It was noted that up to 70% of the sulphur in the metal could be removed. The really important facet of Saniter's work, however, was the introduction of such a treatment as an integral part of Basic Open Hearth steelmaking. Operations of this type formed the subject matter of three patent applications in 1892,[15] the key to their success being quite plainly stated in the original patent, No. 8612:

> "... I have found that, with liquid slags containing up to 45 per cent of lime it is quite impossible to extract the sulphur and I believe there is no instance on record of the percentage of sulphur being reduced by this means. I have found, however, that, if the slag be kept so basic as to contain 50 to 60 per cent of lime, in which state it is in a very thick viscous condition, the addition of a chloride or even a fluoride of an alkaline earth has the property of causing the sulphur to leave the iron and join the slag."

The secret, therefore, was the provision of a fluid, exceedingly basic slag, with up to 60% lime content. This was achieved in the basic Open Hearth furnace by charging with the metal and scrap a much larger amount of limestone, about 2 cwt excess to the ton. When the charge was melted, the feeding in of calcium chloride was commenced to fluidise this slag; addi-

[13] J. E. Stead: Presidential Address, J.I.S.I., 1920, Pt. 1.

[14] E. H. Saniter: "A New Process for the Purification of Iron and Steel from Sulphur", J.I.S.I., 1892, Pt. 2, pp. 216-222.

[15] E. H. Saniter: British Patent 8,612 (6 May 1892): "Improvements in or Relating to Purification of Steel or of Iron during its Conversion into Steel"; British Patent 17,692 (4 Oct 1892): "Improvements in or Relating to the Purification of Iron or Steel from Sulphur"; British Patent 23,534 (21 Dec 1892): "Improvements in the Purification of Iron".

tions were made at intervals until some 50 lb. or so per ton of steel had been introduced. He pointed out that the driest chloride obtainable still contained 30% of water and that this should be further dried off in a reverberatory hearth held at 220°C before use. The procedure was then as normal, through to tapping the steel into the ladle and casting the ingots.

It is clear that this was no small scale trial; at the time of the presentation of a paper in 1892, some two thousand tons of steel had been made, 75% of each charge being what was referred to as "sulphur iron", the product having been sold for all purposes for which Basic Open Hearth steel was normally used – wire, hoops, rivets, tin bars and so on – and had been found to be fully equal in quality to that made from pure pig iron. No great care was necessary in the selection of the raw materials, the only objectionable constituents being silicon and silicates. Subject to this reservation, the commonest descriptions of iron, scrap and ore could be used. The yield of ingots was normal. The cost of the treatment was negligible at around a shilling per ton of ingot make, this being far more than offset by the relative cheapness of the raw materials used. Neither the brickwork of the furnace or of the regenerators had suffered any ill effects from the uses of calcium chloride over a six month period.

A table showing the quality of iron used and the results obtained may be quoted as follows:[16]

	Pig Iron used				Mean S% in Charge	Steel Made				
	Si	S	P	Mn		C	Si	S	P	Mn
1.	.04	.76	1.30	.18	.58	.22	tr	.081	.027	.68
2.	.10	.45	2.10	.50	.35	.20	tr	.072	.052	.75
3.	.10	.45	2.10	.50	.35	.19	tr	.048	.054	.59
4.	.04	.25	2.60	1.00	.20	.08	tr	.048	.025	.43
5.	.20	.23	2.60	1.00	.19	.17	tr	.048	.045	.57
6.	.20	.22	2.60	1.00	.18	.15	tr	.063	.042	.73
7.	.20	.22	2.60	1.00	.18	.39	tr	.053	.045	.61
8.	.40	.22	2.50	1.30	.18	.18	tr	.018	.034	.40
9.	.20	.17	2.60	1.20	.14	.15	tr	.038	.040	.58
10.	.18	.16	3.10	1.50	.13	.13	tr	.032	.040	.20
11.	.44	.15	3.50	1.50	.13	.75	tr	.042	.040	.60
12.	.20	.15	2.60	1.00	.13	.15	tr	.038	.040	.58
13.	.65	.13	3.30	1.50	.11	.16	tr	.025	.035	.68
14.	.56	.05	.05	3.00	.05	.12	tr	.016	.010	.12

16 It should be noted that No. 14 in this table was the result of a trial with haematite pig iron in an attempt to produce ingot iron of the quality of Swedish bar iron.

These figures show a range of sulphur removal efficiencies from 65% to 90% with a mean of 78%; this certainly is impressive. Confirmation of such results was also provided by Dr. Stead subsequent to a visit to the Wigan works. His report provides a further table:[17]

	C%	Si%	S%	P%	Mn%
Calculated Analysis of Charge	1.67	0.15	0.37	1.67	0.46
7.20 am: After Melting	0.40	tr	0.36	1.25	0.22
8.35 am: After adding $CaCl_2$	0.33		0.093	1.11	
10.10 am	0.22		0.082	0.34	
11.30 am:	0.15		0.058	0.065	
Steel	0.15	tr	0.047	0.056	0.59

This shows a total sulphur removal of 87%. The lime content of the slag before addition of the calcium chloride was 58.5%.

The following year Saniter presented a further series of results,[18] pointing out that some 16,000 tons of steel had by then been made from high sulphur iron at Wigan, considerable quantities having been treated at other works in the country with excellent results. He also gave details of a further twenty five melts of Basic Open Hearth steel made using white haematite pig iron treated by the process to produce ingot iron to compete with the best Swedish bar iron; these carried from 0.005% to 0.024% sulphur and from 0.002% to 0.020% phosphorus. This material had been tested in Sheffield as a substitute for Swedish bar iron in making the highest class cutlery and tool steels, with marked success. It had also been found almost equal to the best charcoal iron for conductivity purposes; it welded splendidly and a typical sample gave a breaking strain of 24 tons per sq. in., with 40% elongation and 77% reduction in area, classed as most extraordinary and equal to or better than most test results given by Swedish bars.

Typical test results on specimens with an 8 in. gauge length taken from a variety of steels made in the Basic Open Hearth furnace using the Saniter process at Wigan were quoted as follows:

[17] J. E. Stead: "On the Elimination of Sulphur from Iron", J.I.S.I., 1892, Pt. 2, pp. 260-265. It should be noted that this report on the Saniter process formed a small part of a far-ranging survey of the current state of the art.

[18] E. H. Saniter: "A Supplementary Paper on a New Process for Desulphurising Iron and Steel", J.I.S.I., 1893, Pt. 1 pp. 73-77.

C %	Si %	S %	P %	Mn %	Breaking Stress t.s.i.	Elongation %	Reduction of Area %
0.10	tr	0.049	0.050	0.57	23.4	27.8	60.4
0.115	tr	0.053	0.059	0.60	24.4	30.1	52.6
0.135	tr	0.050	0.053	0.54	25.1	29.6	52.7
0.155	tr	0.051	0.052	0.57	25.9	28.5	51.1
0.205	tr	0.041	0.049	0.56	27.0	27.1	47.2
0.245	tr	0.053	0.062	0.56	29.2	28.2	53.6
0.27	tr	0.044	0.035	0.49	29.3	23.5	34.7
0.315	tr	0.062	0.070	0.70	33.2	23.0	31.9
0.34	tr	0.043	0.050	0.69	33.4	26.2	49.0
0.37	tr	0.054	0.065	0.70	33.1	23.3	33.8
0.54	tr	0.056	0.045	0.58	39.1	21.8	41.8
0.62	tr	0.040	0.055	0.70	48.8	14.0	22.8

Trials had also been made with the process in the Basic Bessemer converter by the Lilleshall Company, using a somewhat larger lime addition than usual, together with 40 lb. of calcium chloride per ton of metal; the blowing proceeded in the normal fashion. The results showed an average sulphur removal of 78% on the treated casts, as against 45% in normal charges – sufficient, the author stated, to make the difference between bad steel and good steel.

An important statement in this report was that some trials had been made substituting calcium fluoride for part of the calcium chloride; this was more infusible and took longer to react but, apparently, was eventually quite effective.

G. J. Snelus appended a report on the operation of the Saniter process at Wigan, having witnessed charge C1134 on 21st February 1893.[19] He provided the following details:

12 tons white cinder pig iron (0.38% S and 2.57% P) charged, together with 3 tons steel scrap (0.06% S); 1 ton 10 cwt limestone and 15 cwt purple ore charged with the metal.

1 ton 7 cwt 2 qr lime, 1 ton limestone, 16 cwt 1 qr pottery mine and 7 cwt of 70% calcium chloride used for feeding.

15 cwt common No. 3 pig (1.5%P) used for pigging.

1 cwt 1 qr 20 lb 80% ferromanganese added to steel ladle.

Ingots made: 14 tons 15 cwt.

[19] G. J. Snelus: "Report on the Saniter Desulphurising Process", J.I.S.I.1893, Pt. 1 pp. 77-83.

Furnace full 11.25 am; Furnace tapped 10.35 pm.

The metal had been all melted down for some time when No. 1 sample (melted sample) was withdrawn at 5.0 pm.

5.20 pm. 2 cwt dry calcium chloride charged, then more lime and a further 2 cwt 2 qr calcium chloride.

6.10 pm. No. 2 samples of metal and slag were drawn. More limestone and about 12 cwt pottery mine were then charged.

7.15 pm. No. 3 sample drawn.

8.15 pm. No. 4 sample drawn and about 4 cwt phosphoric No. 3 pig iron added.

8.20 pm. More lime added, then further 2 cwt 2 qr calcium chloride.

8.30 pm. 4 cwt No. 3 pig added. Sample tested for sulphur.

9.10 pm. No. 5 sample drawn.

9.15 pm. More pig charged. Both samples tested for sulphur.

10.00 pm. Sample tested for sulphur.

10.35 pm. Sample No. 6 drawn and the furnace tapped.

Metal Analyses:	1	2	3	4	5	6	Final Steel
Carbon, %	0.200	0.090	0.070	0.060	0.060	0.055	0.100
Silicon, %	0.060		0.030	0.006	trace	trace	
Sulphur, %	0.320	0.181	0.140	0.129	0.093	0.041	0.040
Phosphorus, %	1.075		0.591	0.163	0.054	0.027	0.030
Manganese, %	0.200						0.504

Slag Analyses:	1	2	3	4	5	6
Silica, %	18.30	15.00	14.20	12.80	11.60	10.80
Ferrous oxide, %	5.39	3.59	3.79	4.11	5.65	7.19
Ferric oxide, %	3.50	2.60	2.60	2.60	2.80	2.60
Lime, %	49.24	49.60	50.00	50.94	55.64	57.00
Sulphur, %	0.315	0.576	0.507	0.480	0.659	0.645
Phosphoric anhydride, %	11.75	16.10	19.20	22.49	18.74	
Alumina, %		6.00		4.00		
Calcium chloride %		5.87				3.09

Snelus also witnessed the rolling of ingots from this cast. The ingots were 12 in. square, weighing 16 cwt, and were reheated and rolled, with twelve passes in the cogging rolls and nine in the finishing rolls, to produce 2 in. square billets. He reported:

> I was struck with the remarkably fine quality of the steel. I could not detect the slightest flaw or sign of red-shortness in any of the blooms or in the

> finished billets, although I examined them minutely. In fact, I never saw any steel roll cleaner than this ... clearly this process is most effectual in desulphurising the commonest kind of pig iron, to such an extent that the highest class of steel can be made from it. The materials employed are not costly and there is no difficulty in carrying out the process and in producing uniformly good results.

Evidence of the interest shown in the Saniter process comes from a report of trials carried out the following year by the Krupp Steelworks in Essen.[20] The first series of tests was made on charges of 7.5 tonnes. The raw materials were relatively low in sulphur but it was desired to reduce this to a minimum; the charges were one third pig with 3% manganese and two thirds steel rail scrap: 0.3 to 0.4 tonnes of limestone were used and the addition of 70% calcium chloride was 0.15 to 0.20 tonnes; the calcium chloride was heated to drive off the bulk of the moisture and was added to the charge in the molten state. Prior to the addition, the sulphur contents ranged from 0.026 to 0.050%; the final analyses were from 0.012 to 0.024%; sulphur removal varied from 35% to 65%, with a mean of 51%. Whilst this was clearly less efficient than with higher sulphur contents, it was significant that normal practice brought about negligible removal of sulphur at these levels. Similar trials with charges containing 0.10% sulphur gave sulphur reductions of 60% to 70%.

Further trials were carried out using charges consisting entirely of high sulphur pig irons in a 10-tonne furnace. These confirmed the trend already demonstrated that the degree of sulphur removal was higher with higher initial sulphur contents. These was also some indication that some sulphur reversion could occur towards the end of a melt, as shown in the following example, which, nevertheless, demonstrates an 82% reduction in sulphur content:

TRIAL 4: The furnace was charged with 600 kg. limestone, 220 kg. burnt lime, 280 kg. calcium chloride, 500 kg. Bilbao ore and 10 tonnes of a high sulphur Luxembourg pig iron. When the whole was well molten and the slag well formed sample (I) was taken. Lime was then added, the slag removed, lime and ore added and sample (II) taken two hours after meltdown. 140 kg further calcium chloride was then added and sample (III) taken. Lime and ore were now added to decarburise the charge and sample (IV) was taken. 100 kg. of ferrosilicon were then added and the charge tapped. The analytical results were as follows:

[20] These results were published in Stahl und Eisen, Vol. xv., pp. 616-617 and were abstracted under the title "Desulphurising in the Basic Open Hearth" in J.I.S.I., 1895, Pt. 2, pp. 511-515.

Sample	C%	Si%	Mn%	P%	S%
Original charge	3.28	0.18	0.17	1.60	0.40
I	2.31		0.15	0.99	0.26
II	1.79		0.28	0.48	0.074
III	1.60		0.27	0.18	0.040
IV	0.054		0.14	0.010	0.063
Final steel	0.093	0.023	0.15	0.024	0.071

The subsequent history of the Saniter process is not documented. It was clearly considered sufficiently important to receive attention in metallurgical textbooks of the early twentieth century[21] but there had been an obvious change in that calcium fluoride had largely, if not totally, been substituted for the calcium chloride. With hindsight, this surely was not unreasonable, since the care required on account of the tendency of calcium chloride to absorb moisture and convert itself into a sticky mess must have been irksome on the furnace stage, whilst calcium fluoride was much more convenient; the relatively longer fluxing period required was of little importance when the overall process time in the Open Hearth furnace is considered. It seems, however, that the use of either the chloride or fluoride had a detrimental effect on the solubility of the phosphate content of the slag in citric acid, thus detracting from its use as a fertiliser. Significantly, however, an investigation done on this topic in 1903 does not mention the use of chloride. The report commenced:[22]

> Fluor Spar, although having its advantages in acting as a flux, producing a thinner and more liquid slag as assisting in the desulphurising and desphosphorising of the pig-iron, has a decided disadvantage in the effect it produces on the basic slag made.

It then proceeded to demonstrate the effect of addition of various proportions of fluorspar (calcium fluoride) to the slag in 18-ton charges of Basic Open Hearth steel, providing the following results, comparative tests on slags produced without the addition of fluorspar having given citric acid solubilities of 83.9% to 90.0% of the phosphoric acid present:

[21] See, for instance, F. W. Harbord and J. W. Hall: "The Metallurgy of Steel", 3rd. ed., London, 1907 , pp. 187-189 and H. H. Campbell: "The Manufacture and Properties of Iron and Steel", 4th. ed., New York, 1907, pp. 202-204.

[22] W. G. Burman, Iron and Coal Trades Review, 9 Jan 1903, p. 94

Fluorspar Addition, cwt.	Total Phosphoric Acid	Soluble Phosphoric Acid	% Solubility
1.5	15.81	8.44	53.3
2.0	17.66	7.42	42.0
5.0	15.84	3.82	24.1
6.0	15.81	3.45	21.8
7.0	15.37	2.56	16.6
10.0	13.43	0.78	5.8

The differences in solubility were attributed to the formation of an insoluble fluor-apatite in the slags treated with calcium fluoride.

It is a matter of record that the Basic Open Hearth process as conducted between 1920 and 1960, when it was responsible for the bulk of steel made world-wide, relied on the use of fluorspar as slag fluidiser; when the use of calcium chloride was abandoned is not clear but would seem to have been within a few years of Saniter's work. Nevertheless, the principle of the provision of a highly fluid, highly basic slag for desulphurisation, as postulated by Saniter in 1892, had its full part to play – and still has wherever basic steelmaking is carried out, particularly in the electric arc furnace.

Recarburisation of Steel in the Ladle

In the original form of the Bessemer process, control of the carbon content of the final steel had been relatively straightforward; either the carbon content was "caught" by stopping the blow at the appropriate time or the whole of the carbon was removed and then brought back by appropriate additions of spiegeleisen, haematite pig iron, ferromanganese, and so on. The same applied to the Siemens Martin and the original Siemens Open Hearth processes. Whilst the introduction of the basic processes had certainly opened up a much wider production capability, it was soon recognised that the recarburisation of the blown metal presented considerable difficulties if the reversion of the phosphorus was not to occur. Even pouring off virtually the whole of the slag before the addition was made, which reduced the magnitude of the problem, was not a real solution and it was commented that:[23]

> ... the production of a low phosphorus steel high in carbon was not accomplished with that degree of certainty and facility which would lead one to regard the introduction of another, simpler method of carburising as other than highly desirable.

[23] A. Thielen: "On the Darby Process of Recarburisation", Amer. Inst. Min.. Engs., Vol. xix (1890-91), pp. 790-803.

As it so happened, there was an enormous demand for softer steel about the time that the basic processes became viable. As one of the foremost American manufacturers reported in 1893:[24]

> The absence of any consideration on my part of methods of carburisation arises from the fact that in our country the Basic Open Hearth produces almost exclusively low steels, below 0.20 per cent carbon.

There were needs elsewhere, however, for producing a range of carbon contents from the basic Open Hearth furnace and the problem was solved in a most straightforward manner at the Brymbo Steel Works. Here it was that J. H. Darby, one of the pioneers of the process in Britain, found that the addition of suitable carbonaceous matter to the metal stream, as it issued from the furnace into the ladle, could be used successfully to provide the desired carbon content in the ingots from more or less completely decarburised metal. The method was patented in Britain,[25] originally specifying the use of graphite or wood charcoal as the carburising agent. Information on the use of the method, however, comes from Germany.[23] Original trials at Ruhrort had been made by feeding powdered coke through a worm screw turned by hand into an intermediate tundish between the ladle and the mould.[26] During these trials it was observed that the coke was almost instantaneously taken into solution and the following figures indicate the effectiveness of the method:

Ingot No.	Desired C%	Actual C%		
		Bottom	Middle	Top
1	0.25	0.255	0.260	0.251
2	0.25	0.244	0.248	0.254
3	0.30	0.285	0.311	0.306
4	0.35	0.330	0.324	0.340
5	0.35	0.350	0.357	0.352
6	0.45	0.461	0.455	0.470

[24] H. H. Campbell produced a magisterial survey on "The Open Hearth Process", Trans. Amer. Inst. Min. Engs., Vol. xxiii (1893-94), pp 345-510. In the author's reply to discussion on this paper the above comments were made (p. 696). During the same discussion (p. 691), it is to be noted that Dr. Wedding pointed out that these low carbon steels would not be referred to in Germany as "Stahl" (Steel) but as "flusseisen" (melted iron), the term "Stahl" being reserved for a material which hardened sensibly on quenching whilst "flusseisen" was an ingot form of "schweisseisen" or wrought iron.

[25] J. H. Darby, British Patent 418, 10 January 1888.

[26] It is interesting to note that the feed rate of the powdered coke was reduced as the metal level neared the top of the mould, to offset the normal higher carbon content, due to segregation effects, which had been noted in the head of the ingot.

The next stage was to fit a chute, protected by a refractory coating, above the ladle and to feed coke into the metal stream issuing from a basic converter, having run off as much slag as possible previously and holding back the small amount of residual slag. The results reported on this trial were as follows:

Heat No.	Carbon %			Phosphorus %	
	On tapping	First ingot	Last ingot	First ingot	Last ingot
1	0.073	0.285	0.290	0.040	0.040
2	0.070	0.268	0.266	0.052	0.070
3	0.082	0.293	0.290	0.045	0.054
4	0.078	0.301	0.320	0.038	0.046
5	0.080	0.274	0.276	0.049	0.049
6	0.091	0.315	0.320	0.074	0.085
7	0.083	0.298	0.290	0.063	0.072
8	0.095	0.274	0.277	0.048	0.049
9	0.072	0.284	0.297	0.055	0.067
10	0.083	0.269	0.266	0.042	0.050
11	0.089	0.300	0.312	0.044	0.056
12	0.077	0.323	0.320	0.073	0.075

The same procedure used on Basic Open Hearth metal gave similar results:

Heat Number	% Carbon Required	% Carbon on Tapping	% Carbon First Ingot	% Carbon Last Ingot
1	0.50	0.075	0.510	—
2	0.60	0.072	0.592	0.592
3	0.65	0.069	0.668	0.660
4	0.70	0.076	0.708	0.700
5	0.80	0.070	0.804	0.804

On the basis of such figures, the process was accounted a satisfactory one and had been put into use whenever higher carbon materials were required in the works. Comparisons were also quoted as to the economies in alloy additions brought about by this process. With 10 ton charges in either case, the figures quoted were as follows:

	10-12% Mn Spiegeleisen	60-65% Ferromanganese	Powdered Coke
Basic Converter:			
Normal Practice	1320 lb.	176 lb.	Nil
Carburising Practice	Nil	176 lb.	132 lb.
Basic Open Hearth:			
Normal Practice	550 lb.	176 lb.	Nil
Carburising Practice	Nil	26 lb.	100 lb.

The loss of carbon from the addition by combustion was said to be of the order of 15-20% using graphite and somewhat greater using coke; obviously this feature was taken into account when the additions were calculated.

In the discussion of the paper from which this data has been taken, Gilchrist reported that he and Arthur Cooper from the North Eastern Steel Works had visited Ruhrort to see the recarburisation being practised and, within a week of their return, the process was being applied with complete success in the works at Middlesbrough.

A variant of the process was reported from Dudlingen in Luxembourg in 1893 in which coal dust and hydrated lime were mixed together with sufficient water to enable them to be moulded into blocks. These were then placed in the bottom of the ladle before the steel was tapped. The result was said to be steel with the desired carbon previously calculated from the weight of carbon used; it was also suggested that the evolution of steam from the blocks assisted in the dissemination of the carbon by its stirring action.[27]

It should be pointed out that it was not long before the British steelworks adopted the used of crushed anthracite coal as the carburising addition and this was being added to ladles of Basic Open Hearth steel for the next fifty years or more – indeed, until the process became obsolete.

The Bertrand Thiel Process

The difficulties when working with blast furnace metal high in silicon in the Basic Open Hearth furnace led to the development of a most interesting process at the Kladno Works near Prague in 1894. This consisted of prior treatment of the metal in one Open Hearth furnace, whereby the bulk of the silicon and some of the phosphorus could be removed, with the production of a slag high in phosphorus pentoxide, followed by passing this metal to another Open Hearth furnace to finish the metal under more ideal conditions. This became known as the Bertrand Thiel process, from its co-inventors.

[27] H. Wedding, in discussion of the paper on "Open Hearth Process", Trans. Amer. Inst. Min. Engs., Vol xxiii (1893-94), p. 691.

TABLE VIII CHEMICAL CHANGES DURING THE HOESCH OR BERTRAND–THIEL PROCESS

Time	Analysis of Metal					Analysis of Slag								Remarks
	C	Si	Mn	S	P	FeO	MnO	Al_2O_3	CaO	MgO	P_2O_5	S	SiO_2	
12.10	3.28	0.32	0.96	0.132	1.86									Charge Pig 51,310 lb, Lime 4,144 lb Swedish Ore 7,582 lb, Mill Scale 1,697lb. Pig run in.
1.10	2.47	tr.	0.17	0.102	0.59									
1.30	1.90	tr.	0.22	0.098	0.47	10.25	5.03	1.36	41.56	4.32	22.85	0.069	12.20	
1.40	1.65	tr.	0.22	0.098	0.37	7.06	4.96	1.38	45.48	4.00	22.36	0.124	11.80	
2.30	1.46	tr.	0.34	0.082	0.26	4.67	3.93	1.60	48.86	4.00	22.13	0.138	11.40	
3.00														End of First period; metal tapped and taken to second furnace which held: Spathic ore 4,804 lb, scrap 11,928lb, lime 2,843 lb.:
3.50	0.385	tr.	0.29	0.100	0.090	19.64	13.49	3.17	33.64	6.92	6.25	0.110	14.20	
4.10	0.205	tr.	0.23	0.089	0.050	15.78	12.08	3.00	35.78	6.70	6.70	0.165	15.40	Charged 210 lb mill scale
4.30	0.090	tr.	0.26	0.90	0.045	15.83	8.88	2.32	43.88	6.00	5.50	0.289	13.20	
4.45	0.075	tr.	0.26	0.080	0.035	14.13	10.19	2.21	43.79	6.10	5.57	0.206	14.20	Charged 352 lb lime; at 4.55 453 lb added.
5.05	0.058	tr.	0.27	0.078	0.035	16.23	8.32	2.20	45.30	5.90	5.55	0.275	14.00	
5.10	0.045	tr.	0.25	0.077	0.030	17.20	7.67	2.00	46.19	6.12	5.15	0.316	13.10	440 lb. ferromanganese
5.15	0.080	tr.	0.47	0.067	0.040	17.03	10.25	1.90	46.28	5.92	5.00	0.344	12.40	added Furnace tapped.

Source: D. Carnegie and S. C. Gladwyn: "Liquid Steel: Its Manufacture and Cost", 2nd. ed., London, 1920, p. 269

It was pointed out by Bertrand[28] that, since complete elimination of the phosphorus was not intended in the primary furnace, less lime might be added than would otherwise be considered necessary and the quantity of slag to be melted was therefore materially diminished; being in a thinner layer, the flame could act with greater intensity on the metal. In practice, using the two furnaces, the high silicon phosphoric pig would be charged to the primary furnace and nearly all the scrap into the finishing furnace, adding such quantities of ore, lime and so on as circumstances and experience demanded. It was elsewhere[29] stated that originally about six heats could be obtained in this way every 24 hours.

Some work was carried out in Britain at the Round Oak Steelworks and the fully developed procedure later applied in a specially built plant at Brymbo Steelworks. On the continent it was adopted on a major scale by the works at Hoesch – so much so, that this method of working was commonly referred to as the Hoesch process. At Hoesch they regularly made up to ten heats in 24 hours and had no difficulty in meeting 0.015% phosphorus maximum in soft steel and 0.020-0.040% in hard steel, the furnace life being in excess of 240 heats. It does seem that there was a tendency to bottom trouble in the second furnace, however. Details of a typical operation at Hoesch may be found in Table VIII; it will be seen that, after oxidation, the primary heat was tapped into a ladle, the slag poured off and the metal then charged to the second furnace which had already received a charge of lime, ore and scrap; the final metal was treated with ferromanganese and then tapped. One rather important feature of the process was that the bulk of the oxidised phosphorous was concentrated in the first slag, which was high in lime and low in iron oxide, and, therefore, eminently suitable for use as agricultural fertiliser.

The Talbot Process

One of the main economies of the Bessemer process resulted from the use of liquid metal from the blast furnace, thus obviating the necessity for the remelting of pig iron in cupolas; this had been made even more acceptable by the use of the mixer which provided a material of more regular composition to the converter. There was clearly a similar advantage to be gained if liquid blast furnace metal could be fed to the Open Hearth furnace. It was with this in mind that a series of trials were made at the Wishaw Steel Works, in 1898, using liquid blast furnace metal in a 25 ton Acid Open Hearth furnace and it was reported that no problems were found; the rate of working

[28] E. Bertrand: "The Practice of the Combined Open Hearth Process of Bertrand and Thiel", J.I.S.I., 1897, Pt 1, pp. 115-124.

[29] D. Carnegie and S. C. Gladwin: "Liquid Steel: Its Manufacture and Cost", 2nd. ed., London, 1920, pp. 267-269.

on the furnaces was improved (a weekly output if 340 ingot tons appears to indicate the product of 14-15 heats), the yield of metal from the furnace was somewhat higher than usual (presumably due to oxidation with iron ore rather than with air during the melt down period of the ordinary procedure) and no furnace bottom problems were encountered over a prolonged period during which 16,000 ingot tons were made.

It seems, however, that general experience in trials in the Basic Open Hearth furnace, with blast furnace metal poured into the furnace already charged with some lime, was not favourable, the hearth not standing up to the erosion which ensued.[31] This gave rise to what became known as the Talbot Process.[32] This called on the recent improvements of the tilting furnace and the mixer but was never completely emptied; some steel was always left in the furnace to act as a cushion for the addition of the next charge of blast furnace metal. In addition, the slag was made very basic and very oxidising before the addition of the blast furnace metal, so that a vigorous reaction occurred. Furthermore, all the oxidation of the impurities in the blast furnace metal was carried out by slag reaction and not by air oxidation and this automatically introduced extra iron into the bath from the ore used, giving a higher yield of metal. The initial trials were carried out at the Pencoyd Works in Pennsylvania, using a 75 ton tilting furnace. Talbot reported on them as follows:

> The pig iron to be converted has an approximate composition of:

C	Si	Mn	S	P
3.75	1.00	0.40	0.06	0.90

> Owing to the absence of blast furnaces, this pig iron has to be melted in cupolas. The furnace should be charged on Sunday evening with about 50 per cent molten cupola metal and 50 per cent scrap, and this first (or filling) heat is worked down to steel in the usual way. When the bath is good finished steel, about one third of it is poured off into a ladle and cast into ingots. No slag is run off with the steel. After tapping off this one third of the charge, oxide of iron in a fairly fine state of division is added to the slag and as soon as this is melted about 20 tons of molten cupola metal are run in to replace the steel tapped out. An immediate reaction takes place, during the continuance of which the gas is cut off from the furnace. The reaction has all the characteristics of the Bessemer blow during the elimination of the carbon, a

[30] J. Riley: "The Use of Fluid Metal in the Open Hearth Furnace", J.I.S.I., 1900, Pt. 1, pp. 22-32.

[31] Harbord and Hall, op. cit., p. 193.

[32] B. Talbot: "The Open Hearth Continuous Steel Process", J.I.S.I., 1900, Pt. 1, pp. 33-61.

large volume of carbon monoxide being given off, which immediately ignites and burns with an intense flame, the heat thus produced partly raising the temperature of the bath and partly being absorbed by the regenerators. After the metal has boiled vigorously for some ten to fifteen minutes, the slag, which is now almost deprived of iron oxide, is partly poured off and the bath worked down into finished steel by the help of fresh additions of iron ore and lime. When the bath is ready, one third (or about 20 tons) of steel is cast, fresh slag additions are made and another 20 tons of molten cupola metal is run in as before. These operations are continued during the whole of the week, the furnace being completely emptied only on the Saturday. The tap hole is some few inches below the fore-plate level, so that no slag is taken in the steel ladle. This is arranged by stopping the hole sufficiently to prevent metal or slag working into it. When the furnace is ready to tap, it is tilted slightly so that the metal is above the tapping hole, a bar is plunged through and metal always runs first. As soon as the desired quantity is obtained, it is tilted back over towards the charging side. This causes the level of the bath to fall well beneath the inside of the taphole, which makes it possible for the furnace helper to dry and clean it rapidly and plug it up without delay. Whilst this is being done, the slag line is examined, gas is turned on and whatever repairs are necessary are made by throwing raw dolomite, or limestone, mixed with 5 per cent pitch or rosin, on the banks at a space of from six to nine inches above the level of the reduced bath. The surface of the bath is very useful since it prevents the repairing material, in a great measure, from rolling down. It acts as a floor or foundation and tends to hold the dolomite in place. Whatever does roll down becomes incorporated in the slag and is beneficial. About three hours forty minutes is the period of time between two successive casts at Pencoyd and ... starting on Sunday night ... from 32 to 34 would readily be obtained... per week. It is not possible to give the actual consumption of fuel, as the furnace is worked from the main gas tube, which supplies several other furnaces. It is interesting to note that from 25 to 33 per cent less deoxidisers are required for the steel from this continuous furnace than for the other furnaces to give the same percentage of manganese (from 0.40 to 0.50 per cent) in the finished steel. Many hundreds of heats and thousands of tons of steel have been made by this method with very satisfactory results, all grades of metal having been produced, from dead soft up to 0.40 carbon steel. The question of excessive wear on the hearth, about which many doubts were entertained, and upon which the practicability of the method depended, has never given the slightest trouble, the hearth being practically in as good a condition as when put in eight months ago. This proves that a heavy surface reaction of short duration is confined to the bath and does not affect the bottom of the hearth. ...

The author considers that the method of working described in this paper offers advantages over the ordinary practice in many particulars. Amongst the chief of these may be cited:

1. The cost and delay of charging cold material is avoided.

2. A saving of fuel in charging molten pig iron and also through not cooling the furnace by charging cold material.

3. The demand for a large and regular supply of good scrap, so important in ordinary practice, is wholly dispensed with.

4. A regular supply of steel to the mills in any wished for quantity and at frequent intervals is insured.
5. An increase in output.
6. An increase in yield.
7. Less repairs to furnace.
8. Saving in labour charges, due to far less skilled labour being required per given quantity of steel.
9. The possibility of using very large furnaces, with consequent reduction in cost of production, without the necessity for very large cranes and ladles.

In a summary of six weeks working, starting with fluid metal on Monday night each week, the following average figures were given:

No. Heats per Week	27
Time per heat, excluding filling and emptying heats	3.8 hours
Weight of Metal Charged per week	488 long tons
Weight of Ingot and Scrap per week	517 long tons
Yield of Ingot and Scrap per 100 tons metal charged	105.94 tons

The announcement of this work came right at the end of the century but it had considerable impact. Within two years the process was in operation at the Frodingham Steel Works in Scunthorpe, using a 100-ton furnace.[33] The developments at Frodingham after the initial steelmaking operations in 1890 are of interest. By 1895 the output had risen to 670 tons per week; in addition to the original two 15-ton furnaces, A and B, two 20-ton furnaces, C and D, had been commissioned. In 1896 output rose to 842 tons per week following the introduction of E furnace of 30 tons capacity, to be followed by the 30-ton F furnace in 1898, with two 40-ton furnaces, G and H, being added in 1900. In 1901 Joseph Cliff and Maximilian Mannaberg visited America and saw the Pencoyd operations and on their return the old A and B furnaces were demolished and preparations made for the installation of a 100-ton tilting furnace for the operation of the Talbot process, for which the building was modified to include an overhead crane, foundations put down for the enlarged regenerators and the rocker stands required for the Wellmann furnace and a railway installed on the stage between the charger and the

[33] The earlier history of steelmaking at Frodingham can be found in Chapter 13. Details of the installation of the Talbot Process at Frodingham quoted here are derived from G. R. Walshaw and C. A. J. Behrendt: "The History of Appleby Frodingham", published privately by United Steel Companies, pp. 69-70. Within ten years there were some fourteen tilting furnaces, of from 100 to 250 tonnes capacity, operating the Talbot Process in Britain. In 1910, Jones and Laughlin in America had thirteen 200 ton furnaces on continuous working by the process. (Carnegie and Gladwin, op. cit., p. 390). It may be of interest to note that, when the author began his metallurgical studies at Sheffield University in 1938, the details of the Hoesch and the Talbot processes were an integral part of the first year lectures on "Iron and Steel Production".

furnace to accommodate the charging bogies; two large mechanically operated gas producers were provided. This was the first installation of its type in Europe and Benjamin Talbot personally spent much time at Frodingham in the early stage of the installation. The output of this furnace alone was over 600 tons per week; it was, indeed, so successful that a second furnace was commissioned within five years.

Views on the Solidification of Ingots

One of the most interesting features which a modern steelmaker discovers in studying the reports of this period, almost a century ago, is the large amount of investigatory work which was being carried out, particularly with regard to the causes for the formation of blowholes and contraction cavities in ingots. A most prestigious American text, published in 1891, devotes many thousands of words to the topic, citing numerous sources.[34] As far as blowhole formation is concerned, it is obvious that ideas about "deoxidation" were still a little uncertain at this date, despite the publicity given to the work at Terrenoire by Holley several years earlier. The method of solidification of liquid steel was clearly understood, including the formation of a central cavity, or "pipe", by the contraction of the metal on solidification, since the density of the liquid metal was less than that of the solid which crystallised from it. There is, indeed, in this connection, a most apt description:[35]

> Immediately the molten steel is poured into it [that is, the ingot mould] thin skins of solidified metal form on the cold sides of the mould and a thinner skin on the surface of the metal which is exposed to the air. As the fluid metal contracts, this upper crust loses its support and sinks into the form of a saucer, while the skin on the sides keeps thickening by the addition of successive solidified layers. All this time the level of the fluid in the centre is falling steadily so that the walls which are thin at the top, where first formed, increase in thickness downwards as more time is allowed them in which to collect material from the fluid core, until there is at last no fluid left. The result is that a hole having the form of an inverted cone, with a thin skin above it, is formed at the upper end of the ingot.
>
> In any case, the feeding of the lower central portion of the ingot cannot continue after the metal has attained to a plastic condition, but as the contraction still continues, the central portions which cool last find themselves contained in a rigid shell which, when they have fallen in temperature, is far too large for them, so that the particles are stretched out in all directions

[34] H. M. Howe: "The Metallurgy of Steel", 2nd. ed., New York, 1891, pp. 146-163. This volume is notable for its weightiness in more senses than one; the page size is 13.5" x 10.5" and the paper is quite substantial. The binding, however, did not stand up to the wear and it is unusual to find a copy in which the pages are still attached to the covers or each other.

[35] Harbord and Hall, op. cit., p. 285.

> to fill it. The first portion of this contraction occurs when the metallic crystals are just forming in the sluggish fluid, in which state the steel somewhat resembles moist brown sugar, the crystals of which are held together by the treacle it contains. At this critical point, when the action of gravity is insufficient to keep the semi-fluid particles in contact and metallic cohesion has hardly begun, steel has practically no tensile strength and is easily ruptured by its own contraction of the parts around it.

The methods which were used by the forgemasters for the production of sound ingots at this date are not publicised – this was probably part of the "mystique". There are vague references to the technique used by Krupp at Essen[36] which imply the lining of the top of the mould with refractory material or the addition of a refractory powder to the top of the metal after casting, or both.[37] The only detailed information available on ingot mould design[38] clearly indicates that, at this date, one piece cast iron moulds, open ended and set on flat bottom plates, were invariably employed. The only exceptions were the semi-closed top moulds still used in some plants for the extra soft steels. Almost without exception, the evidence is that ingots were cast large end down, presumably for ease of stripping. The bottom pouring of rolling mill ingots in clusters appears to have been standard practice.

As far as the behaviour of the metal during teeming is concerned, an American paper dating from 1891 clearly shows the state of the art as regards the provision of rolling mill ingots and sheds an interesting light on general steelworks philosophy at that time:[39]

> The action of gases during teeming has much to do in determining the marketable fitness of the product. In grades of soft steel to be rolled, the gases

[36] Howe, op. cit., p. 150.

[37] This would form a "sink head" or "hot top", akin to the "dozzle" invented by R. F. Mushet, British Patent 1310, 23 May 1861. In this way a reservoir of hot liquid steel was available to feed the contraction below, so concentrating the pipe in the head position and giving solid metal below. The addition of an insulating refractory powder to the top of the metal in the head would assist in maintaining the reservoir of liquid metal for a longer time.

[38] Harbord and Hall, op. cit., pp. 30-38. Specific dimensions for nearly one hundred different ingot moulds are quoted, including squares, rectangular slabs and octagons.

[39] The paper by H.D. Hibbard appeared in Iron Age, 2 July 1891. The quotation given here is taken from H. H. Campbell: "The Open Hearth Process", Trans. Amer. Inst. Min. Engs., Vol. xxiii (1893-94), pp. 488-490. The argument is vaguely familiar to the author, who, whilst his day-to-day work was with fully killed alloy and special steels, became involved in a collaborative exercise in 1948 in which a series of ingots from a basic Open Hearth cast of rimming steel was treated with increasing amounts of aluminium in the moulds. From the initial rimming ingot, he remembers being surprised that very small additions of aluminium caused a more violent evolution of gas and the production of an ingot where the metal rose in the mould quite dramatically, whereas a slightly larger additions provided fully killed and piped ingots. (See P. Macnair: "Ingot structures in a Series from Rimmed to Killed Steels made from the Same Cast", J.I.S.I., 1948, No. III, pp. 151-163.)

should evolve freely during teeming and setting. In hard steels and steel castings they should be kept in solution. The middle ground between these extremes is full of danger and unsound steel is likely to result. At many works especial difficulty is met in making some one grade of carbon, all others, higher or lower, being easily produced. Their practice, particularly in reference to degrees of oxidation maintained, but casting temperatures as well, finds a carbon at which the gases neither escape freely nor are all kept in solution and the steel, as cast, is unsound. Under this head comes for consideration "piping" and "rising" steel. We find in soft steel one kind and in hard steel another kind of pipe. In soft, very low carbon metal, with rather pronounced oxidation and absence of appreciable amounts of silicon and aluminium, the gases are so freely evolved in minute bubbles that they may form half the volume of a mould apparently full of steel at the end of teeming. Later, when this gas has escaped, the metal sinks down and leaves a hollow shell or pipe of metal which has chilled next to the mould. Basic steel is prone to do this. Hard steel which pipes does so because of the shrinkage of the metal during setting and cooling. The gases are kept in solution by ingredients having that effect and the loss of volume of the steel in setting for the whole ingot is concentrated in the pipe. Piping or incipient piping steel is to be desired. In soft steel a slight tendency to pipe is aimed at. This steel will rise a little just before the last of it has solidified if allowed to do so, but it can be prevented and good ingots obtained.

Rising steel is that which increases in bulk after teeming, the result of the formation of bubbles in its interior which do not rise and escape. The casting temperature of such steel is too high and this brings us to the consideration of the effect of temperature on steel. The trouble, when this is too high, is well known, but not the reason why. It seems to be chiefly from the effect on the gases. It is not easy to see how the heat per se can work injury, but by increasing the solvent power of iron for gases to the point that they are evolved so slowly in the cooling steel that they cannot escape, being entangled, as it were, in the plastic steel, but not yet held completely in solution, high heat can and does exert a positive influence in making bad steel. ...

To make steel requiring the free solution of gas in the mould, the temperature must be as low as may be without giving a scull in the ladle. With oxidation rather high, the limit of permissible temperature is raised, but red-shortness lies in wait to head off the melter who would use this agent freely to avoid close watch of the heat of the bath.

High carbon piping steel should not be too hot, as this increases the pipe. For steel castings, high temperature is needed to insure filling all parts of the mould; the pipe must be guarded against by the use of large risers or sink-heads. ...

The weight of charges has a bearing on the resulting quality. When the weight is below 10,000 pounds [4.5 tons], trouble is found in having metal hot enough to cast properly without melting the furnace. The difficulty of this increases about inversely as the square of the decrease in weight. Below 3,000 pounds [1.3 tons] a ladle cannot be used satisfactorily but the metal must be run directly from the furnace to the mould. On the other hand large heats cast into small ingots engender bad results somewhere. If the casting temperature is right at the beginning of teeming it will be too low at the end and, if right at the end, too hot at the beginning.

> The time of making a heat may or may not affect the quality. Where abnormally long, due to bad furnace work, which may come from bad gas, badly designed furnace, bad draught, or ignorance, the quality is pretty apt to be injured. A melter can hardly have had enough experience with this class of heats to know exactly what to do to keep the quality up to his best, though it may pass inspection.

There is much practical experience embodied in this paper. It remains clear, however, that the underlying reasons for gas evolution and blowhole formation were not clearly understood, nor were the functions of elements such as silicon and, particularly, aluminium (which was only available in quantity and at a reasonable cost from about 1890) in suppressing gas evolution. Indeed, the usual terminology was that the presence of such elements increased the solubility of the metal for gas, not that the oxide of the metals reacted so slowly with carbon at the solidification temperature that gas production was negligible.[40]

Locational Variations in Steelmaking Processes

The overall position in world steelmaking in 1900, as far as the major bulk processes is concerned, is summarised in Table IX.[41] It is unfortunate that more detailed information on steelmaking in Russia and France has not survived. Certain interesting regional variations in production can be deduced from the more complete figures from the United States, Great Britain, Germany and Austria-Hungary as follows, the figures being expressed as percentages of the total bulk steel output:

	United States	Great Britain	Germany	Austria Hungary	TOTAL
Bessemer:					
Acid	66.3	25.6	3.4	5.9	36.4
Basic	Nil	10.0	63.6	21.4	21.5
TOTAL	66.3	35.6	67.0	27.3	57.9
Open Hearth:					
Acid	8.5	58.4	2.3	3.1	17.2
Basic	25.2	6.0	30.7	69.6	24.9
TOTAL	33.7	64.4	33.0	72.7	42.1

40 See, for example, Howe, op. cit., p. 162.

41 A. Cooper, Presidential Address, J.I.S.I., 1912, Pt. 1, p. 48. The figures given in the text are calculated by the author from those in Table IX.

TABLE IX WORLD OUTPUT OF STEEL INGOTS IN 1900

	United States	Great Britain	Germany	Austrian Empire	France	Russia	Spain	Sweden	Others*	TOTAL
Acid Bessemer	6,884,770	1,253,903	223,063	68,056						
Basic Bessemer		491,104	4,141,587	245,145						
Bessemer Total	**6,684,770**	**1,745,007**	**4,364,650**	**313,201**	**919,283**	**675,106**	**91,586**	**91,065**		
Acid Open Hearth	853,044	2,862,566	147,800	34,583						
Basic Open Hearth	2,545,091	293,485	1,997,765	797,860						
Total Open Hearth	**3,398,135**	**3,156,051**	**2,145,645**	**832,443**	**645,881**					
TOTAL	10,082,905	4,901,058	6,510,215	1,145,654	1,565,164	2,186,399	150,734	298,483	1,185,668	28,035,280

* The total marked **Others** includes unclassified totals as under:
Belgium 654,827 tons
Italy 115,887 tons
Canada 23,954 tons
together with an unspecified total of 400,000 tons from elsewhere.

Source: A. Cooper, Presidential Address, J.I.S.I., 1912, Pt. 1, p. 48.

	United States	Great Britain	Germany	Austria Hungary	TOTAL
Acid Steel:					
Bessemer	66.3	25.6	3.4	5.9	36.4
Open Hearth	8.5	58.4	2.3	3.1	17.2
TOTAL	74.8	84.0	5.7	9.0	53.6
Basic Steel:					
Bessemer	Nil	10.0	63.6	21.4	21.5
Open Hearth	25.2	6.0	30.7	69.6	24.9
TOTAL	25.2	16.0	94.3	91.0	46.4

This table indicates that, taking these four producing areas, between them responsible for around four fifths the world output of steel in 1900, there was an overall 58/42 split between Bessemer and Open Hearth processes and a 54/46 split between Acid and Basic steelmaking. Considering the countries separately, however, America and Great Britain show a strong bias towards Acid Steel whilst the continental European countries an equally strong bias in favour of Basic Steel; this can be put down in the cases of America, Germany and Austria to the type of domestic ores available; in the case of America, incidentally, it has been suggested that Andrew Carnegie purchased the basic steelmaking rights in America from Gilchrist Thomas so as to restrict the competition.[42] The situation in Britain, however, merits a more extended examination.

Whilst the provision of a basic lining to the Bessemer converter was a British development, it was the British steelmakers who stood to lose most. Cumberland and Furness were clearly threatened, particularly the massive works at Barrow. The other steelmaking centres had relied in the early days on pig iron from Cumberland but had latterly made their own arrangements to import haematite ores, as was the case in South Wales (where iron from imported haematite ores could be produced more cheaply than phosphoric iron from local ores), in Sheffield (where John Brown had set the precedent in self sufficiency) and on Teeside (where the Cleveland pig had proved useless in the original Bessemer practice and, indeed, despite the Basic Bessemer developments in the area, Bolckow Vaughan and Company, as well as the Consett Iron Company, with their vested interests in Spanish ore

[42] The suggestion that possible legal restrictions, subsequent to Andrew Carnegie negotiating extremly favourable terms with Gilchrist Thomas, could have contributed to the poor development of basic converting in the U.S.A. appears in the preface to the translation by W.B. Phillips and E. Prochaska of Herman Wedding's "Basic Bessemer Process", published in 1891 in New York.

mines, continued producing acid steel as well that from more recently installed basic capacity right through to the end of the century). Nevertheless, the basic steel from Cleveland became accepted for a great variety of applications: for rails, wire, tube, girders and particularly for ship plate and angles in the Scottish yards, so much so that two new Basic Bessemer plants were erected in the Glasgow area, at Glengarnock and Wishaw. There was, however, a period of disillusionment with this steel to follow. Although Lloyds had given a partial clearance to the basic steel alongside acid Bessemer in 1884, in view of disastrous results in its use in the shipbuilding yards of the North East, where it was said the very mention of basic Bessemer was like a red rag to a bull, this authorisation was subsequently withdrawn. After the development of the basic Open Hearth material, this was separately assessed; by 1889, one British supplier had been accepted by Lloyds' assessors: surprisingly ten continental sources were also approved and it was commented that the foreign production was generally made from purer materials than the British practice. This could well have been true since, whereas the British material was largely produced from highly phosphoric pig iron, the continental material was derived largely from scrap. Most British shipbuilders, in any case, still refused to use anything other than acid Open Hearth steel for their major construction work. Sheffield produced the engineering forgings: the crankshafts, propellor shafts, rotating axles, as well, of course, as ordnance and armour plate. All these came in acid Open Hearth steel, whilst the cutting tools were still supplied, in the main, from the old fashioned crucible steel. All this may have had a certain amount of prejudice in favour of what was known and tried and a reluctance to change, not only on the part of the user but also on the part of the manufacturer on account of his existing capital investment. The nett effect, however, can be seen by extracting the figures for steel production in Britain:

	1890	**1900**
Acid Bessemer	45.1%	25.6%
Basic Bessemer	11.2%	10.0%
Acid Open Hearth	40.9%	58.4%
Basic Open Hearth	2.8%	6.3%

A catalogue of furnace installations in Germany in 1901 has survived;[43] the details would be tedious at this stage but the information which may be derived is summarised in Table X. The spread of installations at some of the works previously covered is, however, of interest. Krupp at Essen still

[43] "Gemeinfassliche Darstellung des Einenhuttenwesens", 4te Auflage, 1901, pp. 136-139. A copy of this was kindly supplied by F. Toussaint, Verein Deutscher Eisenhuttenleute.

TABLE X GERMAN FURNACE INSTALLATIONS, 1901

	WESTPHALIA	LOTHRINGEN, SAAR & LUXEMBOURG	OTHERS	TOTAL
Acid Bessemer:				
Number of Converters	22	Nil	6	28
Total Capacity, tons	132	...	43	175
Mean Converter Size, tons	6.00	...	7.17	6.25
Basic Bessemer:				
Number of Converters	35	52	11	98
Total Capacity, tons	463	640	146	1249
Mean Converter Size, tons	13.23	12.31	13.27	12.74
Acid Open Hearth:				
Number of Furnaces	20	2	1	23
Total Capacity, tons	238	30	8	276
Mean Furnace Size, tons	11.90	15.00	8.00	12.00
Basic Open Hearth:				
Number of Furnaces	156	14	48	218
Total Capacity, tons	2659	242	767	3668
Mean Furnace Size, tons	17.04	16.13	15.98	16.82

Source: Gemeinfassliche Darstellung des Eisenhuttenwesens, 4te Auflage, 1901.

retained an allegiance to Acid steel; they had had fifteen 6 ton Acid Bessemer furnaces, together with nine 10 ton Acid Open Hearth furnaces; in addition, however, they had thought fit to install eighteen 21 ton Basic Open Hearth furnaces. Similarly, Bochumer Verein still retained three 8 ton Acid converters but were running three 6 ton Basic Bessemer converters and seven 25 ton Basic Open Hearth furnace. The four Basic converters at Hoerde were now 18 ton capacity; there were also eight 18 ton and two 7 ton Open Hearth furnaces, all basic lined except for one of the larger furnaces. Phoenix at Ruhrort had one 12 ton Open Hearth furnace; their three 12 ton converters and five 20 ton Acid Open Hearth furnaces, however, were all basic lined. Hoesch was entirely basic, with three 11 ton converters and four 18 ton Open Hearth furnaces. The most heavy concentration of basic steelmaking was in Lothringen (formerly Lorraine), the Saar and Luxembourg, all working on the Lorraine ores which, up to the Franco-Prussian war, had been in French hands and, up to 1879, had not been considered suited to steelmaking.

The Basic Bessemer process was most important in Germany and a catalogue of the technical improvements which had led to the immense success of the Basic Bessemer process in Germany in the early years of the twentieth century must be of considerable interest:[44]

1. The introduction of various types of mixer.
2. The application of gas engines for blowing.
3. The gradual increase in size of the converters, up to charges of 25 tons, the depth of the bath being kept low, thus lessening mechanical projection.
4. The improvement in the application of the blast, by distributing it over a larger number of smaller holes in the bottom, thus lessening waste of metal.
5. The use of magnesite tuyere bottoms to increase their durability.
6. The increase of pressure up to 500 atmospheres in the moulding of the bricks.
7. The application of compressed air for back ramming the basic converter lining.
8. The use of a mechanical device for rapidly clearing away incrustations in the mouth of the converter.
9. The introduction of transportable pouring ladles.
10. The casting of the steel in ingot moulds set on movable bogies.

Lining life in modern converters in Germany varied between 200 and 350 heats; magnesite tuyere bottoms could last from 120 to 250 heats. In one particular instance at the Rothe Erde works, 236 heats were blown from one vessel, whereupon the lower part was relined and a further 246 charges

[44] This information is derived from an article from Aachener Hutten Verein, published in "Stahl and Eisen", Vol. xxix, pp. 1486-1487, and abstracted in J.I.S.I., 1909, Pt. 2, pp. 455-456.

blown. The bottom part was again relined, still retaining the upper lining and 220 additional charges blown. Thus 720 charges, totalling 14,700 tons of steel, were blown in the one vessel before a complete reline, giving a calcined dolomite consumption of rather less than 20 lb. per ton of steel produced.

Although it is outside the period under survey, it is of interest to provide information based on the statistics for 1910 [45] in which year some 58 million tons of steel were produced. Here again, the four countries on which the detailed statistics were available produced around four fifths of the total world steel.

	United States	Great Britain	Germany	Austria Hungary	TOTAL
Total, ingot tons	25,917,281	6,014,684	13,315,437	2,186,399	47,433,801
Bessemer:					
Acid, %	36.3	18.9	1.3	1.6	22.7
Basic, %	Nil	10.7	60.3	14.0	18.9
TOTAL, %	36.3	29.6	61.6	15.6	41.6
Open Hearth:					
Acid, %	4.7	44.1	1.1	1.3	8.5
Basic, %	59.0	26.3	37.3	83.1	49.9
TOTAL, %	63.7	70.4	38.4	84.4	58.4
Acid Steel:					
Bessemer, %	36.3	18.9	1.3	1.6	22.7
Open Hearth, %	4.7	44.1	1.1	1.3	8.5
TOTAL, %	41.0	63.0	2.4	2.9	31.2
Basic Steel:					
Bessemer, %	Nil	10.7	60.3	14.0	18.9
Open Hearth, %	59.0	26.3	37.3	83.1	49.9
TOTAL, %	59.0	37.0	97.6	97.1	68.6

It will be noted that the ten years had shown an accentuation of the trend which was beginning to make itself clear in 1900 – that of the development of the Basic Open Hearth process as the major source of bulk steel. In America, for instance, the Acid Bessemer process, which had been the foundation of that country's phenomenal rise to that of the major steel producer in the world, had been superseded to a large extent by 1910. Even in Britain the proportion of Basic Open Hearth steel had risen from 6% in

[45] Cooper, op. cit., p. 49. Again the figures in the text are calculated by the author.

1900 to 26%; the Acid Open Hearth process was still the premier method in Britain, however. The other relevant comment would be to draw attention to the fact that the Bessemer process was losing ground against the Open Hearth process; this was less noticeable in Germany than elsewhere but was, nevertheless, the case here also.

Postscript

This, then, concludes the survey of steelmaking in the last half of the nineteenth century. The methods of steelmaking that were used in 1850 had been small scale, highly specialised operations. In the main, they had survived and were still operating on a similar scale at the end of the century but they were, by then, considered too insignificant to be included in the official statistics.[46] The Bessemer and Open Hearth processes, in their Acid and Basic variants, had stolen the stage and they have, therefore, legitimately been the main subjects for discussion. They were, however, in their turn, to be superseded. The Basic Bessemer process became obsolete in Britain in the 1920s; it was, however, resurrected as the most suitable process for the new tube making operations at Corby in the mid 1930s and continued in operation there into the 1970s.[47] The Acid Bessemer process had already died in the 1920s, apart from the operations in the haematite region of the Lake District and here the operations survived at Workington until 1974.[48] The Basic Open Hearth by 1950 was providing the bulk of the world's steel but it too was overtaken by what came to be known as the "basic oxygen processes", which, paradoxically, owed something to Bessemer; there are now none of the old furnaces remaining in Britain and only a few, obsolescent, small scale, isolated furnaces are still operating elsewhere. The Acid Open Hearth furnace survived as long as anywhere in Sheffield, producing forging ingots for which it was ideally suited; its demise was mainly due to lack of suitable raw materials at an economic price and the last furnace was put out of commission in the early 1960s. Thus, in spite of the enormous strides made in steelmaking in the last half of the nineteenth century, the story of the modern steelmaking processes themselves, together with the development of the vast field of alloy engineering steels, belongs to the twentieth century. Nevertheless, this later development was grounded firm-

[46] Further information on these may be found in the author's "Steelmaking Before Bessemer", Vol. II - "Crucible Steel: The Growth of Technology", London., 1984. Despite its title, it carries forward the information on the Crucible Process to its demise in the 1960s.

[47] F. Scopes: "The Development of Corby Works", published privately by Stewarts and Lloyds, 1968.

[48] J. Y. Lancaster and D. R. Wattleworth: "The Iron and Steel Industry of West Cumberland", published privately by British Steel Corporation, 1977. It should be noted that one of the final operating pair of converters from Workington is now situated as a static exhibit at the entrance to the Kelham Island Industrial Museum in Sheffield.

ly on the foundation laid by those who worked and strove so effectively in the operations which have been described in these pages. It should not be forgotten, however, that the pioneers whose names recur in the story could never have succeeded without that now unrecorded host of men whose skill and craftsmanship alone could convert their inventions into viable steel-making techniques, the product of which made possible the enormous advances in engineering throughout the world in the second half of the nineteenth century.[49]

[49] It seems appropriate here to quote the dedication which appears on the flyleaf of the authoritative text by the American metallurgist, H. H. Campbell: "The Manufacture and Properties of Iron and Steel", 4th. ed., New York, 1907:

> "To all those, famous or obscure, who, by the Furnace, in the Shop or at the Desk, are joining hand and brain to solve the problems of the Metallurgic Art, this volume is fraternally dedicated."

Bibliography

UNPUBLISHED MATERIAL

The following collections have been consulted:

Doncaster Archives (held at Kelham Island Museum)
Firth Archives (part held privately; part at Kelham Island Museum)
Jessop Records (held privately)
Percy Papers (held by the Institute of Metals)
Sheffield City Library Archives

PUBLISHED MATERIAL

Official Publications

Ministère de l'Agriculture, du Commerce et des Travaux Publiques: "Statistique de l'Industrie Minerale ... , 1853-1859", Paris, 1861.
Riksdag Bilaga, 1853-4.
Royal Commission on the Depression of Trade, 2nd. report, 1886

PATENT SPECIFICATIONS

British Patents:

1784	1420	H. Cort	Manufacturing Iron
1800	2447	D. Mushet	Manufacturing Metals
1853	6837	C. Schafhautl	Manufacturing Malleable Iron
1839	8021	J. M. Heath	Manufacture of Iron and Steel
	8129	W. Vickers	Manufacture of Cast Steel
1846	11515	A. Chenot	Treatment of Metallic Oxides
1850	12950	E. Riepe	Manufacture of Steel
1854	246	C. B. A. Chenot	Gases for Metallurgical Purposes
	658	C. B. A. Chenot	Manufacture of Steel
	1001	J. Nasmyth	Puddling Iron
	1730	S. Lucas	Manufacturing Steel
1855	629	I. Rogers	Treating Iron Ores
	2082	J. G. Martien	Manufacture of Iron and Steel

1855	2189	F. Uchatius	Manufacturing Cast Steel
	2768	H. Bessemer	Manufacture of Iron
1856	630	H. Bessemer	Manufacturing Iron and Steel
	851	W. Newton	Manufacturing Steel
	1292	H. Bessemer	Manufacturing Iron and Steel
	1587	A. & E. Chenot	Extracting Steel Sponges
	1588	A. & E. Chenot	Sorting Ores
	1589	A. & E. Chenot	Machinery for Compression
	1590	A. & E. Chenot	Reducing Metallic Oxides
	1679	F. Gurlt	Manufacture of Iron and Steel
	2219	R. F. Mushet	Manufacture of Iron and Steel
	2861	F. Siemens	Arrangement of Furnaces
1861	167	W. & F. Siemens	Furnaces
	262	I. Rogers	Furnace for Treating Iron Ores
	1310	R. F. Mushet	Casting Steel Ingots
	2900	G. Parry	Manufacture of Iron and Steel
1862	1473	C. Attwood	Manufacture of Steel
1863	114	H. Bessemer	Manufacture of Iron and Steel
	708	W. E. Newton	Manufacture of Iron and Steel
1864	2031	R. A. Brooman	Manufacture of Cast Steel
1865	2137	R. A. Brooman	Manufacture of Cast Steel
1867	546	A. L. Holley	Casting Ingots
	2395	C. W. Siemens	Improvements in Furnaces
1868	1892	C. W. Siemens	Manufacture of Cast Steel
1869	186	H. A. Bonneville	Reducing Metallic Ores
	616	G. Snelus	Reducing Iron Ores
	908	J. H. Johnson	Furnace Linings
1871	2548	W. & J. Ireland	Casting Steel Ingots
1872	429	W. & J. Ireland	Casting Steel Ingots
	908	G. Snelus	Linings for Furnaces
1873	263	W. & J. Ireland	Casting Steel Ingots
	1574	A. Browne	Metallic Alloys
	1788	B. D. Healey	Casting Ingots
	2819	J. H. Johnson	Mechanical Puddling Furnaces
	4291	J. H. Johnson	Furnaces for Steel
1874	1317	J. H. Johnson	Manufacture of Steel
	1997	J. H. Johnson	Furnaces for Steel and Iron
	3062	J. Whitworth	Compression of Metals in the Fluid State
1875	43	C. W. Siemens	Furnaces for Iron and Steel
	120	A. Browne	Manufacture of Steel
1876	370	A. Browne	Manufacture of Chromium Steel
1877	3370	C. W. Siemens	Furnaces for Iron and Steel
	4422	S. G. Thomas	Lining Bessemer Converters

1878	289	S. G. Thomas	Furnaces for Making Steel
1878	908	S. G. Thomas	Furnaces for Making Steel
	3975	S. G. Thomas	Basic Refractory Materials
	4063	S. G. Thomas	Refractory Bricks
1879	131	S. G. Thomas	Basic Refractory Materials
	257	S. G. Thomas	Furnace Linings
	1079	S. G. Thomas	Making Steel
	1313	S. G. Thomas	Making Bessemer Steel
1885	8269	T. Nordenfelt	Castings in Iron and Steel
1887	630	J. H. Darby	Making Steel
1892	8612	E. H. Saniter	Desulphurisation of Steel or Iron
	17692	E. H. Saniter	Desulphurisation of Steel or Iron
	23534	E. H. Saniter	Desulphurisation of Iron

French Patents:

1869	84,735	E. Muller	Pour la Construction de Garnitures

United States Patents:

1856	16,082	H. Bessemer	Manufacture of Iron and Steel
1857	17,389	R. F. Mushet	Manufacture of Iron and Steel
	17,628	W. Kelly	Manufacture of Iron and Steel
1865	49,495	J. Baur	Manufacture of Chromium Steel

Periodicals, etc.

American Iron and Steel Institute: Statistical Reports.
British Iron and Steel Federation: Collected Statistics
Engineering: 25 October 1867; 8 December 1876
Gemeinfassliches Darstellung des Eisenhuttenwesens, 4te Auflage, 1901
"Iron": 1 December 1884
Iron and Steel Institute: Statistical Summaries in J.I.S.I.
"The Times": 14 August 1856; 8 September 1866; 14 March 1879; 2 September 1880

Private Publications

Anon: "Ett Svenskt Jernverk: Sandviken, 1862–1937", Sandvik Steels, 1937.
Anon: "Thomas and Gilchrist – Bolckow and Vaughan – 1879-1929", Bolckow, Vaughan and Company, 1929.
A. Borthwick: "Hallside: One Hundred Years, 1873-1973", British Steel Corporation, Special Steels Division, 1973.

A. L. Holley: Private Reports:
No. 1. Siemens Direct Ore Process. London, 1874
No. 2. The Value of Manganese. Brooklyn, 1875
Supplement to No. 2. The Effect of Manganese, Carbon, Phosphorus and Other Ingredients on Iron and Steel. New York, 1876
No. 3. The Siemens Steel Manufacture from Pig and Ore. New York 1874
No. 4. West Cumberland Iron and Steel Works. London, 1875
No. 5. The Pernot Revolving Hearth Furnace. Brooklyn, 1875
No. 6. Brown, Bayley and Dixon's Steel Works, Sheffield. London, 1875
No. 7. Barrow Haematite Iron and Steel Works. London, 1875
No. 8. Steel Manufacture in France and Belgium. New York, 1876
Second Series:
No. VII. Solid Steel Castings by the Terrenoire Process. New York, 1877
No. IX. The Open Hearth Steel Manufacture in Europe. New York, 1879

G. R. Walshaw and C. A. J. Behrendt: "The History of Appleby Frodingham", United Steel Companies, 1950.

Articles in Journals

R. Akerman: "The Proper Definition of Steel", Iron, Vol. 7, p. 738 (10 June 1876) [Translation from Jernkontorets Annaler]

R. Akerman: "The Bessemer Process as Conducted in Sweden", Trans. Amer. Inst. Mining Engs, Vol XXII (1893), pp. 265-288.

Anon: "Grinding Basic Slag", Iron and Coal Trades Review, Vol. xxxiv, p. 479.

K. C. Barraclough: "Sir William Siemens: 1823-1883", Historical Metallurgy, Vol. 18 (1984), No. 1, pp. 1-7.

H Bauermann: "Iron and Steel at the Universal Exhibition, Paris, 1900", J.I.S.I., 1900, Pt. 2, pp. 160-168.

I. Lowthian Bell: "The Chemical Phenomena of Iron Smelting", J.I.S.I., 1871, p. 85.

I. Lowthian Bell: "On the Separation of Phosphorus from Pig Iron", J.I.S.I., 1878, Pt. 1, pp. 17-34; also discussion, pp. 34-47.

E. Bertrand: "The Practice of the Combined Open Hearth Process of Bertrand and Thiel", J.I.S.I., 1897, Pt. 1, pp. 115-124.

H. Bessemer: "On the Manufacture of Malleable Iron and Steel"; lecture

24 May 1859, reported in "The Mechanics' Magazine", 1859, pp. 397-398 and 405-407.
H. Bessemer: "On the Manufacture of Cast Steel and its Application to Constructive Purposes", Proc. Inst. Mech, Eng., 1861, pp. 133-157.
H. Bessemer: "On Some Early Forms of Bessemer Converter", J.I.S.I., 1886, Pt. 2, pp. 638-649.
H. Bessemer: "Historical and Technical Sketch of the Origin of the Bessemer Process", Trans. Amer. Soc. Mech. Engs., Vol. 18 (1897), pp. 455-475; discussion pp. 480-491.
E. Boman: "Jernfarskning efter Bessemers Metod", Jernkontorets Annaler, 1863, Pt. 3, pp. 184-192; Pt. 4, pp. 193-227.
W. G. Burman: "Use of Fluor Spar in the Basic Open Hearth Furnace and Its Effect on the Solubility of Basic Slag", Iron and Coal Trades Review, 9 Jan 1903, p. 94.
H. H. Campbell: "The Open Hearth Process", Trans. Amer. Inst. Mining Eng., Vol. XXIII, (1893-4), pp. 345-510.
Per Carlberg: "Early Industrial Production of Bessemer Steel at Edsken", J.I.S.I., Vol. 189 (1958), pp. 201-204.
W. Clay: "On the Manufacture of Puddled or Wrought Steel, with an Account of Some of the Uses to which it has been Applied", Journal Society of Arts, Vol. VI (1857-8), pp. 140-148.
A. Cooper: "Metal Mixers as Used at the North Eastern Steel Company", J.I.S.I., 1895, Pt. 1, pp. 66-70.
A. Cooper: Presidential Address, J.I.S.I., 1912, Pt. 1, pp. 31-50.
C.G. Dahlerus: "Undersokningar rorande forloppet vid Bessemerfarskning", Jernkontorets Annaler, 1881, pp. 314-322.
J. H. Darby: "The Manufacture of Open Hearth Steel", J.I.S.I., 1889, Pt. 1, pp. 79-83.
H. Dickmann: "Alfred Krupp – Henry Bessemer", Stahl und Eisen, Vol. 82 (1962), pp. 911-913.
G. Dove: "The Frodingham Iron Field", J.I.S.I., 1876, pp. 318-326.
W. F. Durfee: "An Account of the Experimental Steel Works at Wyanotte, Michigan", Trans. Amer. Soc. Mech. Engs., Vol. 6 (1884-5), pp. 40-59.
W. F. Durfee: "Development of American Industries Since Columbus", Popular Science Monthly, Vol. xxxix (1892), pp. 729-750.
H. Fehland: "Geschichtliches uber die Puddelstahl Fabrikation", Stahl und Eisen, 1886, pp. 224-230.
Delvaux de Fenffe: "De la Fabrication de l'Acier Puddlé en Allemagne", Revue Universelle, Vol. 1 (1857), pp. 59-94.
R. Forsyth: "The Bessemer Plant at South Chicago", Trans. Amer. Inst. Mining Engs., Vol. XII (1883-4), pp. 254-274.

T. Gillott: "The Basic Open Hearth Steel Process", Proc. Inst. Civ.. Engs., Vol. 77, (1883), pp. 297-310.

K. F. Goransson: "The First Successful Bessemer Steel", Metal Progress, Jan 1956, pp. 85-87.

F. Gouvy: "La Maison Gouvy depuis sa Foundation jusqu'a Nos Jours", Le Fer a Travers les Ages, Nancy, 1956, pp. 351-352.

E. Grateau: "Memoire sur la Fabrication de l'Acier Fondu par le Procédé Chenot", Revue Universalle, Vol. 6 (1859), pp. 1-62.

A. Greiner: "Les Installations Bessemer Americaines Adaptées au Procédé Basique d'apres les Plans de Monsieur Holley", Revue Universelle, Tome ix, 2me Serie (1882), pp. 1-14.

L. Gruner: "Notes Diverses concernant l'Acier Puddlé et l'Acier de Forge", Annales des Mines, Vol. xv (1859), pp. 291-336.

L. Gruner: "Note sur les Sens des Mots FER et ACIER", Revue Univer-elle, Tome 40 (1876), pp. 437-440.

W. Hackney: "The Manufacture of Steel", Proc. Inst. Engs., Vol. xlii (1874-5), pp. 2-68.

F. W. Harbord: "The Thomas Gilchrist Basic Process, 1879-1937", J.I.S.I., 1937, Vol. II, pp. 77-95.

J. Hartshorne: "The Basic Bessemer Plant at the Pottstown Iron Company", Trans. Amer. Inst. Mining Engs., Vol. XXI (1892-3), pp. 743-765.

A. P. Head: "Tilting Open Hearth Furnaces", J.I.S.I., 1899, Pt. 1, pp. 69-75.

B. D. Healey: "On a New Mode of Casting Ingots", J.I.S.I., 1873, Pt. 1, pp. 183-185.

C. B. Holland and A. Cooper: "On the Manufacture of Bessemer Steel and Ingot Iron from Bessemer Pig", J.I.S.I., 1880, Pt. 1, pp. 79-89.

A. L. Holley: "Setting Bessemer Converter Bottoms", J.I.S.I., 1874, pp. 368-370.

A. L. Holley: "What is Steel?", Iron, Vol. 7, pp. 102-103 (22.1.1876) and pp. 165-166 (2.2.1876)

A. L. Holley: "The Chemical and Physical Analysis of Phosphoric Steel", Proc. Inst. Civ. Engs., Vol liii (1877-8), Pt. 3, pp. 3-28.

A. L. Holley: "An Adaptation of the Bessemer Plant to the Basic Process", Trans. Amer. Soc. Mech. Engs., Vol. 1 (1880), pp. 130-132.

H. M. Howe: "The Nomenclature of Iron", Trans. Amer. Inst. Mining Engs., Vol V (1876-7), pp. 515-528.

R. W. Hunt: "A History of the Bessemer Manufacture in America", Trans. Amer. Inst. Mining Engs., Vol. V (1876-7), pp. 201-215.

R. W. Hunt: "The Original Bessemer Plant at Troy ", Trans. Amer. Inst. Mining Engs., Vol V (1876-7), pp. 60-70.

M. Janoyer: "Memoire sur la Fabrication du Fer a Grains et de l'Acier Naturel aux fours a Puddler", Annales des Mines, Vol. xv (1859), pp. 146-167.

S. Jordan: "The Iron and Steel Manufacture in France", J.I.S.I., 1889, Pt II, pp. 20-22.
S. Kern: "The Manufacture of Siemens Martin Steel in Russia", Chemical News No. 41, p. 278 (abstracted J.I.S.I., 1880. Pt. 2, p. 777)
S. Kern: "On the Manufacture of Steel and Steel Plates in Russia", J.I.S.I., 1881, Pt. 1, pp. 80-83.
I. Krulis-Randa: "The Introduction of British Metallurgical Techniques into Austria", J.I.S.I., Vol. 194 (1960), pp. 43-45.
P. Kupelwesier: "On the Basic Bessemer Process", J.I.S.I., 1881, Pt. 2, pp. 397-406.
E. F. Lange: "Bessemer, Goransson and Mushet: A Contribution to Technical History", Manchester Memoirs, Vol. lvii (1913), No. 7. pp. 1-44.
J. F. Lundin: "Branstålsugn for Sagspan", Jernkontorets Annaler, 1862, No. 2, pp. 115ff.
P. Macnair: "Ingot Structures in a Series from Rimmed to Killed Steels made from the Same Cast", J.I.S.I., 1948, No. III, pp. 151-163.
J. Massenez: "Dephosphorising in the Converter", J.I.S.I., 1880, Pt. 2, pp. 475-487.
J. Massenez: "History of the Basic Process in Germany", J.I.S.I., 1909, Pt. 2, pp. 452-454 (abstracted from an article in "Stahl und Eisen", Vol. xxix, pp. 1465-1477)
G. W. Maynard: "A Biographical Note on Sidney Gilchrist Thomas", Trans. Amer. Inst. Mining Engs., Vol. XIII (1884-5), pp. 785-791.
M. Millard: "Old Methods of Ironmaking", Proc. Staffordshire Iron and Steel Inst., Vol. xxvii (1911-2), pp. 188-191.
E. G. Odelstjerna: "The Manufacture of Open Hearth Steel in Sweden", Trans. Amer. Inst. Mining Engs., Vol. XXIV (1894), pp. 288-315.
G. Parry: "On Puddled Steel", Proc. South Wales Inst. of Engs., Vol iii (1862-3), pp. 74-81.
R. Pink: "On the Dephosphorisation of Iron in the Bessemer Converter", J.I.S.I., 1880, Pt. 1, pp. 57-67.
F. le Play: "Memoire sur la Fabrication de l'Acier en Yorkshire", Annales des Mines, Serie 4me, Tome III (1943), pp. 583-714.
E. Windsor Richards: Presidential Address, Proc. Cleveland Inst. Engs. 1880-1881, pp. 15-52.
E. Riley: "On a Ready Means of Moulding Lime and Making Lime or Basic Bricks and Linings for Furnaces and Converters", J.I.S.I., 1879, Pt. 1, pp. 144-149.
J. Riley: "The Use of Fluid Metal in the Open Hearth Furnace", J.I.S.I., 1900, Pt. 1, pp. 22-32.
H. E. Roscoe: "On the Spectrum Produced by the Flame Evolved in the Manufacture of Steel by the Bessemer Converter", Philosophical Magazine, Vol. xxv, pp. 318ff.

H. E. Roscoe: "Spectrum Analysis", J.I.S.I., 1871, pp. 38-62.
F. Rowan: "On the Examination of the Flame of the Bessemer Converter", Chem. News., Vol. xix, pp. 170ff.
C. Sahlin: "De Svenska Degelstålsverken", Med Hammare och Fackla, Vol. IV (1932), pp. 35-134.
E. H. Saniter: "A New Process for the Purification of Iron and Steel from Sulphur", J.I.S.I., 1892, Pt. 2, pp. 216-222.
E. H. Saniter: "A Supplementary Paper on a New Process for Desulhurising Iron and Steel", J.I.S.I., 1893, Pt. 1, pp. 73-77.
C. W. Siemens: "On a Regenerative Gas Furnace as Applied to Glashouses, Puddling and Heating", Proc. Inst. Mech, Eng., 1862, pp. 21-45.
C. W. Siemens: "On the Regenerative Gas Furnace as Applied to the Manufacture of Cast Steel", J. Chem. Soc., 1868, pp. 279-310.
C. W. Siemens: "On Smelting Iron and Steel", J. Chem. Soc., 1873, pp. 661-678.
J. M. Silliman: "On the Examination of the Bessemer Flame with Coloured Glasses and with the Spectroscope", Philosophical Magazine, Vol. xli, pp. 1ff.
G. Snelus: "On the Removal of Phosphorus and Sulphur during the Besemer and Siemens Martin Processes of Steel Manufacture", J.I.S.I., 1879, Pt, 1, pp. 144-149.
G. Snelus: "Report on the Saniter Desulphurisation Process", J.I.S.I., 1893, Pt. 1, pp. 77-83.
T. Spencer: "On the Manufacture of the Uchatius Cast Steel", Proc. Inst. Mech. Engs., August 1858, pp. 146-154.
J. E. Stead: "On the Elimination of Sulphur from Iron", J.I.S.I., 1892, Pt. 2, pp. 223-278.
J. E. Stead, Presidential Address, J.I.S.I., 1920, Pt. 1, pp. 33-109.
B. Talbot: "The Open Hearth Continuous Steel Process", J.I.S.I., 1900, Pt. 1, pp. 33-61.
M.A. Tamm: "Undersokningar rorande sammansattingen af fram en Bessemerugn under blasningen gaserna", Jernkonorets Annaler, 1875, Pt. 5, pp. 257-273.
M.A. Tamm: "Undersokningar sammansattningen af de fran en Bessemerugn under blasningen bortgaserna af dervid anvandt tackjern och erhallna producter", Jernkontorets Annaler, 1878, pp. 444 ff.
M.A.Tamm: "Undersokningar rorande Bessemergases sommansattning", Jernkontorets Annaler, 1880, pp. 462 ff.
A. Thielen: "On the Darby Process of Recarburisation", Trans. Amer. Inst. Mining Engs., Vol. XXIII (1893-4), pp. 790-803.
S. G. Thomas and P. C. Gilchrist: "On the Elimination of Phosphorus",

J.I.S.I., 1879, Pt. 1, pp. 135-143.
B. Walker: "Development of Latest Improvements in Appliances for the Manufacture of Bessemer Steel", J.I.S.I., 1874, pp. 374-383.
K. Warren: "The Sheffield Rail Trade, 1861-1930: An Episode in the Locational History of the British Steel Industry", Trans. Inst. British Geography, June 1964, pp. 131-155.
M. W. Watts: "On the Spectrum of the Bessemer Flame", Philosophical Magazine, Vol. xxxiv, pp. 318ff.
H. Wedding: "Progress of German Practice in Metallurgy", J.I.S.I., 1890, Pt. 2, pp. 491-560.
D. Weeks: "The Invention of the Bessemer Process", Trans. Amer. Inst. Mining Engs., Vol. XXVI (1897), pp. 980-991.
S. T. Wellman: "The Early History of Open Hearth Steel Manufacture in the United States", Trans. Amer. Soc, Mech, Engs., Vol xxiii (1901), pp. 78-98.

Other Publications

J. K. Almond: "Making Steel from Phosphoric Iron in the Converter" – Chapter 7 in "Cleveland Iron and Steel", British Steel Corporation, 1979.
L. Ansiaux et L. Masion: "Traité Pratique de la Fabrication du Fer et de l'Acier Puddlé, Liege, 1861
K. C. Barraclough: "Steelmaking Before Bessemer", – Volume I: Blister Steel, London, 1984
K. C. Barraclough: "Steelmaking Before Bessemer" – Volume II: Crucible Steel, London, 1984.
B. Baylis: "On Puddling – by a Practical Puddler", London, 1855.
L. Beck: "Die Geschichte des Eisens", Vol. 5, Brunswick, 1903.
H. Bessemer: "An Autobiography", London, 1905. (Republished in facsimile, 1989, Institute of Metals).
H. Brearley: "Steelmakers", Sheffield, 1933.
G. Broling: "Anteckningar under en Resa i England aren 1797, 1798 och 1799", Vol. III, Stockholm, 1816.
H. H. Campbell: "The Manufacture and Properties of Iron and Steel", 4th. ed., New York, 1907.
D. Carnegie and S. Gladwin: "Liquid Steel: Its Manufacture and Cost", 2nd. ed., London, 1920.
J. C. Carr and W. Taplin: "History of the British Steel Industry", Oxford, 1962.
W. Crookes and E. Rohrig: "A Practical Treatise on Metallurgy", New York, 1870.
W. K. V. Gale: "The Black Country Iron Industry", London, 1966.

A. J. Grant: "Steel and Ships", London, 1949.
S. Griffiths (ed. W. K. V. Gale): "Guide to the Iron Trade of Great Britain, 1871", Newton Abbot, 1967.
L. E. Gruner et C. Lan: "L'Etat Present de la Metallurgie du Fer en Angleterre", Paris, 1862.
F. W. Harbord and J. W. Hall: "The Metallurgy of Steel", 3rd. ed., London, 1907.
A. S. Hewitt: "The Production of Iron and Steel in its Economic and Social Relations", Washington, 1868.
J. Hunter (ed. A. Gatty): "The History and Topography of the Parish of Sheffield in the County of York", Sheffield, 1869.
J. S. Jeans: "Notes on the Northern Industries", London, 1878.
J. S. Jeans: "Steel: Its History, Manufacture, Properties and Uses", London, 1880.
W. T. Jeans: "Creators of the Age of Steel", 2nd. ed., London, 1885.
O. Johannsen: "Geschichte des Eisens", Dusseldorf, 1953.
S. Jordan: "Revue de L'Industrie du Fer", Vol. 4, Paris, 1871.
J. Y. Lancaster and D. R. Wattleworth: "The Iron and Steel Industry of West Cumberland", Workington, 1977.
D. S. Landes: "The Unbound Prometheus", Cambridge, 1969.
H. C. Landrin (trans. A. A. Fesquet): "A Treatise on Steel", Philadelphia, 1868.
W. Lewis ("Lewys Afan"): "The Tinman's Companion", Pontypridd, 1876.
A. C. Marshall and H. Newbould: "The History of Firth's, 1842-1918", Sheffield, 1924.
J. McHugh: "Alexander Holley and the Makers of Steel", Baltimore, 1980.
F. G. C. Muller: "Krupp's Steel Works", being "an authorised translation from the German", London, 1898.
F. M. Osborn: "The Story of the Mushets", London, 1952.
J. B. Pearse: "A Concise History of Iron Manufacture in Pennsylvania", New York, 1876.
J. Percy: "Metallurgy: Iron and Steel", London, 1864.
W. Pole: "The Life of Sir William Siemens", London, 1888.
F. Scopes: "The Development of Corby Works", Portsmouth, 1968.
J. D. Scott: "Vickers: A History", London, 1962.
J. M. Swank: "History of the Manufacture of Iron in All Ages", 2nd. ed., Philadelphia, 1892.
L. G. Thompson: "Sidney Gilchrist Thomas: An Invention and Its Consequences", London, 1940.

G. H. Thurston: "Pittsburgh and Allegheny in the Centennial Year", Pittsburgh, 1876.
W. Truran: "The Iron Manufacture of Great Britain", London, 1855.
H. Wedding, trans. W. B. Phillips and E. Prochaska: "Basic Bessemer Process", New York, 1891.

Index